PRINCIPLES OF LASERS AND OPTICS

Principles of Lasers and Optics describes both the fundamental principles of lasers and the propagation and application of laser radiation in bulk and guided wave components. All solid state, gas and semiconductor lasers are analyzed uniformly as macroscopic devices with susceptibility originated from quantum mechanical interactions to develop an overall understating of the coherent nature of laser radiation.

The objective of the book is to present lasers and applications of laser radiation from a macroscopic, uniform point of view. Analyses of the unique properties of coherent laser light in optical components are presented together and derived from fundamental principles, to allow students to appreciate the differences and similarities. Topics covered include a discussion of whether laser radiation should be analyzed as natural light or as a guided wave, the macroscopic differences and similarities between various types of lasers, special techniques, such as super-modes and the two-dimensional Green's function for planar waveguides, and some unusual analyses.

This clearly presented and concise text will be useful for first-year graduates in electrical engineering and physics. It also acts as a reference book on the mathematical and analytical techniques used to understand many opto-electronic applications.

WILLIAM S. C. CHANG is an Emeritus Professor of the Department of Electrical and Computer Engineering, University of California at San Diego. A pioneer of microwave laser and optical laser research, his recent research interests include electro-optical properties and guided wave devices in III–V semiconductor heterojunction and multiple quantum well structures, opto-electronics in fiber networks, and RF photonic links.

Professor Chang has published over 150 research papers on optical guided wave research and five books. His most recent book is *RF Photonic Technology in Optical Fiber Links* (Cambridge University Press, 2002).

PRINCIPLES OF LASERS AND OPTICS

WILLIAM S. C. CHANG

Professor Emeritus
Department of Electrical Engineering and Computer Science
University of California San Diego

CAMBRIDGE UNIVERSITY PRESS
Cambridge, New York, Melbourne, Madrid, Cape Town, Singapore, São Paulo

Cambridge University Press
The Edinburgh Building, Cambridge CB2 8RU, UK

Published in the United States of America by Cambridge University Press, New York

www.cambridge.org
Information on this title: www.cambridge.org/9780521642293

First published 2005
This digitally printed version 2007

A catalogue record for this publication is available from the British Library

Library of Congress Cataloguing in Publication data
Chang, William S. C. (William Shen-chie), 1931–
Principles of lasers and optics / William Shen Chie Chang.
p. cm.
Includes bibliographical references and index.
ISBN 0 521 64229 9 (alk. paper)
1. Lasers. 2. Photonics. 3. Optical wave guides. 4. Quantum optics. I. Title.
TK1675 C485 2005
621.36′6 – dc22 2004054604

ISBN 978-0-521-64229-3 hardback
ISBN 978-0-521-64535-5 paperback

Contents

Preface

When I look back at my time as a graduate student, I realize that the most valuable knowledge that I acquired concerned fundamental concepts in physics and mathematics, quantum mechanics and electromagnetic theory, with specific emphasis on their use in electronic and electro-optical devices. Today, many students acquire such information as well as analytical techniques from studies and analysis of the laser and its light in devices, components and systems. When teaching a graduate course at the University of California San Diego on this topic, I emphasize the understanding of basic principles of the laser and the properties of its radiation.

In this book I present a unified approach to all lasers, including gas, solid state and semiconductor lasers, in terms of "classical" devices, with gain and material susceptibility derived from their quantum mechanical interactions. For example, the properties of laser oscillators are derived from optical feedback analysis of different cavities. Moreover, since applications of laser radiation often involve its well defined phase and amplitude, the analysis of such radiation in components and systems requires special care in optical procedures as well as microwave techniques. In order to demonstrate the applications of these fundamental principles, analytical techniques and specific examples are presented. I used the notes for my course because I was unable to find a textbook that provided such a compact approach, although many excellent books are already available which provide comprehensive treatments of quantum electronics, lasers and optics. It is not the objective of this book to present a comprehensive treatment of properties of lasers and optical components.

Our experience indicates that such a course can be covered in two academic quarters, and perhaps might be suitable for one academic semester in an abbreviated form. Students will learn both fundamental physics principles and analytical techniques from the course. They can apply what they have learned immediately to applications such as optical communication and signal processing. Professionals may find the book useful as a reference to fundamental principles and analytical techniques.

1

Scalar wave equations and diffraction of laser radiation

1.1 Introduction

Radiation from lasers is different from conventional optical light because, like microwave radiation, it is approximately monochromatic. Although each laser has its own fine spectral distribution and noise properties, the electric and magnetic fields from lasers are considered to have precise phase and amplitude variations in the first-order approximation. Like microwaves, electromagnetic radiation with a precise phase and amplitude is described most accurately by Maxwell's wave equations. For analysis of optical fields in structures such as optical waveguides and single-mode fibers, Maxwell's vector wave equations with appropriate boundary conditions are used. Such analyses are important and necessary for applications in which we need to know the detailed characteristics of the vector fields known as the modes of these structures. They will be discussed in Chapters 3 and 4.

For devices with structures that have dimensions very much larger than the wavelength, e.g. in a multimode fiber or in an optical system consisting of lenses, prisms or mirrors, the rigorous analysis of Maxwell's vector wave equations becomes very complex and tedious: there are too many modes in such a large space. It is difficult to solve Maxwell's vector wave equations for such cases, even with large computers. Even if we find the solution, it would contain fine features (such as the fringe fields near the lens) which are often of little or no significance to practical applications. In these cases we look for a simple analysis which can give us just the main features (i.e. the amplitude and phase) of the dominant component of the electromagnetic field in directions close to the direction of propagation and at distances reasonably far away from the aperture.

When one deals with laser radiation fields which have slow transverse variations and which interact with devices that have overall dimensions much larger than the optical wavelength λ, the fields can often be approximated as transverse electric and magnetic (TEM) waves. In TEM waves both the dominant electric field and the dominant magnetic field polarization lie approximately in the plane perpendicular to the direction of propagation. The polarization direction does not change substantially within a propagation distance comparable to wavelength. For such waves, we usually need only to solve the scalar wave equations to obtain the amplitude and the phase of the dominant electric field along its local polarization direction. The dominant magnetic field can be calculated directly from the dominant electric field. Alternatively, we can first solve the scalar equation of the dominant magnetic field, and the electric field can be calculated from the magnetic field. We have encountered TEM waves in undergraduate electromagnetic field courses usually as plane waves that have no transverse amplitude and phase variations. For TEM waves in general, we need a more sophisticated analysis than plane wave analysis to account for the transverse variations. Phase information for TEM waves is especially important for laser radiation because many applications, such as spatial filtering, holography and wavelength selection by grating, depend critically on the phase information.

The details with which we normally describe the TEM waves can be divided into two categories, depending on application. (1) When we analyze how laser radiation is diffracted, deflected or reflected by gratings, holograms or optical components with finite apertures, we calculate the phase and amplitude variations of the dominant transverse electric field. Examples include the diffraction of laser radiation in optical instruments, signal processing using laser light, or modes of solid state or gas lasers. (2) When we are only interested in the propagation velocity and the path of the TEM waves, we describe and analyze the optical beams only by reference to the path of such optical rays. Examples include modal dispersion in multimode fibers and lidars. The analyses of ray optics are fairly simple; they are discussed in many optics books and articles [1, 2]. They are also known as geometrical optics. They will not be presented in this book.

We will first learn what is meant by a scalar wave equation in Section 1.2. In Section 1.3, we will learn mathematically how the solution of the scalar wave equation by Green's function leads to the well known Kirchhoff diffraction integral solution. The mathematical derivations in these sections are important not only in order to present rigorously the theoretical optical analyses but also to allow us to appreciate the approximations and limitations implied in various results. Further approximations of Kirchhoff's integral then lead to the classical Fresnel and Fraunhofer diffraction integrals. Applications of Kirchhoff's integral are illustrated in Section 1.4.

Fraunhofer diffraction from an aperture at the far field demonstrates the classical analysis of diffraction. Although the intensity of the diffracted field is the primary concern of many conventional optics applications, we will emphasize both the amplitude and the phase of the diffracted field that are important for many laser applications. For example, Fraunhofer diffraction and Fourier transform relations at the focal plane of a lens provide the theoretical basis of spatial filtering. Spatial filtering techniques are employed frequently in optical instruments, in optical computing and in signal processing.

Understanding the origin of the integral equations for laser resonators is crucial in allowing us to comprehend the origin and the limitation of the Gaussian mode description of lasers. In Section 1.5, we will illustrate several applications of transformation techniques of Gaussian beams based on Kirchhoff's diffraction integral, which is valid for TEM laser radiation.

Please note that the information given in Sections 1.2, 1.3 and 1.4 is also presented extensively in classical optics books [3, 4, 5]. Readers are referred to those books for many other applications.

1.2 The scalar wave equation

The simplest way to understand why we can use a scalar wave equation is to consider Maxwell's vector wave equation in a sourceless homogeneous medium. It can be written in terms of the rectangular coordinates as

$$\nabla^2 \underline{E} - \frac{1}{c^2}\frac{\partial^2 \underline{E}}{\partial t^2} = 0,$$

$$\underline{E} = E_x \underline{i_x} + E_y \underline{i_y} + E_z \underline{i_z},$$

where c is the velocity of light in the homogeneous medium. If $\underline{E}$ has only one dominant component $E_x \underline{i_x}$, then E_y, E_z, and the unit vector $\underline{i_x}$ can be dropped from the above equation. The resultant equation is a scalar wave equation for E_x.

In short, for TEM waves, we usually describe the dominant electromagnetic (EM) field by a scalar function U. In a homogeneous medium, U satisfies the scalar wave equation

$$\nabla^2 U - \frac{1}{c^2}\frac{\partial^2}{\partial t^2} U = 0. \qquad (1.1)$$

In an elementary view, U is the instantaneous amplitude of the transverse electric field in its direction of polarization when the polarization is approximately constant (i.e. $|U|$ varies slowly within a distance comparable to the wavelength). From a different point of view, when we use the scalar wave equation, we have implicitly assumed that the curl equations in Maxwell's equations do not yield a sufficient magnitude of electric field components in other directions that will affect significantly the TEM characteristics of the field. The magnetic field is calculated directly from the dominant electric field. In books such as that by Born and Wolf [3], it is shown that U can also be considered as a scalar potential for the optical field. In that case, electric and magnetic fields can be derived from the scalar potential.

Both the scalar wave equation in Eq. (1.1) and the boundary conditions are derived from Maxwell's equations. The boundary conditions (i.e. the continuity of tangential electric and magnetic fields across the boundary) are replaced by boundary conditions of U (i.e. the continuity of U and normal derivative of U across the boundary). Notice that the only limitation imposed so far by this approach is that we can find the solution for the EM fields by just one electric field component (i.e. the scalar U). We will present further simplifications on how to solve Eq. (1.1) in Section 1.3.

For wave propagation in a complex environment, Eq. (1.1) can be considered as the equation for propagation of TEM waves in the local region when TEM approximation is acceptable. In order to obtain a global analysis of wave propagation in a complex environment, solutions obtained for adjacent local regions are then matched in both spatial and time variations at the boundary between adjacent local regions.

For monochromatic radiation with a harmonic time variation, we usually write

$$U(x, y, z; t) = U(x, y, z)e^{j\omega t}. \tag{1.2}$$

Here, $U(x, y, z)$ is complex, i.e. U has both amplitude and phase. Then U satisfies the Helmholtz equation,

$$\nabla^2 U + k^2 U = 0, \tag{1.3}$$

where $k = \omega/c = 2\pi/\lambda$ and c = free space velocity of light $= 1/\sqrt{\varepsilon_0 \mu_0}$. The boundary conditions are the continuity of U and the normal derivative of U across the dielectric discontinuity boundary.

In this section, we have defined the equation governing U and discussed the approximations involved when we use it. In the first two chapters of this book, we will accept the scalar wave equation and learn how to solve for U in various applications of laser radiation.

We could always solve for U for each individual case as a boundary value problem. This would be the case when we solve the equation by numerical methods. However, we would also like to have an analytical expression for U in a homogeneous medium when its value is known at some boundary surface. The well known method used to obtain U in terms of its known value on some boundary is the Green's function method, which is derived and discussed in Section 1.3.

1.3 The solution of the scalar wave equation by Green's function – Kirchhoff's diffraction formula

Green's function is nothing more than a mathematical technique which facilitates the calculation of U at a given position in terms of the fields known at some remote boundary without explicitly solving the differential Eq. (1.4) for each individual case [3, 6]. In this section, we will learn how to do this mathematically. In the process we will learn the limitations and the approximations involved in such a method.

Let there be a Green's function G such that G is the solution of the equation

$$\nabla^2 G(x, y, z; x_0, y_0, z_0) + k^2 G = -\,\delta(x - x_0, y - y_0, z - z_0) = -\,\delta(\underline{r} - \underline{r_0}). \tag{1.4}$$

Equation (1.4) is identical to Eq. (1.3) except for the δ function. The boundary conditions for G are the same as those for U; δ is a unit impulse function which is zero when $x \neq x_0$, $y \neq y_0$ and $z \neq z_0$. It goes to infinity when (x, y, z) approaches the discontinuity point (x_0, y_0, z_0), and δ satisfies the normalization condition

$$\iiint_V \delta(x - x_0, y - y_0, z - z_0)\, dx\, dy\, dz = 1 = \iiint_V \delta(\underline{r} - \underline{r_0})\, dv, \tag{1.5}$$

where $\underline{r} = x\underline{i_x} + y\underline{i_y} + z\underline{i_z}$, $\underline{r_0} = x_0\underline{i_x} + y_0\underline{i_y} + z_0\underline{i_z}$ and $dv = dx\, dy\, dz = r^2 \sin\theta\, dr\, d\theta\, d\phi$. V is any volume including the point $(x_0,\ y_0,\ z_0)$. First we will show how a solution for G of Eq. (1.4) will let us find U at any given observer position $(x_0,\ y_0,\ z_0)$ from the U known at some distant boundary.

From advanced calculus [7],

$$\nabla \cdot (G\nabla U - U\nabla G) = G\nabla^2 U - U\nabla^2 G.$$

Applying a volume integral to both sides of the above equation and utilizing Eqs. (1.4) and (1.5), we obtain

$$\iiint_V \nabla \cdot (G\nabla U - U\nabla G)\, dv = \iint_S (G\underline{n} \cdot \nabla U - U\underline{n} \cdot \nabla G)\, ds = \iiint_V \left[-k^2 GU + k^2 UG + U\delta(\underline{r} - \underline{r_0})\right] dv = U(\underline{r_0}). \tag{1.6}$$

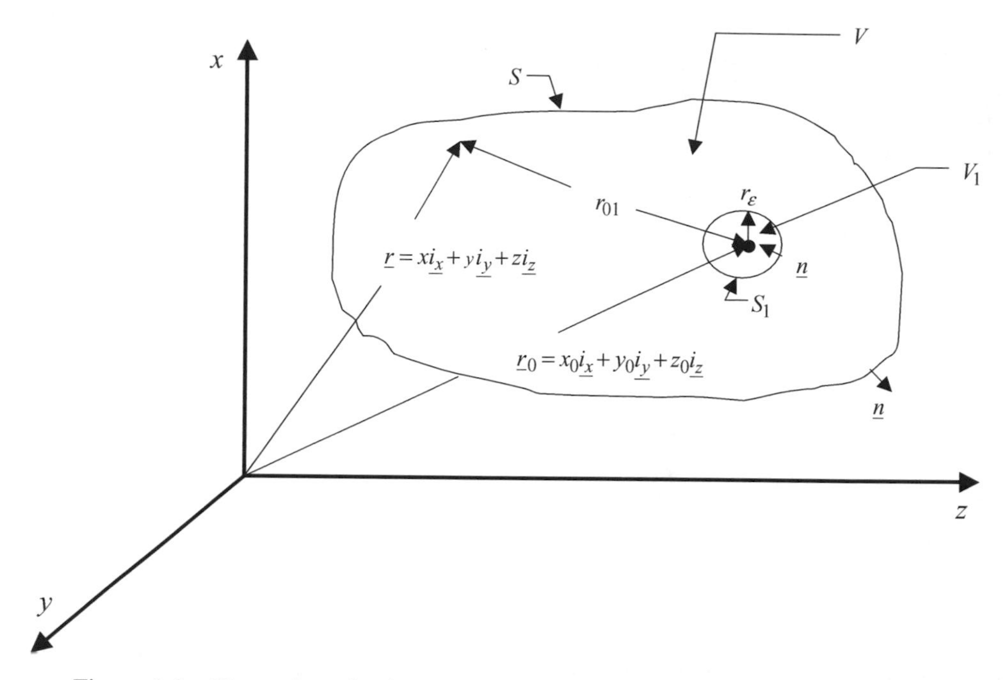

Figure 1.1. Illustration of volumes and surfaces to which Green's theory applies. The volume to which Green's function applies is V', which has a surface S. The outward unit vector of S is $\underline{n}$; $\underline{r}$ is any point in the x, y, z space. The observation point within V is $\underline{r_0}$. For the volume V', V_1 around $\underline{r_0}$ is subtracted from V. V_1 has surface S_1, and the unit vector $\underline{n}$ is pointed outward from V'.

V is any closed volume (within a boundary S) enclosing the observation point $\underline{r_0}$ and n is the unit vector perpendicular to the boundary in the outward direction, as illustrated in Fig. 1.1.

Equation (1.6) is an important mathematical result. It shows that, when G is known, the U at position (x_0, y_0, z_0) can be expressed directly in terms of the values of U and ∇U on the boundary S, without solving explicitly the Helmholtz equation, Eq. (1.3). Equation (1.6) is known mathematically as Green's identity. The key problem is how to find G.

Fortunately, G is well known in some special cases that are important in many applications. We will present three cases of G in the following.

1.3.1 The general Green's function G

The general Green's function G has been derived in many classical optics textbooks; see, for example, [3]:

$$G = \frac{1}{4\pi} \frac{\exp(-jkr_{01})}{r_{01}}, \tag{1.7}$$

where

$$r_{01} = |\underline{r_0} - \underline{r}| = \sqrt{(x - x_0)^2 + (y - y_0)^2 + (z - z_0)^2}.$$

As shown in Fig. 1.1, r_{01} is the distance between $\underline{r_0}$ and $\underline{r}$.

This G can be shown to satisfy Eq. (1.4) in two steps.

(1) By direct differentiation, $\nabla^2 G + k^2 G$ is clearly zero everywhere in any homogeneous medium except at $\underline{r} \approx \underline{r_0}$. Therefore, Eq. (1.4) is satisfied within the volume V', which is V minus V_1 (with boundary S_1) of a small sphere with radius r_ε enclosing $\underline{r_0}$ in the limit as r_ε approaches zero. V_1 and S_1 are also illustrated in Fig. 1.1.

(2) In order to find out the behavior of G near $\underline{r_0}$, we note that $|G| \to \infty$ as $r_{01} \to 0$. If we perform the volume integration of the left hand side of Eq. (1.4) over the volume V_1, we obtain:

$$\lim_{r_\varepsilon \to 0} \iiint\limits_{V_1} [\nabla \cdot \nabla G + k^2 G]\, dv = \iint\limits_{S_1} \nabla G \cdot \underline{n}\, ds$$

$$= \lim_{r_\varepsilon \to 0} \int_0^{2\pi} \int_{-\pi/2}^{\pi/2} \left[-\frac{e^{-jkr_\varepsilon}}{4\pi\, r_\varepsilon^2} \right] r_\varepsilon^2 \sin\theta\, d\theta\, d\phi = -1.$$

Thus, using this Green's function, the volume integration of the left hand side of Eq. (1.4) yields the same result as the volume integration of the δ function. In short, the G given in Eq. (1.7) satisfies Eq. (1.4) for any homogeneous medium.

From Eq. (1.6) and G, we obtain the well known Kirchhoff diffraction formula,

$$U(\underline{r_0}) = \iint\limits_{S} (G \nabla U - U \nabla G) \cdot \underline{n}\, ds. \tag{1.8}$$

Note that we need only to know both U and ∇U on the boundary in order to calculate its value at $\underline{r_0}$ inside the boundary.

1.3.2 Green's function, G_1, for U known on a planar aperture

For many practical applications, U is known on a planar aperture, followed by a homogeneous medium with no additional radiation source. Let the planar aperture be the surface $z = 0$; a known radiation U is incident on the aperture Ω from $z < 0$, and the observation point is located at $z > 0$. As a mathematical approximation to this geometry, we define V to be the semi-infinite space at $z \geq 0$, bounded by the surface S. S consists of the plane $z = 0$ on the left and a large spherical surface with radius R on the right, as $R \to \infty$. Figure 1.2 illustrates the semi-sphere.

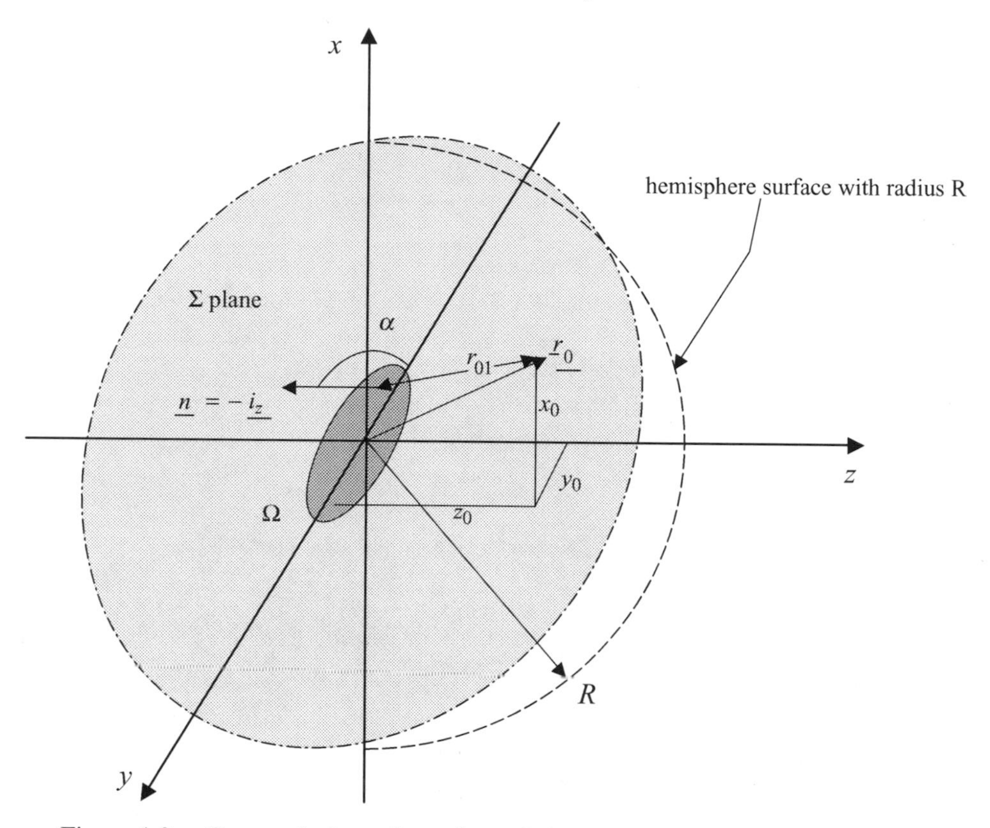

Figure 1.2. Geometrical configuration of the semi-spherical volume for the Green's function G_1. The surface to which the Green's function applies consists of Σ, which is part of the *xy* plane, and a very large hemisphere that has a radius *R*, connected with Σ. The incident radiation is incident on Ω, which is an open aperture within Σ. The outward normal of the surfaces Σ and Ω is $-i_z$. The coordinates for the observation point r_0 are x_0, y_0 and z_0.

The boundary condition for a sourceless U at $z > 0$ is given by the radiation condition at very large R; as $R \to \infty$ [8],

$$\lim_{R\to\infty} R\left(\frac{\partial U}{\partial n} + jkU\right) = 0. \tag{1.9}$$

The radiation condition is essentially a mathematical statement that there is no incoming wave at very large R. Any U which represents an outgoing wave in the $z > 0$ space will satisfy Eq. (1.9).

If we do not want to use the ∇U term in Eq. (1.8), we like to have a Green's function which is zero on the plane boundary (i.e. $z = 0$). Since we want to apply Eq. (1.8) to the semi-sphere boundary S, Eq. (1.4) needs to be satisfied only for $z > 0$. In order to find such a Green's function, we note first that any function F in the form $\exp(-jkr)/r$, expressed in Eq. (1.7), will satisfy $\lfloor \nabla F + k^2 F = 0 \rfloor$ as

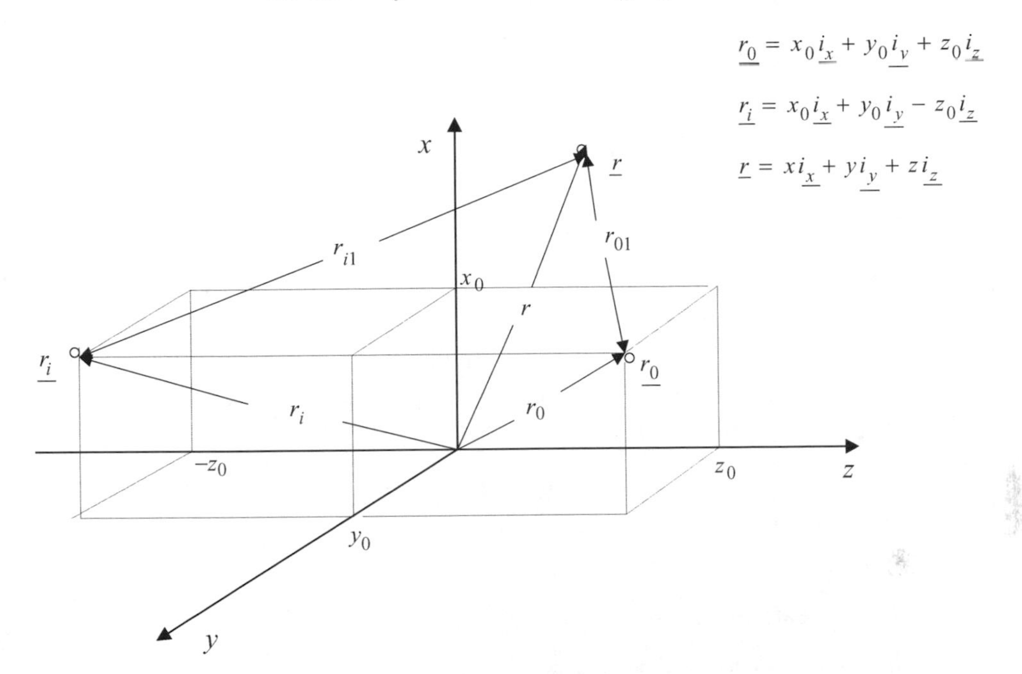

Figure 1.3. Illustration of $\underline{r}$, the point of observation $\underline{r_0}$ and its image $\underline{r_j}$, in the method of images. For G, the image plane Σ is the xy plane, and $\underline{r_i}$ is the image of the observation point $\underline{r_0}$ in Σ. The coordinates of $\underline{r_0}$ and $\underline{r_i}$ are given.

long as r is not allowed to approach zero. We can add such a second term to the G given in Eq. (1.7) and still satisfy Eq. (1.4) for $z > 0$ as long as r never approaches zero for $z > 0$. To be more specific, let $\underline{r_i}$ be a mirror image of (x_0, y_0, z_0) across the $z = 0$ plane at $z < 0$. Let the second term be $e^{-jkr_{i1}}/r_{i1}$, where r_{i1} is the distance between (x, y, z) and $\underline{r_i}$. Since our Green's function will only be used for $z_0 > 0$, the r_{i1} for this second term will never approach zero for $z \geq 0$. Thus, as long as we seek the solution of U in the space $z > 0$, Eq. (1.4) is satisfied for $z > 0$. However, the difference of the two terms is zero when (x, y, z) is on the $z = 0$ plane. This is known as the "method of images" in electromagnetic theory. Such a Green's function is constructed mathematically in the following.

Let the Green's function for this configuration be designated as G_1, where

$$G_1 = \frac{1}{4\pi}\left[\frac{e^{-jkr_{01}}}{r_{01}} - \frac{e^{-jkr_{i1}}}{r_{i1}}\right], \tag{1.10}$$

where $\underline{r_i}$ is the image of $\underline{r_0}$ in the $z = 0$ plane. It is located at $z < 0$, as shown in Fig. 1.3. G_1 is zero on the xy plane at $z = 0$. When G_1 is used in the Green's identity,

Eq. (1.8), we obtain

$$U(\underline{r_0}) = \iint_{\Omega} U(x, y, z=0)\,\frac{\partial G_1}{\partial z}\,dx\,dy. \tag{1.11}$$

Here, Ω refers to the xy plane at $z = 0$. Because of the radiation condition expressed in Eq. (1.9), the value of the surface integral over the very large semi-sphere enclosing the $z > 0$ volume (with $R \to \infty$) is zero.

For most applications, $U \neq 0$ only in a small sub-area of Σ, e.g. the radiation U is incident on an opaque screen that has a limited open aperture Ω. In that case, $-\partial G_1/\partial z$ at $z_0 \gg \lambda$ can be simplified to obtain

$$-\nabla G_1 \cdot \underline{i_z} = 2\cos\alpha\,\frac{e^{-jkr_{01}}}{4\pi\, r_{01}}\,(-jk),$$

where α is as illustrated in Fig. 1.2. Therefore, the simplified expression for U is

$$U(\underline{r_0}) = \frac{j}{\lambda}\iint_{\Omega} U\frac{e^{-jkr_{01}}}{r_{01}}\cos\alpha\,dx\,dy. \tag{1.12}$$

This result has also been derived from the Huygens principle in classical optics.

Let us now define the paraxial approximation for the observer at position (x_0, y_0, z_0) in a direction close to the direction of propagation and at a distance reasonably far from the aperture, i.e. $\alpha \approx 180^\circ$ and $|r_{01}| \approx |z| \approx \rho$. Then, for observers in the paraxial approximation, α is now approximately a constant in the integrand of Eq. (1.12) over the entire aperture Ω, while the change of ρ in the denominator of the integrand also varies very slowly over the entire Ω. Thus, U can now be simplified further to yield

$$U(z \approx \rho) = \frac{-j}{\lambda\rho}\iint_{\Omega} U e^{-jkr_{01}}\,dx\,dy. \tag{1.13}$$

Note that $k = 2\pi/\lambda$ and ρ/λ is a very large quantity. A small change of r_{01} in the exponential can affect significantly the value of the integral, while the ρ factor in the denominator of the integrand can be considered as a constant in the paraxial approximation.

Both Eqs. (1.8) and (1.13) are known as Kirchhoff's diffraction formula [3]. In the case of paraxial approximation, limited aperture and $z \gg \lambda$, Eq. (1.8) yields the same result as Eq. (1.13). However, Eq. (1.13) is more commonly used in engineering.

1.3.3 Green's function for ∇U known on a planar aperture, G_2

The Green's function for calculating $U(x_0, y_0, z_0)$ from just the derivative of U on the plane aperture is also known. In this case,

$$G_2 = \frac{1}{4\pi}\left[\frac{e^{-jkr_{01}}}{r_{01}} + \frac{e^{-jkr_{i1}}}{r_{i1}}\right]. \tag{1.14}$$

Clearly, $\partial G_2/\partial z$ is zero on the $z = 0$ plane. According to Eq. (1.8), the value of U calculated from G_2 now depends only on ∇U on the boundary, i.e. the $z = 0$ plane. However, this Green's function is seldom used. Therefore, we will not discuss it further.

It is most important to note that, in principle, if we substitute the true U and ∇U into any one of the integrals using G, G_1 or G_2, we should get the same answer. However, we do not know the true U and ∇U because we have not yet solved Eq. (1.3). For Eqs. (1.12) and (1.13), it is customary to use just the incident U in optics without considering the electromagnetic effects involving the aperture Ω. For example, when we used the incident radiation U as the U in the aperture, we ignored the induced currents near the edge of the aperture. This is an approximation. In this case, we will obtain the same result from the three different Green's functions only in the paraxial approximation, i.e. for $z \gg \lambda$, for an observer located at a relatively small angle from the z axis and for a limited aperture size Ω. In the paraxial approximation, no information concerning the fringe field at small z values or at large angles of observation can be obtained. See ref. [9] for a more detailed discussion.

1.3.4 The expression for Kirchhoff's integral in engineering analysis

Equation (1.13) is usually presented in a different format for engineers. Let

$$\frac{-j}{\lambda\rho}\exp(-jkr_{01}) = h\left[(x - x_0), (y - y_0), (z - z_0)\right].$$

Then we obtain

$$U(\underline{r_0}) = \iint_{\Omega} U(x, y, 0)h\left[(x - x_0), (y - y_0), -z_0\right] dx\, dy. \tag{1.15}$$

This is the well known transform relation between the U's at two different planes, $z = 0$ and $z = z_0$. The expression $h[(x - x_0), (y - y_0), (z - z_0)]$ now has the same format as the electrical impulse response in electrical circuit and system analysis.

This implies that the h function is the response of the optical system for any unit impulse excitation at (x, y, $z = 0$). $U(x, y, 0)$ is just the excitation at the $z = 0$ plane. For large Ω, h essentially determines completely the $U(x_0, y_0, z_0)$ from $U(x, y, 0)$. For small Ω, the position and shape of Ω are also important. Equation (1.15) is the foundation of many pattern recognition, optical computing and optical signal processing techniques. Many theoretical techniques, such as superposition, convolution theory, sampling theory, spatial filtering method and spatial Fourier transform, can be applied to Eq. (1.15). However, strictly speaking, mathematical techniques used in electrical engineering circuit and system analyses usually apply only to integrals with $-\infty$ and $+\infty$ limits of integration, while the limits of integration in Eq. (1.15) are determined by the aperture size and position. Nevertheless, much can be learned from those techniques, especially when the aperture is large. Furthermore, the integral in Eq. (1.15) can also be regarded as a unit impulse integral of the product of the $U(x, y, 0)$ and a unit step function of x and y representing Ω, with limits of integration at ∞. In Sections 1.4 and 1.5, we will discuss some examples of these techniques. See ref. [10] for many examples illustrating the importance of this transform relation.

1.3.5 Fresnel and Fraunhofer diffraction

Before applying Eq. (1.13) or Eq. (1.15), we note that the binomial expansion may be applied to simplify ρ further:

$$\begin{aligned}\rho &= (z_0 - z)\sqrt{1 + \frac{(x_0 - x)^2 + (y_0 - y)^2}{(z_0 - z)^2}} \\ &= d\left[1 + \frac{1}{2d^2}\left(x_0^2 + y_0^2 - 2xx_0 - 2yy_0 + x^2 + y^2\right) + \text{higher order terms}\right].\end{aligned} \tag{1.16}$$

Here, $d = z_0 - z$, and, in the paraxial approximation, $d \gg |x_0 - x|$ and $|y_0 - y|$.

If d is sufficiently large that we can drop the higher order terms, we obtain from Eq. (1.13) the following:

$$U(\underline{r_0}) = \frac{-j}{\lambda d} e^{-jkd} e^{-jk\frac{x_0^2+y_0^2}{2d}} \iint_{\Omega} \left[U(z=0)\, e^{-j2\pi\frac{x^2+y^2}{2\lambda d}}\right] e^{+j2\pi\frac{xx_0}{\lambda d}} e^{+j2\pi\frac{yy_0}{\lambda d}}\, dx\, dy. \tag{1.17a}$$

This is known as the Fresnel diffraction integral, which describes near field diffraction effects.

If d is so large that the term involving $(x^2 + y^2)$ can also be neglected, then we obtain an even simpler diffraction integral,

$$U(\underline{r_0}) = \frac{-j}{\lambda d} e^{-jkd} e^{-jk\frac{x_0^2+y_0^2}{2d}} \iint_{\Omega} U(z=0)\, e^{+j2\pi\frac{xx_0}{\lambda d}} e^{+j2\pi\frac{yy_0}{\lambda d}}\, dx\, dy. \quad (1.17b)$$

This is known as the Fraunhofer diffraction integral of the far radiation field. Note that the U as a function of x_0 and y_0 is approximately a Fourier transform of U as a function of x and y.

So far, we have presented primarily the mathematical derivations to obtain the results given in Eqs. (1.13), (1.15) and (1.17). We have learned two things from these derivations: (1) the approximations employed in Fresnel and Fraunhofer diffraction formulae; (2) the significance of the radiation condition and the paraxial approximation involved. Whenever the field can be analyzed by the scalar wave equation, and whenever the limitations and approximations used in Eqs. (1.17a) and (1.17b) are acceptable, Kirchhoff's formula can be used to solve practical problems, as demonstrated in Section 1.4.

1.4 Applications of the analysis of TEM waves

Equations (1.15), (1.17a) and (1.17b) are used in many applications of laser radiation. Examples include holography, addressing of laser radiation by diffraction, micro-optics, wavelength selection by grating diffraction, optical signal processing, computing, etc. To discuss all these applications is beyond the scope of this book. Extensive discussions on holography, transformation optics, grating, interference, etc., are already available in other books [10, 11, 12]. However, we will not understand clearly the significance of the analyses of TEM waves, including Eqs. (1.13), (1.15), (1.17a) and (1.17b), without demonstrating some practical applications. Therefore, in this section we will present four applications of the analysis of TEM waves. The applications will not only illustrate the significance of this method, but will also lead to results which are basic to diffraction, transformation optics and laser modes that will be discussed in later chapters.

1.4.1 Far field diffraction pattern of an aperture

Far field diffraction from radiation U incident on a rectangular aperture is the simplest example to illustrate the power of Eqs. (1.17a) and (1.17b). The result is

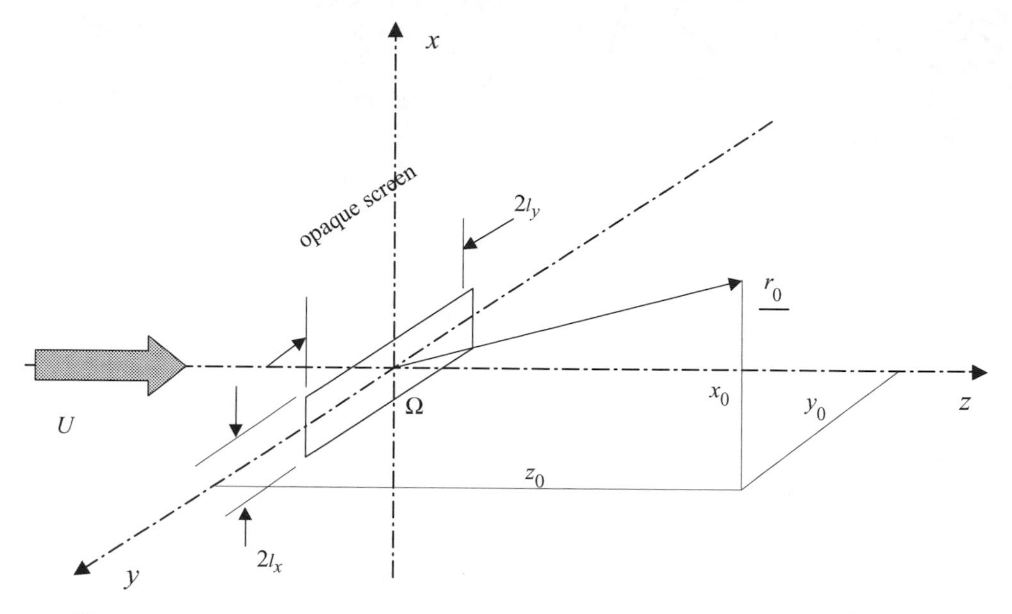

Figure 1.4. Geometrical configuration of a rectangular aperture. The radiation U is incident on a rectangular aperture Ω on an opaque screen, which is the xy plane. The size of Ω is $2l_x \times 2l_y$. For a far field, r_0 is far away with large z_0 coordinate. In the paraxial approximation, $|z_0| \gg |x_0|$ and $|y_0|$.

also very useful for subsequent discussions. Let the radiation U be a plane wave with amplitude A normally incident on an opaque screen at $z = 0$ with a rectangular open aperture with dimensions $2l_x$ and $2l_y$ in the x and y directions, respectively; i.e.

$$U(x, y, z = 0) = A \operatorname{rect}\left(\frac{x}{l_x}\right) \operatorname{rect}\left(\frac{y}{l_y}\right),$$

where

$$\begin{aligned} \operatorname{rect}(\chi) &= 0 \quad \text{for} \quad |\chi| > 1, \\ \operatorname{rect}(\chi) &= 1 \quad \text{for} \quad |\chi| \leq 1. \end{aligned}$$

Figure 1.4 illustrates the geometric configuration. Substituting into Eq. (1.17a), we obtain

$$\begin{aligned} U(x_0, y_0, d) = &\frac{-je^{-jkd}e^{-j\frac{k}{2d}(x_0^2+y_0^2)}}{\lambda d} \\ &\times \iint_{\Omega} \left[Ae^{-jk\frac{x^2+y^2}{2d}}\right] e^{j2\pi\left(\frac{x_0}{\lambda d}\right)x} e^{j2\pi\left(\frac{y_0}{\lambda d}\right)y} dx\, dy. \end{aligned}$$

In the far field case, d is very large, such that

$$k\frac{l_x^2 + l_y^2}{2d} \ll 1, \tag{1.18}$$

and then

$$\exp\left(-jk\frac{x^2 + y^2}{2d}\right) \approx 1.$$

Since

$$\int_{-l_x}^{l_x} e^{j2\pi\left(\frac{x_0}{\lambda d}\right)x}\, dx = \frac{e^{j2\pi\left(\frac{x_0}{\lambda d}\right)l_x} - e^{-j2\pi\left(\frac{x_0}{\lambda d}\right)l_x}}{j\,2\pi\left(\frac{x_0}{\lambda d}\right)} = \frac{j\,2\,\sin\left[2\pi\left(\frac{x_0}{\lambda d}\right)l_x\right]}{j\,2\pi\left(\frac{x_0}{\lambda d}\right)},$$

we obtain the far field U from Eq. (1.17b) as

$$U(x_0, y_0, d) = \frac{4je^{-jkd}e^{-j\frac{k}{2d}(x_0^2+y_0^2)}}{\lambda d} A\, l_x l_y \,\mathrm{sinc}\left(\frac{2l_x x_0}{\lambda d}\right) \mathrm{sinc}\left(\frac{2l_y y_0}{\lambda d}\right), \tag{1.19}$$

where

$$\mathrm{sinc}(\alpha) = \frac{\sin \pi\alpha}{\pi\alpha}.$$

U is the classical Fraunhofer diffraction pattern of the rectangular aperture for a plane wave normally incident on the aperture. Two additional comments are important to note. (1) The Fraunhofer diffraction pattern is ignored in geometric or ray optics because the transverse amplitude and phase variations are ignored. This would be the case in which one is interested only in the U as x_0/d and $y_0/d \to 0$ in Eq. (1.19). (2) The U at the far field has a spherical phase front about the z axis. This is not important for most classical optic applications. Unlike microwaves, the electric field cannot be detected directly in optics. Detectors and films measure the intensity of the radiation. This is the reason why the phase information is not emphasized in conventional optics. However, the phase information becomes very important for a number of laser applications that involve wavelength selection, signal processing, interference and diffraction. The effect of the phase of U is detected from its interference with another U' or from diffraction effects. As a reminder of the importance of phase, we recall that when the laser radiation is used to illuminate an image pattern, there are many speckles created by interference effects due to small irregularities. This is the primary reason why laser light is not used for imaging.

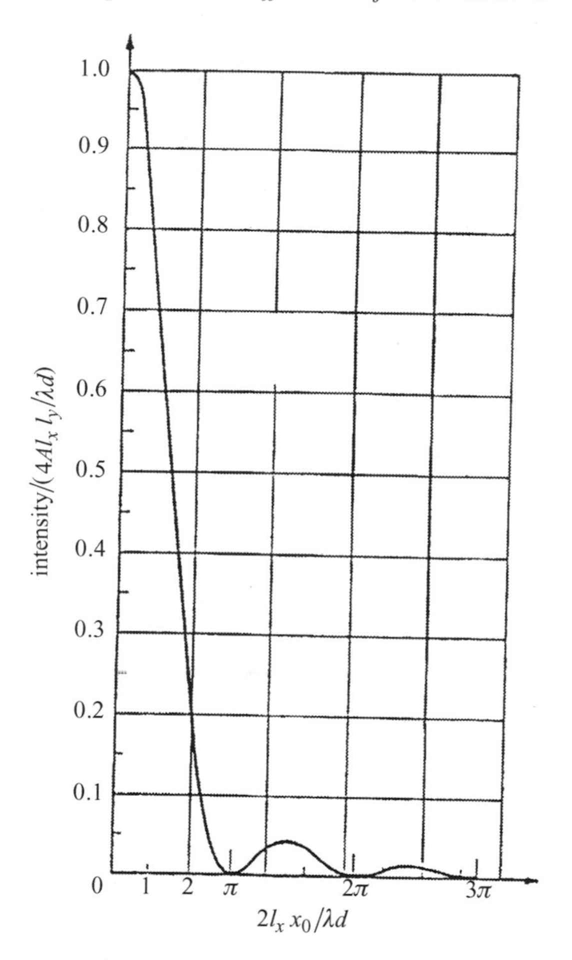

Figure 1.5. The intensity distribution of the Fraunhofer diffraction pattern of a rectangular aperture for a normally incident plane wave. Taken from ref. [3].

All detectors convert the optical power into electrical current. In electromagnetic field theory, we learn that $I = (1/2)\,|E|^2 / \sqrt{\mu_0/\varepsilon}$, where E is the transverse electric field. In optics, U is usually normalized (i.e. $|U|$ is just proportional to the magnitude of the transverse electric field) such that UU^* is the intensity. Thus, the intensity I at x_0 and y_0 is

$$I(x_0, y_0) = UU^* = \left[\frac{4Al_x l_y}{\lambda d}\,\mathrm{sinc}\left(\frac{2l_x x_0}{\lambda d}\right)\mathrm{sinc}\left(\frac{2l_y y_0}{\lambda d}\right)\right]^2. \tag{1.20}$$

Figure 1.5 illustrates the intensity I as a function of x_0 when y_0 is zero. Clearly, I is inversely proportional to d^2, as expected in a divergent wave, and it has a major radiation loop directed along the z axis. In optics, the minimum diffraction beam-width of the major loop is defined as the angle θ between the direction of propagation (i.e. the z axis) and the first zero of I. Thus, for a rectangular aperture,

$$\theta_x = \lambda/2\,l_x \tag{1.21a}$$

and

$$\theta_y = \lambda/2\,l_y. \tag{1.21b}$$

I also has minor radiation loops in directions $x_0/d = (3/2)\lambda/l_x$, $(5/2)\lambda/l_x$, etc., and in directions $y_0/d = (3/2)\lambda/l_y$, $(5/2)\lambda/l_y$, etc.

The preceding discussion clearly demonstrates the characteristics of the diffracted far field without complex mathematics. This is why we chose to present the diffraction from a rectangular aperture as an example. Similar results have been described for circular apertures with radius r' in classical optics books using cylindrical coordinates and Bessel functions [13, 14]. In that case the beam-width of the main radiation loop is given by [3]

$$\theta_c = 0.62\lambda/r'. \tag{1.22}$$

As a result, Eq. (1.22) is commonly used to specify the angular resolution of a telescope.

The Fraunhofer (or far field) diffraction pattern is a favorable way to describe the output radiation from many instruments. For example, (1) the output from a laser is frequently described in trade brochures by its far field radiation pattern; and (2) for communication among distant stations, the far field pattern is used to specify the angular resolution obtained through a telescope. The difference in the results obtained from the circular or the rectangular aperture is minor. However, we should be very careful about using the far field radiation formula because of the condition specified in Eq. (1.18). For an aperture 1 mm wide and at 1 μm wavelength, or laser output with 1 mm lateral mode size, the Fraunhofer diffraction formula is not strictly valid until the distance of observation is 30 m or more. Such distances are often not available in laboratories. Most often what we observe is the effect of Fresnel diffraction, which is more tedious to calculate. We should again keep in mind that, even in the far field, the laser radiation field has a spherical wave front.

It is also interesting to note that when a plane wave (microwave) is incident on a metal screen with a rectangular opening, the solution of the Maxwell equation for that problem will be the precise solution of the diffraction problem that we have just solved. However, we will include the radiation field caused by the induced

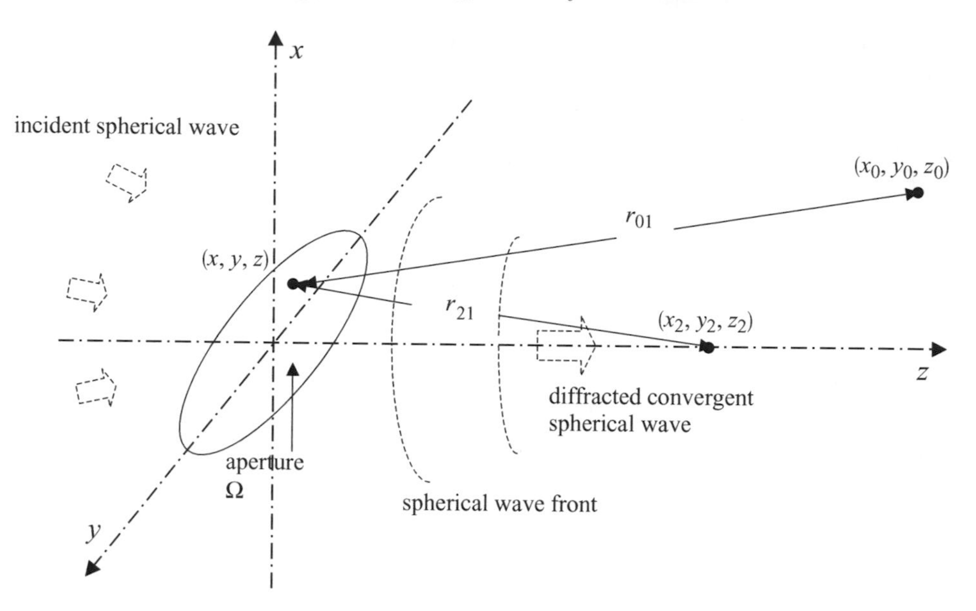

Figure 1.6. Illustration of a spherical wave incident on a plane aperture. The incident wave is a converging spherical wave focused at (x_2, y_2, z_2). It passes through an opening aperture Ω of an opaque screen, which is the xy plane.

current on the edges of the opening. When the opening is small or comparable to the wavelength, the radiation from the induced current is an important part of the total radiation. When the opening is large, the radiation field contribution from the induced current is small for the far field near the axis. From the mathematical point of view, as long as the radiation field can be approximated by a TEM wave, there is nothing wrong with the results expressed in Eq. (1.8) or Eq. (1.13) for far fields near the axis. If the induced current must be taken into account, we simply cannot assume that U in these equations is just the incident U. However, even for large openings, the contribution from the induced current may still be important at distances close to the opening.

1.4.2 Fraunhofer diffraction in the focal plane of a lens

A Fraunhofer diffraction pattern is also obtained near the focus of a lens. This is an important case to study since it is much easier to capture the Fraunhofer diffraction pattern at the focal plane of a lens than at distances far away. Furthermore, the Fourier transform relationship between the incident field and the field at the focal plane will allow us to perform many signal processing functions, such as spatial filtering [15].

Consider the case where the incident wave on the plane aperture is a convergent wave, as shown in Fig. 1.6. A convergent wave is normally produced from a plane wave by a lens in front of the aperture. Let the focal point of the convergent wave be (x_2, y_2, z_2), then the U without any aperture can be expressed for $z_2 \geq z$ as

$$U = A\frac{e^{+jkr_{21}}}{r_{21}}, \tag{1.23}$$

where $r_{21} = |\underline{r_2} - \underline{r}|$. The focal length of the lens is z_2. Note that the + sign in the exponential combined with the $\exp(+j\omega t)$ time variation represents a convergent wave. When an aperture is placed at $z = 0$, Eq. (1.13) can be used to calculate U at any point (x_0, y_0, z_0) for $z_0 > 0$. In this case, the incident U is given by Eq. (1.23) in the aperture. In other words,

$$U(x_0, y_0, z_0) = \frac{-jA}{\lambda z_0 z_2}\iint_{\Omega} e^{-jk(r_{01} - r_{21})}\, dx\, dy.$$

Using paraxial approximation and the binomial expansion, we obtain:

$$\begin{aligned} k(r_{01} - r_{21}) = k\,[z_0 - z_2] + \frac{k}{2}\left[\frac{x_0^2 + y_0^2}{z_0} - \frac{x_2^2 + y_2^2}{z_2}\right. \\ -k\left[\frac{x_0 x + y_0 y}{z_0} - \frac{x_2 x + y_2 y}{z_2}\right] + \frac{k}{2}\left[\frac{x^2 + y^2}{z_0} - \frac{x^2 + y^2}{z_2}\right. \\ +\left\{-\frac{1}{4}\frac{[(x_0 - x)^2 + (y_0 - y)^2]^2}{z_0^3} + \frac{1}{4}\frac{[(x_2 - x)^2 + (y_2 - y)^2]^2}{z_2^3}\right\} \\ \left. + \text{ other higher order terms}\right]. \end{aligned}$$

Under the conditions

$$\left|\frac{k}{8}\frac{\left[(x_0 - x)^2 + (y_0 - y)^2\right]^2}{z_0^3}\right|_{\max} \ll 2\pi$$

and

$$\left|\frac{k}{8}\frac{\left[(x_2 - x)^2 + (y_2 - y)^2\right]^2}{z_2^3}\right|_{\max} \ll 2\pi,$$

the terms in { } and other higher order terms can be neglected. Let us consider the case $x_2 = 0$, $y_2 = 0$, and the higher order terms and the { } are negligible. Then,

$$\begin{aligned} U(x_0, y_0, z_0) = {} & \frac{-jAe^{-jk(z_0 - z_2)}e^{-jk\left(\frac{x_0^2 + y_0^2}{2z_0}\right)}}{\lambda\, z_2 z_0} \\ & \times \iint_{\Omega} e^{j\frac{\pi}{\lambda}\left[\frac{1}{z_0} - \frac{1}{z_2}\right](x^2 + y^2)} e^{j2\pi\left[\frac{x_0}{\lambda z_0}\right]x} e^{j2\pi\left[\frac{y_0}{\lambda z_0}\right]y}\, dx\, dy. \end{aligned} \tag{1.24}$$

When $z_0 = z_2$, the [ ] factor involving $(1/z_0) - (1/z_2)$ in the above integral is zero. Therefore, the radiation in the focal plane of the lens is a Fourier transform with the limits of integration given by the aperture Ω.

Four conclusions can be drawn from this result.

(1) Except for the constant A/z_2, Eq. (1.24) is the same result as that obtained for Fraunhofer diffraction in the far field expressed in Eq. (1.17b).

(2) Let there be a thin plane transparent film placed before the aperture at $z = 0$. The amplitude and phase transmission of the film is given by $t(x, y)$. Then,

$$U(x_0, y_0, z_0) = \frac{-jAe^{-j2\pi\left(\frac{x_0^2+y_0^2}{2\lambda z_0}\right)}}{\lambda z_0^2} \iint_\Omega t(x, y)e^{j2\pi\left(\frac{x_0}{\lambda z_0}\right)x} e^{j2\pi\left(\frac{y_0}{\lambda z_0}\right)y}\, dx\, dy. \quad (1.25)$$

This is an important result. It states that when the limit of integration is large, the U at the focal plane $z_0 = z_2$ is essentially the Fourier transform of t at $z = 0$.

(3) The convergent wave is usually created by a lens with focal length z_2. Thus, for a plane wave incident on a lens followed immediately by a transparent object with transmission t, we obtain approximately the Fourier transform of t at the focal plane of the lens.

(4) For an arbitrary U incident on a lens, we obtain approximately the Fourier transform of the field U at the focal plane of the lens.

Let us consider two practical applications using this result. (1) In the first example, a student wants to measure the far field radiation pattern of a laser. It is not necessary to take the measurement at a distance far away. All the student needs to do is to use a camera focused to infinity. In that case, the image plane is exactly the focal plane of the lens, and the far field pattern on the image plane of the camera is obtained. (2) In a second example, let us consider two optical lenses with focal length f. Let the lenses be placed in series and perpendicular to the optical axis. They are separated from each other by a distance $2f$. If the size of the lens is sufficiently large, then the integration limit in Eq. (1.25) can be approximated by ∞. In other words, we expect to see the spatial Fourier transform of the U incident on the first lens at the focal plane midway between the two lenses. We also expect to see $-U$ following the second lens. Now, consider the optical signal processing setup shown in Fig. 1.7. Let U be a normally incident plane wave. The field at the focal plane of the first lens is now the Fourier transform of the transmission of the transparent film, t, placed in front of the first lens. When this radiation is transmitted through an aperture placed at the focal plane of the first lens, the higher Fourier frequencies are blocked by the opaque portion of the aperture. Thus, the U obtained after the second lens is the $-tU$ filtered through a low pass spatial frequency filter. Such a setup has many applications. For example, when a laser mode passes through optical instruments, it is frequently perturbed because of imperfections

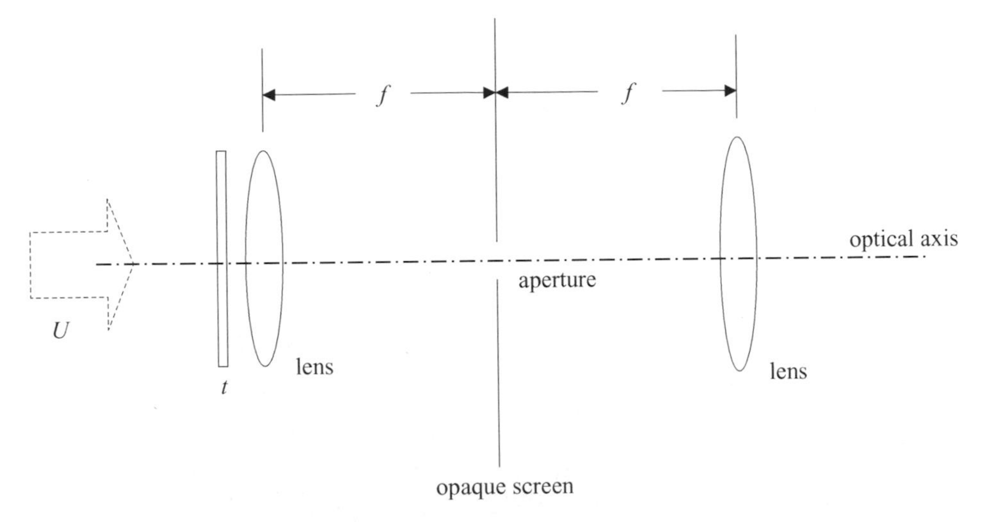

Figure 1.7. Spatial low pass filtering of an optical wave. The opaque screen with an open aperture is at the focal plane of the first lens. The field just before the aperture is approximately the Fourier transform of t times U. The aperture is, in effect, at a low pass filter that cuts out the higher spatial frequency components of tU. A second lens is placed at a focal distance after the aperture. The field emerging after the second lens is $-tU$ filtered through a low pass filter.

or defects in the optical elements. A setup such as that shown in Fig. 1.7 (without the transparent film) is commercially sold as a spatial filter to clean up the effects of perturbations or defects which typically have higher spatial frequencies than the laser mode. More sophisticated spatial filtering examples are presented in ref. [10].

1.4.3 The lens as a transformation element

A simple but very useful example that illustrates a TEM wave analysis other than diffraction is to consider the transmission function t of a thin lens. In order to find the transmission function of a thin lens, we go back to wave propagation to analyze what happens to an optical wave as it propagates through the lens. No diffraction is involved as we are analyzing the changes in the optical TEM wave just before and just after the thin lens. From another point of view, within a short distance from the incident plane, the diffraction effects are insignificant except near the edge.

Let us consider a spherical lens whose geometrical configuration is shown in Fig. 1.8. The right surface of the lens is described by

$$x''^2 + y''^2 + z''^2 = r_1^2.$$

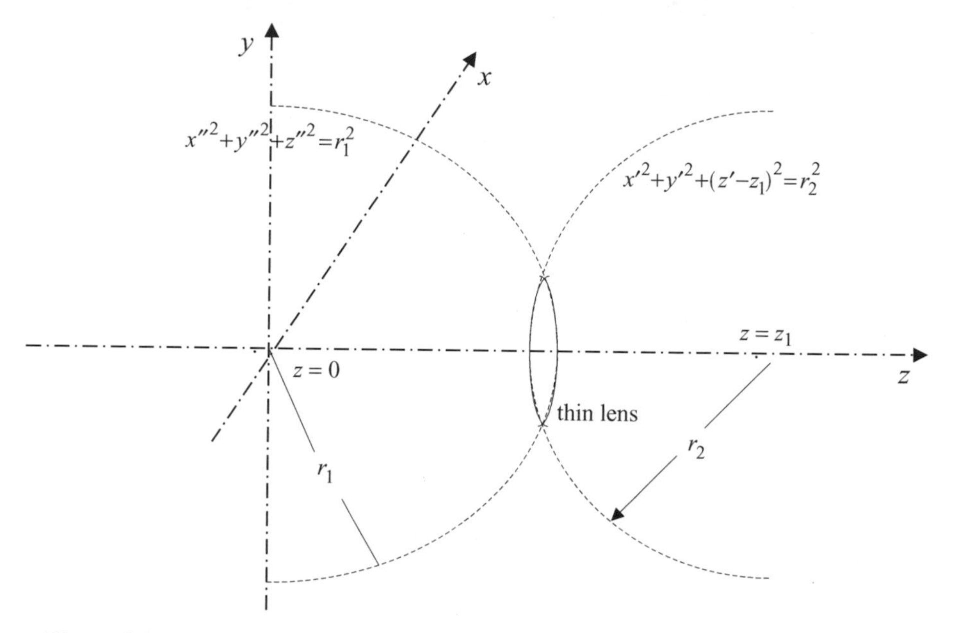

Figure 1.8. The geometrical configuration of a spherical lens. Two spherical surfaces centered about $z = z_1$ and $z = 0$ are shown. The front (left) and back (right) surfaces of a thin spherical lens are made from the interception section of these two spherical surfaces. The lens is made from a material that has refractive index n.

The left surface of the lens is described by

$$x'^2 + y'^2 + (z' - z_1)^2 = r_2^2.$$

Consider an optical ray, i.e. a pencil TEM wave which has a beam size small compared with the size of the lens, propagating in the $+z$ direction. At the transverse position (x, y), it passes through the lens. Its phase at the output will depend on x and y because the ray goes through a higher index region which has thickness $z'' - z'$ at $x = x' = x''$ and $y = y' = y''$. The change in its phase, in comparison with a beam in free space without the lens, is

$$\begin{aligned}\Delta\phi &= -\frac{2\pi}{\lambda}(n-1)(z''-z') \\ &= -\frac{2\pi}{\lambda}(n-1)\left(r_1\left\{1-\frac{x^2+y^2}{r_1^2}\right\}^{1/2} - z_1 + r_2\left\{1-\frac{x^2+y^2}{r_2^2}\right\}^{1/2}\right),\end{aligned}$$

where the refractive index of the lens material is n and $z'' > z'$ for x and y inside the lens. The binomial expansion can be used again for the terms in { }, and only the first-order approximation is used for a thin lens. We then obtain

$$\Delta\phi = \frac{\pi}{\lambda}(n-1)\left(\frac{1}{r_1}+\frac{1}{r_2}\right)(x^2+y^2). \tag{1.26}$$

The focal length of a thin spherical lens is usually given by $1/f = (n-1)(1/r_1 + 1/r_2)$. Thus, for any U passing through a thin lens, we can now multiply the incident U on the lens by a phase function [15],

$$t_l = e^{j\frac{\pi}{\lambda f}(x^2+y^2)}, \tag{1.27}$$

to obtain the U immediately after the lens. This is a very simple result that can be applied to any U passing through a lens. We emphasize that this is a thin lens approximation. Only an ideal lens can be represented by Eq. (1.27). A practical lens will have other higher order terms in the phase shift which are considered as distortions from an ideal lens. Note also that this description of a lens by its effect on the phase of the incoming wave does not address the issue of the amplitude of the wave as it propagates toward the focus of the lens.

Although we have derived this result only for a thin spherical lens, it is used in general to represent any ideal compound lens provided that f is the focal length of the compound lens.

As an illustration of this method, let us consider again the use of a camera focused to infinity to obtain the Fraunhofer diffraction effect given in Eq. (1.17b). If the incident U passes through an ideal thin lens just before the aperture Ω (or passes through the lens right after the aperture), we should multiply U by the phase function given by Eq. (1.27). We then apply the diffraction formula to the transmitted U. On the focal plane of the lens, $d = f$. In the diffraction integral, the two exponential spherical factors cancel each other out, and we obtain the result given in Eq. (1.17b) without satisfying the far field condition expressed in Eq. (1.18).

Note that if we multiply a plane wave ($U = A$) by the phase shift given in Eq. (1.27), we obtain a convergent spherical wave focused at a distance f after the lens. Combining this result with the result derived in Section 1.4.2, we conclude that if a lens is placed immediately after or before the transparent film with transmission function $t(x, y)$, we obtain the Fourier transform of t at the focal plane of the lens. We assume here that the size of the lens is large (so that the effect of the finite size of the lens can be ignored). This is the foundation of transformation optics.

The lens transformation formula is particularly convenient when we analyze resonant modes of complex cavities containing lenses in Chapter 2.

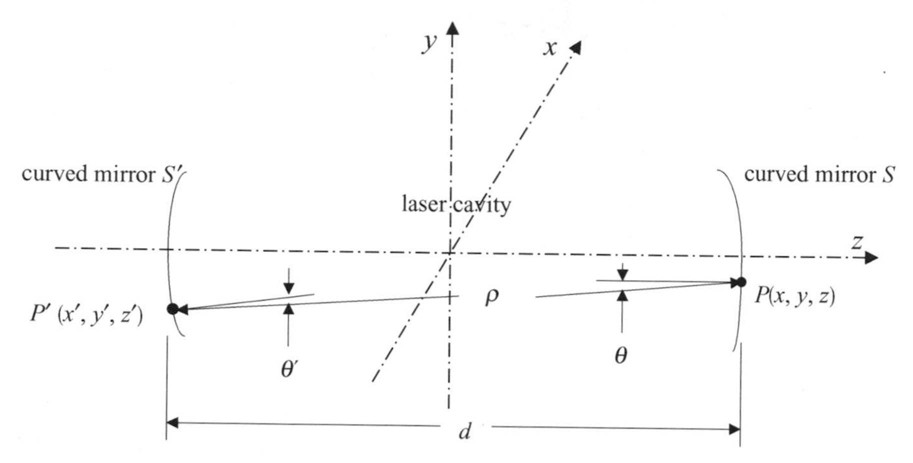

Figure 1.9. Illustration of the multiple diffraction in a laser cavity. A solid state or gas laser cavity consists of two curved mirrors at S and S'. The mirrors are separated by a distance d, and ρ is the distance between a point P on P and a point P' on S'.

1.4.4 Integral equation for optical resonators

In optical resonators that have transverse dimensions much larger than the optical wavelength, the resonator modes are TEM modes that are the solutions of the integral equations derived from Kirchhoff's diffraction formula. This is the case for solid state, surface emitting semiconductor and gas lasers, but not for waveguide semiconductor lasers, which do not have TEM modes. Consider a typical laser cavity as shown in Fig. 1.9. Let the y-polarized electric field on the S' mirror be $E'_y(x', y')$ and the electric field on S be $E_y(x, y)$. The diffracted electric field $E''_y(x, y)$ on the S mirror can be calculated by the Green's function G from the E'_y on S'. Similarly, the diffracted E'''_y on S' can be calculated from E_y on S:

$$E''_y(x, y) = \int_{S'} \frac{jk(1 + \cos\theta')}{4\pi\rho} e^{-jk\rho} E'_y(x', y')\, ds'$$

and

$$E'''_y(x', y') = \int_{S} \frac{jk(1 + \cos\theta)}{4\pi\rho} e^{-jk\rho} E_y(x, y)\, ds.$$

ρ is the distance between P and P', $\rho = \sqrt{(x - x')^2 + (y - y')^2 + (z - z')^2}$. If we have a symmetric pair of mirrors and if the cavity supports a stable mode, then

E_y and E'_y must eventually reproduce each other except for a complex constant γ; i.e.

$$\gamma E_y = E''_y \quad \text{and} \quad \gamma E'_y = E'''_y$$

$$\gamma E_y(x, y) = \int_{S'} \frac{jk(1 + \cos\theta')}{4\pi\rho} e^{-jk\rho} E'_y(x', y')\, ds', \tag{1.28a}$$

$$\gamma E'_y(x', y') = \int_{S} \frac{jk(1 + \cos\theta)}{4\pi\rho} e^{-jk\rho} E_y(x, y)\, ds. \tag{1.28b}$$

E_y and E'_y should have the same x and y function. Note that G_1 should not be used here since the integration is not performed over a planar aperture.

Any stable resonant mode of the cavity must satisfy Eqs. (1.28). Conversely, any solution of Eqs. (1.28) is a resonant mode of the cavity. The field pattern of the resonant mode of the laser was found first by Fox and Li [16]. They calculated the diffraction integral numerically on a computer, starting from an assumed mode pattern on S. The resultant electric field pattern on the opposite mirror S' is then used as the E_y in the diffraction integral to calculate the field on S after a round trip. This process is iterated back and forth many times. Eventually, stabilized mode patterns (i.e. mode patterns that differ from each other only by a complex constant after one diffraction) were found. We will discuss solutions of the integral equation in detail in Chapter 2.

1.5 Superposition theory and other mathematical techniques derived from Kirchhoff's diffraction formula

In practical applications, we frequently need to simplify complex problems to some basic problems that we can analyze easily. Since Maxwell's equations are linear differential equations, the superposition theory is always a very useful tool to break down a more complex problem. For example, when there are two radiation sources at the same frequency, such as two beams split from the same laser, the total diffracted field is the superposition of the U from the individual sources. However, superposition theory is not generally applicable unless the boundary conditions remain the same for both cases. Fortunately, for those optical problems that can be solved via Kirchhoff's integral in Eq. (1.15), we have more powerful techniques to use. These techniques are derived from the mathematical properties of Eqs. (1.15) and (1.17). Five techniques are listed here.

(1) Any integration over the aperture Ω is equivalent to the summation of integrals over different apertures, the sum of which is Ω.

(2) Kirchhoff's integral gives zero contribution to $U(\underline{r_0})$ from any opaque portion of the screen, where both U and ∇U are considered to be zero. In other words, we can change the geometrical configuration of the opaque portion of the screen as long as U and ∇U remain zero there.

(3) When the position of the aperture is moved, the new Kirchhoff integral may be related to the old integral simply by a change of coordinates.

(4) When there is more than one transmission function, t, at the aperture, then the results obtained from Eqs. (1.17b) and (1.25) will be in the form of a Fourier transform of the product of all the t functions over a finite integration limit defined by Ω.

(5) The convolution theory of the Fourier transform of the product of two functions may be applied when the limits of integration are large.

We will illustrate in the following examples how to use these techniques.

(1) As the first example, let us consider the diffraction of U by an aperture which is a rectangular opening on the $z = 0$ plane, from $y = h - l_y$ to $h + l_y$ and from $x = -l_x$ to $+l_x$.

The Fraunhofer diffraction integral in Eq. (1.17b) can now be explicitly written as follows:

$$U(\underline{r_0}) = \frac{-j}{\lambda d} e^{-jkd} e^{-jk\frac{x_0^2+y_0^2}{2d}} e^{+j2\pi\left(\frac{y_0}{\lambda d}\right)h}$$
$$\times \int_{-l_x}^{+l_x} \int_{h-l_y}^{h+l_y} U(x, y, z = 0) e^{+j2\pi\left(\frac{x_0}{\lambda d}\right)x} e^{+j2\pi\left(\frac{y_0}{\lambda d}\right)y} dx\, dy$$
$$= \frac{-j}{\lambda d} e^{-jkd} e^{-jk\frac{x_0^2+y_0^2}{2d}} e^{+j2\pi\left(\frac{y_0}{\lambda d}\right)h} \left[\int_{-\infty}^{+\infty} \int_{-\infty}^{+\infty} \mathrm{rect}\left(\frac{x}{l_x}\right) \mathrm{rect}\left(\frac{y'}{l_y}\right) \right.$$
$$\left. \times\, U(x, y' + h, z = 0) e^{+j2\pi\left(\frac{x_0}{\lambda d}\right)x} e^{+j2\pi\left(\frac{y_0}{\lambda d}\right)y'} dx\, dy' \right]. \tag{1.29}$$

A change of coordinates from y to $y' = y - h$ has been carried out. Equation (1.29) allows us to find the U in terms of the diffraction pattern of a rectangular aperture centered on $x = y = 0$, which we have already discussed.

(2) As the second example, let the incident U in the above example be an optical wave with complex functional variation instead of a simple plane wave; then the integral given in Eq. (1.29) is the Fourier transform, F, of the product of two functions, $RR = \mathrm{rect}\,(x/l_x)\,\mathrm{rect}\,(y'/l_y)$ and $U(x, y' + h, z = 0)$.

Let the Fourier frequencies be

$$f_x = \frac{x_0}{\lambda d} \quad \text{and} \quad f_y = \frac{y_0}{\lambda d}.$$

Let

$$F_{RR}(f'_x, f'_y) = \int_{-\infty}^{+\infty}\int_{-\infty}^{+\infty} \text{rect}\left(\frac{x}{l_x}\right)\text{rect}\left(\frac{y'}{l_y}\right) e^{+j2\pi f'_x x} e^{+j2\pi f'_y y'} dx\, dy',$$

$$F_U(f'_x, f'_y) = \int_{-\infty}^{+\infty}\int_{-\infty}^{+\infty} U(x, y' + h, z = 0) e^{+j2\pi f'_x x} e^{+j2\pi f'_y y'} dx\, dy'.$$

Then, according to the convolution theory,

$$U(\underline{r_0}) = \frac{-j}{\lambda d} e^{-jkd} e^{-j2\pi f_y h} e^{-jk\frac{x_0^2+y_0^2}{2d}} \times \int_{-\infty}^{+\infty}\int_{-\infty}^{+\infty} F_{RR}(f'_x, f'_y)\, F_U(f_x - f'_x, f_y - f'_y)\, df'_x\, df'_y. \quad (1.30)$$

F_{RR} and F_U are likely to be results that we already have. Thus we can obtain U at $\underline{r_0}$ from the known results.

(3) As the third example, let us consider the diffraction pattern of a double slit, from $y = h - l_y$ to $y = h + l_y$ and from $y = -h - l_y$ to $y = -h + l_y$. The incident U is a plane wave ($U = A$) propagating in the $+z$ direction.

Using the superposition theory and the results obtained in Eq. (1.29), we immediately obtain the diffraction pattern as

$$U(\underline{r_0}) = \frac{-2Aj}{\lambda d} e^{-jkd} e^{-jk\frac{x_0^2+y_0^2}{2d}} \cos\left[2\pi\left(\frac{y_0}{\lambda d}\right) h\right] \times \int_{-\infty}^{+\infty}\int_{-\infty}^{+\infty} \text{rect}\left(\frac{x}{l_x}\right)\text{rect}\left(\frac{y'}{l_y}\right) e^{+j2\pi\left(\frac{x_0}{\lambda d}\right)x} e^{+j2\pi\left(\frac{y_0}{\lambda d}\right)y'} dx\, dy'. \quad (1.31)$$

The cosine function expresses the interference effect of the double slit diffraction.

(4) As the fourth example, let us consider the diffracted field when the screen is an opaque obstacle such as a finite sized disk. We note that, for any very large open aperture Ω', we will get back the incident U at $\underline{r} = \underline{r_0}$. We can express any opaque aperture Ω as $\Sigma - (\Sigma - \Omega)$, where Σ is the entire $z = 0$ plane. $\Sigma - \Omega$ is the complementary aperture of Ω. Therefore, we can rewrite Eq. (1.17b) as follows:

$$U(\underline{r_0}) = U_{\text{inc}}(\underline{r} = \underline{r_0}) - \left[\frac{-j}{\lambda d} e^{-jkd} e^{-jk\frac{x_0^2+y_0^2}{2d}} \times \iint_{\Sigma-\Omega} U_{\text{inc}}(z = 0) e^{+j2\pi\frac{xx_0}{\lambda d}} e^{+j2\pi\frac{yy_0}{\lambda d}} dx\, dy\right].$$

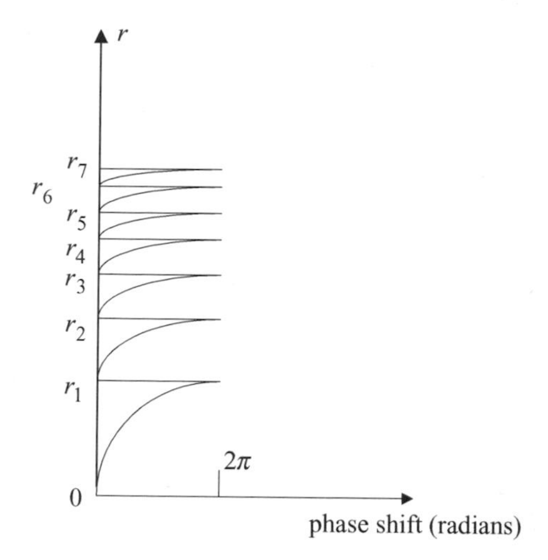

Figure 1.10. Phase shift in a Fresnel lens. A refractive Fresnel lens has a circular symmetric sectional continuous variation of its thickness. The phase shift of optical light transmitted within each circular section varies as a function proportional to r^2, from $2n\pi$ to $2(n+1)\pi$. The phase shift as a function of r for the first few rings is illustrated here for the case of $n = 0$. For a plane wave incident normally on a Fresnel lens, its phase shift varies continuously like an ideal lens, as shown in Eq. (1.27).

(5) As the fifth example, let us consider a refractive Fresnel lens. A refractive Fresnel lens does not have a spherical surface as shown in Fig. 1.8. It has a material structure that has a sectional continuous profile. For the first segment, the surface profile is such that, from the center $r = 0$ to a radius r, the phase shift is described by Eq. (1.27). However, this surface stops at $r = r_1$ when the phase shift is 2π, i.e. when $r_1^2/\lambda f = 2$. A new segment of the surface starts at r_1 with zero phase shift. This second segment of the surface will provide a phase shift proportional to $(r^2/\lambda f) - 2$. The second surface segment stops at r_2 when $r_2^2/\lambda f = 4$. The third segment starts at r_2 with zero phase shift. These segments continue until the shortest length of the segments, $r_j - r_{j-1}$, reaches the resolution limit of the fabrication technology. Figure 1.10 illustrates the phase shifts along the radial direction r.

When one calculates the diffraction pattern of the Fresnel lens, the Kirchhoff integral will be performed separately over each segment of continuous phase shift zone. The sum of all the diffraction integrals gives the $U(\underline{r_0})$. The insertion of $e^{j2n\pi}$ (n = any integer) into any integrand does not change the value of the integral. We can easily show that for any normally incident plane wave, the U given by Kirchhoff's integrals for the Fresnel lens behaves identically to any thin spherical lens with the same focal length. The difference between the spherical lens and the

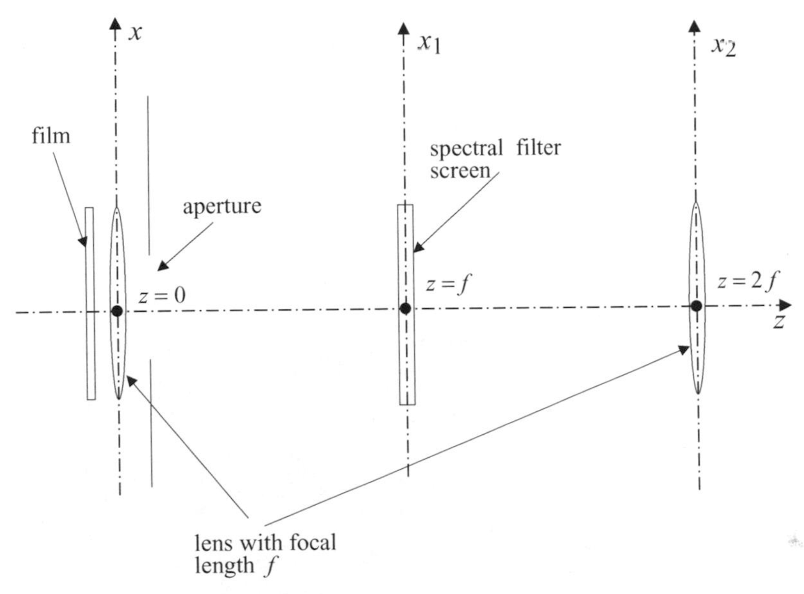

Figure 1.11. Illustration of an example of spatial filtering in the Fourier transform plane. A transparent film with transmission function $t(x, y)$ is placed in a square aperture $(d \times d)$ in front of an ideal lens with focal length f at $z = 0$. A spatial filter is placed at the focal plane at $z = f$. A second lens with focal length f is placed at $z = 2f$ to reconstruct the filtered incident light.

Fresnel lens lies in the higher order terms of the phase shifts which we neglected in the first-order approximation. This difference is important for oblique incident radiation.

(6) As the final example (from ref. [15]), a plane wave with amplitude A is incident normally on a transparent film at $z = 0$, followed immediately by an ideal thin lens with focal length f, as shown in Fig. 1.11. The film is placed in a square aperture $(d \times d)$ centered at $x = y = 0$. The electric field transmission t of the transparent film is

$$t(x, y) = \frac{1}{2}[1 + \cos(2\pi Hx)] \operatorname{rect}\left(\frac{x}{d/2}\right) \operatorname{rect}\left(\frac{y}{d/2}\right),$$

where $H \gg (1/d)$. An optical filter (shown in Fig. 1.12) is placed at the focal plane of this lens. The screen is opaque in two regions: (1) $|x| < l/2$ and $|y| < l/2$ for the inside region, and (2) $|x| > l$ and $|y| > l$ for the outside region. A second lens with focal length f is placed at a distance f behind the screen. To find the diffracted field after the second lens, we proceed as follows.

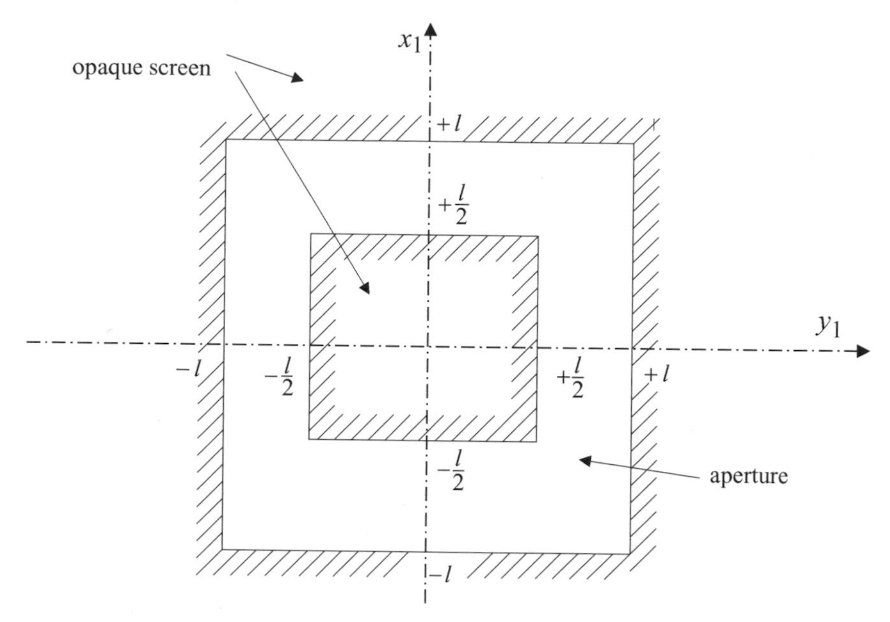

Figure 1.12. The optical filter in Fig. 1.11 The spatial filter in Fig. 1.11 is illustrated here. It is opaque for $|x|$ and $|y| < l/2$ and for $|x|$ and $|y| > l$.

At $z = f$, the incident field on the screen is

$$U(x_1, y_1, f) = \frac{-jAe^{-jkf}e^{-jk\left(\frac{x_1^2+y_1^2}{2f}\right)}}{\lambda f}$$

$$\times \int_{-d/2}^{d/2}\int_{-d/2}^{d/2} \frac{1}{2}\left[1 + \frac{1}{2}e^{j2\pi Hx} + \frac{1}{2}e^{-j2\pi Hx}\right] e^{j2\pi\left(\frac{x_1}{\lambda f}\right)x} e^{j2\pi\left(\frac{y_1}{\lambda f}\right)y}\, dx\, dy$$

$$= \frac{-jAe^{-jkf}e^{-jk\left(\frac{x_1^2+y_1^2}{2f}\right)}}{2\lambda f}\left[\frac{\sin\left(\frac{\pi x_1 d}{\lambda f}\right)}{\pi\frac{x_1}{\lambda f}} + \frac{\sin\left(\pi\left(H + \frac{x_1}{\lambda f}\right)d\right)}{2\pi\left(H + \frac{x_1}{\lambda f}\right)}\right.$$

$$\left. + \frac{\sin\left(\pi\left(H - \frac{x_1}{\lambda f}\right)d\right)}{2\pi\left(H - \frac{x_1}{\lambda f}\right)}\right]\frac{\sin\left(\frac{\pi y_1 d}{\lambda f}\right)}{\pi\frac{y_1}{\lambda f}}.$$

Thus there are three radiation peaks in the x_1 direction, centered at $x_1 = 0$, $x_1 = \lambda f H$ and $x_1 = -\lambda f H$. The width (defined by the first zero of the field) of the peaks in the x direction is $\lambda f/d$ around the centers. All radiation peaks are centered about $y_1 = 0$ with width $\lambda f/d$ in the y direction. However, the x transmission range of the screen at $z = f$ is $l/2 < |x_1| < l$. Thus the peak centered about $x_1 = 0$ is always blocked by the screen. In order for the two side peaks to pass the screen, $l/2 < f\lambda H < l$. In order for the main lobe of the two side peaks to pass through the screen, we need

$$\lambda H f + \frac{\lambda f}{d} < l < 2\lambda H f - \frac{2\lambda f}{d}.$$

Since the peaks are centered in the y direction at $y_1 = 0$, the transmission of the screen is effectively from $y_1 = -l$ to $y_1 = l$ whenever the peak is transmitted in the x direction. If we approximate the transmitted radiation field by deleting the term representing the peak centered about $x_1 = y_1 = 0$, we obtain the diffracted transmitted radiation after the screen:

$$U'(x, y, f) \approx \frac{-Ae^{-jkz}e^{-jk\left(\frac{x^2+y^2}{2(z-f)}\right)}}{2\lambda^2 f(z-f)}$$
$$\times \int_{-l}^{l} \left[\frac{\sin\left(\frac{\pi y_1 d}{\lambda f}\right)}{\pi \frac{y_1}{\lambda f}} e^{-jk\frac{y_1^2}{2f}}\right] e^{-jk\frac{y_1^2}{2(z-f)}} e^{j2\pi\left(\frac{y}{\lambda(z-f)}\right)y_1}\, dy_1$$
$$\times \left\{\int_{l/2}^{l} \left[\frac{\sin\left(\pi\left(\frac{x_1}{\lambda f} - H\right)d\right)}{2\pi\left(\frac{x_1}{\lambda f} - H\right)} e^{-jk\frac{x_1^2}{2f}}\right] e^{-jk\frac{x_1^2}{2(z-f)}} e^{j2\pi\left(\frac{x}{\lambda(z-f)}\right)x_1}\, dx_1\right.$$
$$\left. + \int_{-l}^{-l/2} \left[\frac{\sin\left(\pi\left(\frac{x_1}{\lambda f} + H\right)d\right)}{2\pi\left(\frac{x_1}{\lambda f} + H\right)} e^{-jk\frac{x_1^2}{2f}}\right] e^{-jk\frac{x_1^2}{2(z-f)}} e^{j2\pi\left(\frac{x}{\lambda(z-f)}\right)x_1}\, dx_1\right\}.$$

When this diffracted field passes through the second lens at $z = 2f$, the exponential term in front of the integral, $\exp(-jk[(x^2 + y^2)/2(z - f)])$, is canceled by the quadratic phase change of an ideal lens, $\exp(jk[(x^2 + y^2)/2f])$, for $z \geq 2f$. The integration is quite messy in the general case. However, the answer is simple for the following special case. Let $\lambda f/d$ be small (i.e. the width of radiation peaks is narrow), and let the dimension l be such that at least the main lobe of the two side peaks passes through the screen at $z = f$. Then $\sin(\pi y_1/(\lambda f/d))/(\pi y_1/\lambda f)$ is significant only for $y_1 < \lambda f/d$, and its peak value is proportional to d. Within such a small range of y_1, the three exponential factors in the above y_1 integral can be approximated by constant values at $y_1 \approx 0$. This means that

$$\exp\left(-jky_1^2/2f\right)\exp\left(-jky_1^2/2(z-f)\right)\exp(jky_1/2(z-f)y) \approx 1.$$

Similarly, the three exponential factors in the two x_1 integrals can be approximated by $x_1 = \lambda Hf$ and $x_1 = -\lambda Hf$, respectively.

Therefore, immediately after the second lens at $z = 2f$, we have the following field:

$$U''(x_2, y_2, 2f) \approx \frac{-Ae^{-jk2f}}{2\lambda^2 f^2} \int_{-l}^{l} \frac{\sin\left(\pi \dfrac{2\pi\, y_1 d}{\lambda f}\right)}{\pi \dfrac{y_1}{\lambda f}}\, dy_1$$

$$\times \left\{ e^{-jk\lambda^2 H^2 f} e^{jk\lambda\, Hx_2} \int_{l/2}^{l} \frac{\sin\left[\pi\left(\dfrac{x_1}{\lambda f} - H\right) d\right]}{2\pi\left(\dfrac{x_1}{\lambda f} - H\right)}\, dx_1 \right.$$

$$\left. + \; e^{-jk\lambda^2 H^2 f} e^{-jk\lambda\, Hx_2} \int_{-l}^{-l/2} \frac{\sin\left[\pi\left(\dfrac{x_1}{\lambda f} + H\right)\right]}{2\pi\left(\dfrac{x_1}{\lambda f} + H\right)}\, dx_1 \right\}.$$

If the second lens is sufficiently large, the diffraction effect due to the finite size of the second lens can be neglected. The far field diffraction pattern will be given by two beams, one beam propagating as $\exp(-jkz)\exp(jk\lambda Hx)$ and the second beam propagating as $\exp(-jkz)\exp(-jk\lambda Hx)$. The incident beam propagating along the z axis has been filtered out.

References

1 M. V. Klein, *Optics*, Chapter 2, New York, John Wiley and Sons, 1970
2 M. Born and E. Wolf, *Principles of Optics*, Chapters 3 & 4, New York, Pergamon Press, 1959
3 M. Born and E. Wolf, *Principles of Optics*, Chapter 8, New York, Pergamon Press, 1959
4 J. W. Goodman, *Introduction to Fourier Optics*, Chapters 3 & 4, New York, McGraw-Hill, 1968
5 M. V. Klein, *Optics*, Chapters 7–9, New York, John Wiley and Sons, 1970
6 P. M. Morse and H. Feshback, *Methods of Theoretical Physics*, Chapter 7, New York, McGraw-Hill, 1953
7 W. Kaplan, *Advanced Calculus*, Chapters 2–5, Reading, MA, Addison-Wesley, 1984
8 J. A. Stratton, *Electromagnetic Theory*, Section 9.1, New York, McGraw-Hill, 1941
9 A. Sommerfeld, *Optics*, translated by O. Laporte and P. A. Moldauer, Chapter 5, New York, Academic Press, 1954

10 J. W. Goodman, *Introduction to Fourier Optics*, Chapters 7 & 8, New York, McGraw-Hill, 1968
11 J. B. Develis and G. O. Reynalds, *Theory and Applications of Holography*, Chapter 8, Reading, MA, Addison-Wesley, 1967
12 Francis T. S. Yu, *Optical Signal Processing, Computing, and Neural Networks*, Chapters 1–6, New York, John Wiley and Sons, 1992
13 J. A. Stratton, *Electromagnetic Theory*, Chapter 6, New York, McGraw-Hill, 1941
14 P. M. Morse and H. Feshback, *Methods of Theoretical Physics*, Chapter 11, New York, McGraw-Hill, 1953
15 J. W. Goodman, *Introduction to Fourier Optics*, Chapters 5 & 6, New York, McGraw-Hill, 1968
16 A. G. Fox and T. Li, "Resonant Modes in a Maser Interferometer," *Bell System Technical Journal*, **40**, 1961, 453

2

Gaussian modes in optical laser cavities and Gaussian beam optics

It is well known that basic solid state and gas laser cavities consist of two end reflectors that have a certain transverse (or lateral) shape such as a flat surface or a part of a large sphere. The reflectors are separated longitudinally by distances varying from centimeters to meters. The size of the end reflectors is small compared with the separation distance. All cavities for gaseous and solid state lasers have slow lateral variations within a distance of a few wavelengths (such as the variation of refractive index and gain of the material and the variation of the shape of the reflector). Therefore these cavity modes are analyzed using the scalar wave equation. Laser cavities are also sometimes called Fabry–Perot cavities because of their similarity to Fabry–Perot interferometers. However, Fabry–Perot interferometers have distances of separation much smaller than the size of the end reflectors. The diffraction properties of the modes in Fabry–Perot interferometers are quite different from the properties of modes in laser cavities.

The analysis of the resonant modes is fundamental to the understanding of lasers. Modes of solid state and gas lasers are solutions of Eqs. (1.28a) and (1.28b), known as Gaussian modes. They are TEM modes. The analysis of laser modes and Gaussian beam optics constitutes a nice demonstration of the mathematical techniques presented in Chapter 1. It is also important to learn the mathematical analysis for the following two reasons: (1) we will appreciate the limitations and circumstances in which Gaussian modes can be used; and (2) we will know how to apply these mathematical techniques to new structures.

The propagation properties of Gaussian modes outside the laser cavity are important in many applications such as filtering or matching a laser beam to a specific input beam shape. Gaussian modes are also solutions of Maxwell's equations, just like plane waves, cylindrical waves, etc. When Gaussian waves are propagating through lenses or other optical components, they retain the form of Gaussian modes. In other words, when a Gaussian beam is diffracted by an optical component, it can usually be described as a transformation into another Gaussian beam with different parameters. Such a transformation can be performed simply by a transformation matrix of the parameters without the use of complicated diffraction integrals as we have done in Chapter 1. Therefore, the analyses of many applications of laser beams can be greatly simplified if the optical beam can be considered as a Gaussian beam. This chapter is devoted entirely to Gaussian beams and their propagation. Such a discussion is not available in most classical optics books.

In Chapter 3 we will discuss modes of a different kind, guided wave modes, which arise from material structures with significant index variation within a distance of a wavelength in the lateral direction. Vector wave equations will be used for the analysis of guided waves. The resonant modes of edge emitting semiconductor lasers are resonant modes of guided waves in cavities with flat end reflectors or grating reflectors. These guided wave modes are similar to the modes in microwave structures. They are not TEM modes. For the vertical cavity surface emitting semiconductor lasers (VCSEL) discussed in Chapter 7, there is usually no waveguide structure in the transverse directions. The end reflectors are flat, separated longitudinally by a distance of the order of micrometers. Therefore neither the analysis of the guided wave modes nor the analysis of the Gaussian modes is applicable. Because of the short dimensions in which diffraction effects are negligible, the VCSEL modes are often treated as plane waves.

In this chapter, we will present the discussion on TEM cavity modes and their propagation in the following order.

(1) First we will analyze the modes of laser cavities by solving the integral equations, Eqs. (1.28). Each mode is an independent electromagnetic field solution of the integral equation. All the modes discussed in this chapter are "cold" or "passive" cavity modes, meaning that the gain in the material is zero. The contribution to the susceptibility of the material from the induced quantum mechanical transition is not included. The solutions of Eqs. (1.28a) and (1.28b) are commonly called the Gaussian modes because their lateral amplitude variation has a Gaussian envelope. We will begin our discussion with confocal cavities, because Eqs. (1.28) have analytical solutions for such a cavity. We will then extend the analysis to other cavity configurations. When there is sufficient gain, solid state or gas lasers will oscillate in one or several of these passive cavity modes modified by the gain and susceptibility contributed from the induced quantum mechanical transition. These properties involve laser oscillation, and will be discussed in Chapter 6.

(2) We will show that, when end reflectors are partially transmitting, these modes will continue to propagate as Gaussian beams outside the laser cavity. This is an important result. Since the diffraction of a Gaussian beam yields another Gaussian beam, the propagation of these beams through different optical elements, such as lenses, etc., can be treated simply as a matrix transformation of the parameters of the Gaussian beam, without explicitly carrying out the Kirchhoff integral in each

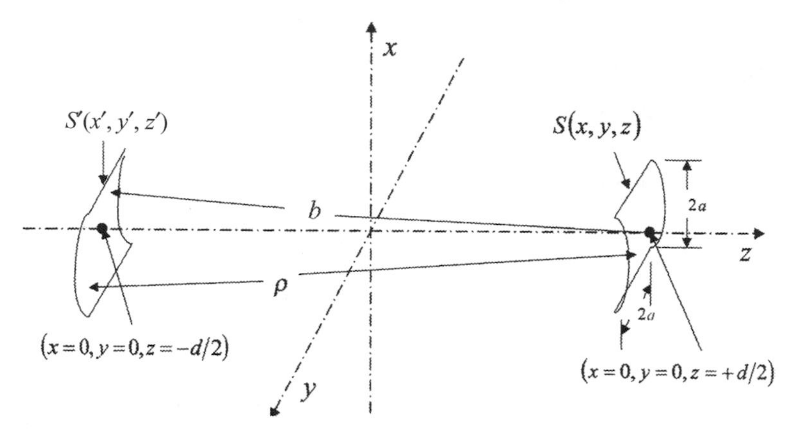

Figure 2.1. Illustration of a confocal cavity. A confocal resonator has two spherical end reflectors, S and S'. The mirrors have a square aperture, $2a \times 2a$. The spherical center of the S surface is at the center of S', with radius b. The spherical center of the S' surface is at the center of S, also with radius b. The focus of both the S mirror and the S' mirror is at the origin; ρ is the distance from a point on the S' mirror to a point on the S mirror.

step. Because of the convenience of Gaussian beam transformation, the fields of other components, such as fibers and waveguides, are frequently approximated by Gaussian modes.

(3) Mathematically, Gaussian modes form a complete orthogonal set, meaning that any radiation field can be expressed as a superposition of these modes. This is important for the analysis of various applications. For example, the excitation of the different "cold" cavity modes by external radiation is usually analyzed by expanding the incident radiation field as a superposition of orthogonal Gaussian cavity modes. The transmission or reflection of any arbitrary external radiation by Fabry–Perot interferometers with curved end reflectors is analyzed by expanding the external radiation in terms of the Gaussian modes of the interferometer.

2.1 Modes in confocal cavities

The rigorous theory of laser modes is based on the analytical solution of the integral equation created by repeated Kirchhoff diffraction in cavities that have confocal reflectors. Consider the confocal resonator shown in Fig. 2.1. There are two identical spherical mirrors, each with radius b, symmetrically placed about the z axis at $z = \pm d/2$ ($d = b$ in confocal cavities). In order to take advantage of the simplicity of mathematical analysis in rectangular coordinates, both mirrors are assumed to have a square shape ($2a \times 2a$ in the transverse dimension). The size of the mirror is small compared with the separation distance, i.e. $d \gg a$. Although the centers of the spherical surfaces are located at $x = y = 0$ and $z = \pm d/2$, the focal point of both mirrors is at $x = y = 0$ and $z = 0$. Therefore it is called a confocal cavity. We will analyze rigorously the confocal cavities following ref. [1].

Note that all dimensions of solid state and gaseous laser cavities are much larger than the wavelength. Since there are many resonant modes in a large cavity, the small lateral dimension of the reflectors is employed so that the losses of higher order resonant modes will be much higher than the loss of the fundamental mode. In Chapter 6, we will show that the difference in losses is necessary to prevent the oscillation of the unwanted higher order modes. Since laser oscillation takes place only in lowest order modes, our discussion in this chapter will focus on lower order modes.

2.1.1 The simplified integral equation for confocal cavities

Since $a \ll d$, $\theta \approx 0$ and $\cos\theta \approx 1$ in Eqs. (1.28). Thus, Eqs. (1.28) for the electric field polarized linearly in the y direction can be simplified as follows:

$$\gamma E_y(x, y)|_{\text{on } S} = \left(\frac{j}{\lambda d}\right) \int_{-a}^{a} \int_{-a}^{a} E_y(x', y')|_{\text{on } S'} e^{-jk\rho} dx' dy'. \tag{2.1}$$

Here, ρ is the distance between P and P' on the S and S' surfaces, respectively. Clearly, the E_y on S and S' must be identical. Thus, Eq. (2.1) is an integral equation for E_y. It is well known mathematically that, like differential equations with appropriate boundary conditions, such an integral equation has independent eigen functions and eigen values [2]. If we can find these independent solutions, we have found the modes of the confocal cavity.

The S' and S surfaces are described by

$$z' - \frac{d}{2} = -\sqrt{d^2 - x'^2 - y'^2} \approx -d + \frac{x'^2 + y'^2}{2d},$$

$$z + \frac{d}{2} = \sqrt{d^2 - x^2 - y^2} \approx d - \frac{x^2 + y^2}{2d}.$$

When $e^{-jk\rho}$ is simplified by a binomial approximation and when the higher order terms are neglected, we obtain

$$\begin{aligned}
\rho &= \sqrt{(z - z')^2 + (x - x')^2 + (y - y')^2} \\
&\approx (z - z') + \left[\frac{1}{2} \frac{(x - x')^2 + (y - y')^2}{(z - z')}\right] \\
&\approx \left(d - \frac{x'^2 + y'^2}{2d} - \frac{x^2 + y^2}{2d}\right) + \left[\frac{(x^2 + x'^2) + (y^2 + y'^2) - 2xx' - 2yy'}{2d}\right] \\
&\approx d - \frac{xx' + yy'}{d}.
\end{aligned}$$

Here, z and z' on S and S' in the preceding expression are used to simplify the $(z - z')$ term, and d is used to approximate the $(z - z')$ term in the denominator. Note that the quadratic terms in the first-order term of the binomial series expansion in the square brackets of the above equation are canceled by the quadratic terms in parentheses created by the spherical surfaces of the confocal resonator. This coincidence gives us a simplified expression for ρ. Neglecting the higher order terms in the binomial expansion is justified when $a^2/b\lambda \ll (b/a)^2$. When higher order terms are neglected, the E_y at (x, y, z) on S is related to E_y at (x', y', z') on S' by a simplified equation:

$$\gamma E_y(x, y, z)|_{\text{on } S} = \left(\frac{j}{\lambda d} e^{-jkd}\right) \int_{-a}^{+a} \int_{-a}^{+a} E_y(x', y', z')|_{\text{on } S'} e^{jk\left(\frac{xx'+yy'}{d}\right)} dx' \, dy'.$$

Let us compare this equation with the diffraction integrals for Fraunhofer diffraction in the focal plane of a lens. We see that the relation between E_y on S and E_y on S' is again a Fourier transform with finite integration limits, $\pm a$. There are known mathematical solutions for such an integral equation. This is really the secret of the confocal cavity and the reason we started the analysis with it.

2.1.2 Analytical solutions of the modes in confocal cavities

If we let the E_y on S be described by $F(x)G(y)$, then the integral equation for F and G is

$$\sigma_l \sigma_m F_l(x) G_m(y) = \int_{-a}^{+a} \int_{-a}^{+a} \frac{je^{-jkb}}{\lambda d} F_l(x') G_m(y') e^{jk\left(\frac{xx'+yy'}{b}\right)} dx' dy',$$

where γ is represented by $\sigma_l \sigma_m$. When we make the following change of variables:

$$\Lambda = \frac{a^2 k}{b}, \quad X = \frac{\sqrt{\Lambda}}{a} x, \quad \text{and} \quad Y = \frac{\sqrt{\Lambda}}{a} y,$$

we obtain

$$\sigma_l \sigma_m F_l(X) G_m(Y) = \frac{je^{-jkb}}{2\pi} \int_{-\sqrt{\Lambda}}^{+\sqrt{\Lambda}} F_l(X') e^{jXX'} dX' \int_{-\sqrt{\Lambda}}^{+\sqrt{\Lambda}} G_m(Y') e^{jYY'} dY'. \quad (2.2)$$

This is a product of two well known identical integral equations, one for X and one for Y. In order for both of them to be satisfied for all X and Y, each integral equation must be satisfied separately. Slepian and Pollak [3] have shown that the

lth independent solution to

$$F_l(X) = \frac{1}{\sqrt{2\pi}\,\chi_l} \int_{-\sqrt{\Lambda}}^{+\sqrt{\Lambda}} F_l(x')e^{jXX'}dX'$$

is

$$F_l(X) = S_{0l}\left(\Lambda, \frac{X}{\sqrt{\Lambda}}\right)'$$

and

$$\chi_l = \sqrt{\frac{2\Lambda}{\pi}}\, j^l R_{0l}{}^{(1)}(\Lambda, 1),$$

where $l = 0, 1, 2, \ldots$, and S_{0l} and $R_{0l}{}^{(1)}$ are, respectively, the angular and radial wave functions in prolate spheroidal coordinates, as defined by Flammer [4]. Thus, the eigen values and eigen functions of Eq. (2.1) are

$$\sigma_l \sigma_m = jX_l X_m e^{-jkb} = \frac{2\Lambda}{\pi} R_{0l}{}^{(1)}(\Lambda, 1) R_{0m}{}^{(1)}(\Lambda, 1) j^{m+l+1} e^{-jkb}, \quad \text{(2.3a)}$$

and

$$E_y = U_{lm}(x, y) = S_{0l}\left(\Lambda, \frac{x}{a}\right) S_{0m}\left(\Lambda, \frac{y}{a}\right), \quad \text{(2.3b)}$$

with $l, m = 0, 1, 2, 3, \ldots$ According to Slepian and Pollak [3], the R and the S functions are real. This confirms that the mirrors are surfaces of constant phase of E_y.

2.1.3 Properties of resonant modes in confocal cavities

Many conclusions can immediately be drawn from the solution for the fields on the reflector surface discussed in Section 2.1.2. Seven properties of the resonant modes of the confocal cavities are presented in the following.

(1) Transverse field pattern

We normally designate the resonant modes as TEM_{lm} modes which have the transverse variation given by U_{lm}. Figure 2.2 illustrates the transverse field distribution of lowest order TEM_{lm} modes in confocal resonators. Note that the lth-order mode will have l nodes or zeros in the x direction, while the mth-order mode will have m nodes in the y direction. This is important information. It allows us to identify experimentally the mode order by examining the nodes in its intensity pattern. We will also show in the third conclusion that you cannot expect to couple a mode

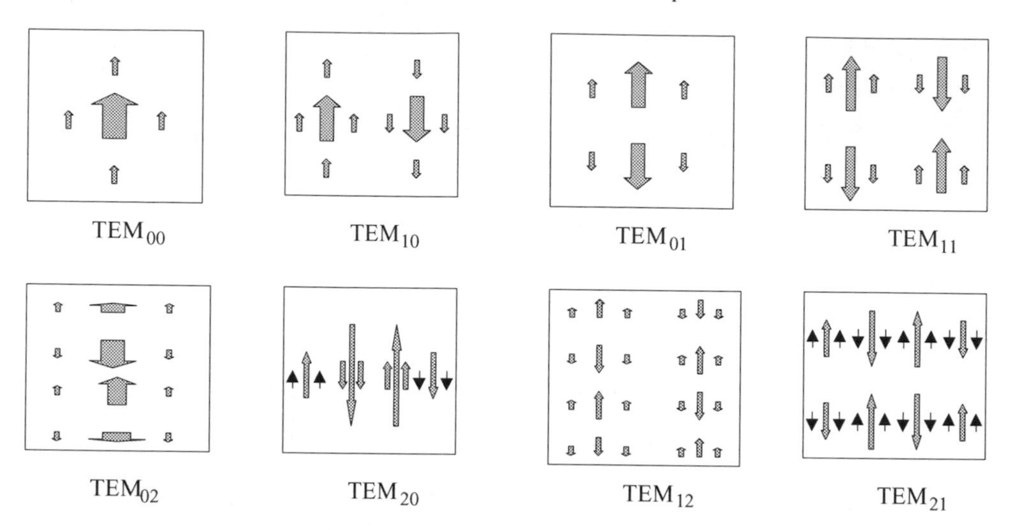

Figure 2.2. Sketch of the transverse field distribution of modes in confocal resonators. The arrows are used to indicate the electric field patterns of various lower order TEM_{lm} modes on the mirror. The direction of the field is shown by the direction of the arrows and the magnitude of the field is indicated by the size of the arrows.

with an odd number of nodes to any other field which has a symmetric mode pattern.

(2) *Resonance frequency*

At resonance, the phase shift after each round trip of propagation in the z direction must be an integer multiple of 2π. Thus, resonance in the z direction occurs only at discrete wavelengths λ_{lmq} that correspond to various values of q multiples of 2π:

$$\left| \pi - \frac{4\pi b}{\lambda_{lmq}} + (m + l)\,\pi \right| = 2q\pi,$$

where m, l and q are all integers. From here on, we designate modes belonging to different l and m as transverse modes and modes belonging to different q as longitudinal modes. However, lower order transverse modes have small integers or zero for m and l, whereas q may be a very large number, up to millions, for long cavities.

In summary, the resonance frequency f_{lmq} for a given order of mode, designated by l, m and q, is

$$f_{lmq} = \frac{c}{4b}\,(2q + l + m + 1)\,, \tag{2.4}$$

where c is the velocity of light in free space.

From Eq. (2.4), we see that the TEM_{lm} modes are degenerate with respect to l and m. Degeneracy means that independent modes with the same $l + m$ value, but different l and m values, will have the same resonance frequency. As we will show in the following section, such degeneracy does not exist in non-confocal cavities. In principle, degenerate modes may oscillate at the same frequency. However, we do not want more than one mode to oscillate because it may cause uncertainty in its total field. Thus we do not like to use cavities with degenerate modes in practical applications. The mode degeneracy is a disadvantage of confocal resonators.

TEM_{lm} modes which have adjacent longitudinal mode orders, i.e. q and $q + 1$, will have resonance frequencies separated by $c/2b$, where $2b/c$ is the round trip propagation time for a wave front to travel around the cavity. Thus, the frequency spacing of the longitudinal modes is controlled by the mirror separation between the reflectors and the velocity of light. For cavities filled with a dielectric that has refractive index n, the resonance frequency separation of the adjacent longitudinal modes will be $2bn/c$.

(3) Orthogonality of the modes

The U_{lm} are orthogonal functions, i.e.

$$\int_{-a}^{a}\int_{-a}^{a} F_m\left(\frac{x\sqrt{\Lambda}}{a}\right) G_n\left(\frac{y\sqrt{\Lambda}}{a}\right) F_{m'}\left(\frac{x\sqrt{\Lambda}}{a}\right) G_{n'}\left(\frac{y\sqrt{\Lambda}}{a}\right) dx\,dy$$
$$= \int_{-a}^{a}\int_{-a}^{a} U_{mn} U_{m'n'}\,dx\,dy = 0,$$

when $m \neq m'$ or $n \neq n'$. Therefore these modes are orthogonal modes. Moreover, it can be shown mathematically that eigen functions of the integral equation of the form given in Eq. (2.2) always form a complete set.

Any arbitrary TEM field polarized in the y direction can be expressed as a superposition of these TEM_{lm} modes, just as we can express them as a superposition of plane waves or cylindrical waves. The selection of the specific form of modal expansion will be based on mathematical convenience.

The orthogonality relation is very helpful in expanding any arbitrary $U(x, y)$ in terms of U_{lm}. For example, let

$$U = \sum_{l,m} a_{lm} U_{lm}.$$

Then, because of the orthogonality relation,

$$a_{lm} = \frac{\int_{-a}^{a}\int_{-a}^{a} U\, U_{lm}\, dx\, dy}{\int_{-a}^{a}\int_{-a}^{a} U_{lm}^2\, dx\, dy}.$$

There are many important applications of the orthogonality properties in modal expansion analysis. For the application of coupling laser radiation to another optical component, such as single-mode optical fibers, modal expansion of the fiber mode in terms of TEM_{lm} modes, or vice versa, is an important technique. For example, the coupling between the TEM_{lm} mode with odd l or m and any cylindrically symmetric mode of the fiber will be zero because the integral of the product of any symmetric and any anti-symmetric function is zero.

(4) Simplified analytical expression for the field

For x and $y \ll a$, U can be approximated by the product of an Hermite polynomial and a Gaussian envelope,

$$U_{lm}(x, y) = \frac{\Gamma[(l/2)+1]\Gamma[(m/2+1)]}{\Gamma(l+1)\Gamma(m+1)} H_l\left(\frac{x\sqrt{\Lambda}}{a}\right) H_m\left(\frac{y\sqrt{\Lambda}}{a}\right) e^{-\pi(x^2+y^2)/\lambda d}. \tag{2.5}$$

Γ is the usual gamma function, and Hermite polynomials are tabulated in many physics and mathematics books [5]:

$$\begin{aligned}
&H_0(x) = 1,\\
&H_1(x) = 2x,\\
&H_2(x) = 4x^2 - 2,\\
&\vdots\\
&H_n(x) = (-1)^n\, e^{x^2} \frac{\partial^n}{\partial x^n} e^{-x^2}.
\end{aligned}$$

For $l = m = 0$, the lowest order Hermite polynomial is just unity. Thus, the TEM_{00} mode is just a simple Gaussian function. An lth-order Hermite polynomial is an lth-order algebraic polynomial function. Thus, it will have l zeros. Even-order modes will be even functions, while odd-order modes will be odd functions. At large x and y values, polynomials are weakly varying functions, and the exponential function dominates the amplitude variation. Thus, the envelope of all TEM_{lm} modes is a Gaussian function which is independent of the mode order, l and m.

(5) Spot size

The radius ω_s at which the exponential envelope term falls to $1/e$ of its maximum value at $x = 0$ and $y = 0$ is the spot size ω_s of the Gaussian modes on the mirror. At this distance from $x = 0$ and $y = 0$, the intensity falls to $1/e^2$ of its maximum value. Therefore, for all TEM_{lm} modes, the spot size on the mirror is

$$\omega_s = \sqrt{b\lambda/\pi}. \tag{2.6}$$

Note that the spot size on the mirror is independent of the mode order, l and m. It is controlled only by the radius of curvature of the confocal mirror.

(6) Diffraction loss

There is a fractional energy loss per reflection γ_D. It is commonly called the diffraction loss per pass (i.e. the loss for propagation of the wave front from one reflector to the second reflector and reflected back by the second reflector) of the TEM_{lm} mode. It means that the diffracted field of the first mirror is only partially captured by the second mirror. Because of this loss, the magnitude of the eigen value χ_m is less than unity. There are two ways to calculate γ_D.

(a)

$$\gamma_D = 1 - |\chi_l \chi_m|^2. \tag{2.7}$$

(b) We can calculate γ_D from the ratio of the energy captured by the mirror to the total energy in the E field at the mirror, i.e.

$$\gamma_D = 1 - \frac{\iint_{\Omega} |E(x, y, z)|^2 \, dx \, dy}{\int_{-\infty}^{\infty}\int_{-\infty}^{\infty} |E(x, y, z)|^2 \, dx \, dy}. \tag{2.8}$$

E will be given in Eq. (2.10), and Ω is the aperture representing the mirror. Figure 2.3 shows the γ_D for several lowest order modes of the confocal resonators obtained by Boyd and Gordon [1], as well as the γ_D obtained by Fox and Li in their numerical calculation for two flat mirrors (see ref. [16], Chapter 1). This is a very important result. (1) Note that TEM_{lmq} and $TEM_{lmq'}$ modes have the same diffraction loss (i.e. the diffraction loss is independent of the longitudinal mode order). The diffraction loss increases, in general, for higher order transverse modes. Note also that the diffraction loss varies rapidly as a function of $a^2/b\lambda$. In lasers, we like to have just a single TEM oscillating mode most of the time. If the diffraction loss is sufficiently high for higher order modes, the lasers will not oscillate. Controlling

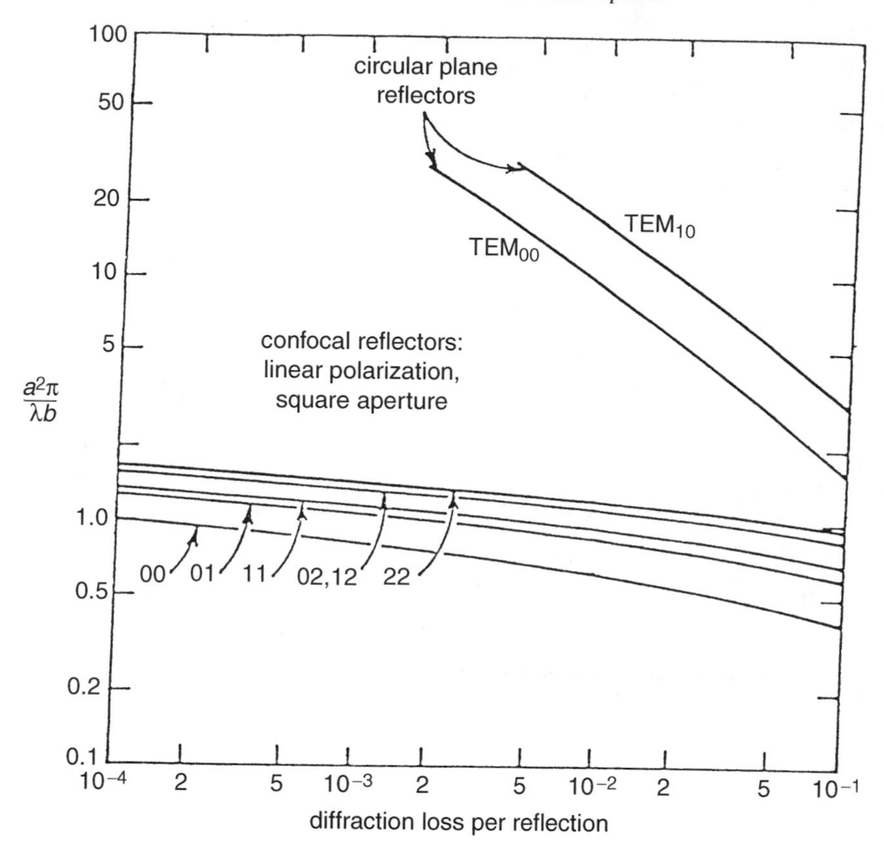

Figure 2.3. Diffraction loss per pass for the lowest order mode of a plane parallel cavity and for low order modes of confocal cavities; a is the mirror radius and b is the mirror spacing. The pairs of numbers under the arrows refer to the transverse mode order l and m of the confocal cavity; n is the refractive index of the material between the reflectors. Taken from ref. [6], by permission of the publisher.

the diffraction loss by the aperture size is a very important technique in laser design. (2) Note that in conventional Fabry–Perot interferometers $a^2/b\lambda$ is much bigger than in laser cavities, shown in Fig. 2.3. Therefore, diffraction loss is insignificant for many modes in Fabry–Perot interferometers. Diffraction loss is also insignificant in microwave cavities and in edge emitting semiconductor lasers.

(7) Quality factor Q

The sharpness of any resonance is commonly described by the quality factor Q. Let us consider the cavity resonance mode as two waves traveling simultaneously in opposite directions inside the cavity. Let the average stored electromagnetic energy per unit volume in the cavity be ρ_E. Since there are two oppositely propagating waves in a cavity, the intensity of one propagating wave is $c\rho_E/2$, where c is the velocity of light. The propagation loss for a round trip is twice the loss of a single pass. Therefore, the power loss due to diffraction loss for a given stored energy per unit volume ρ_{E} is

$$2\gamma_{\mathrm{D}}\int c\frac{\rho_{\mathrm{E}}}{2}\,ds.$$

Q is also the ratio of the stored energy to the power loss times ω_{r}, where ω_{r} is the resonance angular frequency. Thus, the cavity Q of the resonant mode due to diffraction loss alone is

$$Q_{\mathrm{c}}=\frac{\omega_{\mathrm{r}}\displaystyle\int\rho_{\mathrm{E}}\,dv}{\gamma_{\mathrm{D}}c\displaystyle\int\rho_{\mathrm{E}}\,ds}=\frac{2\pi d}{\gamma_{\mathrm{D}}\lambda}. \tag{2.9}$$

Let the stored energy be given by the number of photons in the mode. If we assume that the power loss is causing the decay of the number of photons in the mode, we obtain the photon decay time of the mode to be Q/ω_{r}.

2.1.4 Radiation fields inside and outside the cavity

Inside the cavity, the internal field U can be obtained by applying Kirchhoff's diffraction formula to the U on the mirror. If the mirror is partially transmitting, there will also be a radiation field outside the cavity. Since the transmission is usually low, the outside field will have a much smaller amplitude than the internal field. Since U must be continuous across a partially transmitting surface, the propagation of U outside the cavity can also be calculated by Kirchhoff's diffraction formula from the U on the mirror. The result is as follows:

$$\begin{aligned}E_{y_{lm}}(x,y,z)&=A\frac{2}{1+\xi^2}\frac{\Gamma[(m/2)+1]\Gamma[(l/2)+1]}{\Gamma(m+1)\Gamma(l+1)}\\&\times H_l\left(\frac{x}{\sqrt{\frac{b\lambda}{2\pi}}}\sqrt{\frac{2}{1+\xi^2}}\right)H_m\left(\frac{y}{\sqrt{\frac{b\lambda}{2\pi}}}\sqrt{\frac{2}{1+\xi^2}}\right)\exp\left[-\frac{kr^2}{b(1+\xi^2)}\right]\\&\times\exp\left(-j\left\{k\left[\frac{b}{2}(1+\xi)+\frac{\xi}{1+\xi^2}\frac{r^2}{b}\right]-(1+l+m)\left(\frac{\pi}{2}-\phi\right)\right\}\right),\end{aligned} \tag{2.10}$$

where $r^2=x^2+y^2$, $\xi=2z/b$, $\tan\phi=(1-\xi)/(1+\xi)$ and A is the amplitude.

Equation (2.10) implies that the amplitude spot size at any z is

$$\omega(z) = \sqrt{\frac{b\lambda}{2\pi}(1+\xi^2)}. \qquad (2.11)$$

The intensity of the radiation is proportional to $E_y E_y^*$, and thus the intensity falls to $1/e^2$ of its maximum value at the edge of the spot. Clearly the minimum spot size ω_0 is at $z = 0$:

$$\omega_0 = \sqrt{\frac{b\lambda}{2\pi}}. \qquad (2.12)$$

The Gaussian beam at $z = 0$ is known as the beam waist. Note again that, at large x and y, the amplitude variation will be dominated by the exponential function, instead of any polynomial function dependent on l and m. Thus, the spot size is independent of the order of the mode.

2.1.5 *Far field pattern of the TEM modes*

From Eq. (2.11), we can calculate ω_s/z for very large z. This ratio is the radiation beam-width θ_{rad} of the TEM modes at the far field,

$$\theta_{rad} = \tan^{-1}\left(\frac{\lambda}{\pi\omega_0}\right) \approx \frac{\lambda}{\pi\omega_0} = \sqrt{\frac{2\lambda}{\pi b}}. \qquad (2.13)$$

If we compare this far field beam-width, $\lambda/\pi\omega_0$, with the beam-width of a plane wave incident on a rectangular aperture given in Eqs. (1.21), $\lambda/2l_x$ or $\lambda/2l_y$, we immediately see the similarity between them. The main difference is the constant $\pi/2$. However, in the case of Eqs. (1.21) we defined the radiation intensity beam-width by the first node of the radiation intensity; whereas here we define the radiation beam-width as that point when the intensity falls to $1/e^2$ of its maximum.

2.1.6 *General expression for the TEM$_{lm}$ modes*

We can now rewrite Eq. (2.10) in terms of quantities that have clear physical meanings, as follows:

$$(E_y)_{lm} = E_0 \frac{\omega_0}{\omega(z)} H_l\left[\frac{\sqrt{2}x}{\omega(z)}\right] H_m\left[\frac{\sqrt{2}y}{\omega(z)}\right] \exp\left[-\frac{r^2}{\omega^2(z)}\right]$$
$$\times \exp\left[-jk\left(\frac{r^2}{2R(z)}\right)\right] \exp[-jkz + j(l+m+1)\eta]. \qquad (2.14)$$

Here, E_0 is just the amplitude, a constant, and

$$\omega = \omega_0 \left[1 + \left(\frac{z}{z_0} \right)^2 \right]^{1/2}, \qquad z_0 = \frac{b}{2},$$

$$R(z) = \frac{1}{z} \left[z^2 + z_0^2 \right],$$

$$\eta = \tan^{-1}(z/z_0).$$

The three exponential factors in Eq. (2.14) have important physical meanings. (1) The first exponential factor exhibits the Gaussian envelope amplitude variation at any z. This is the most commonly cited property of TEM cavities. (2) The second exponential factor exhibits the quadratic phase variation (i.e. the spherical wave front) with a specific radius of curvature $R(z)$ at each z value. Note that at $z = \pm d/2$, R is just the curvature of the confocal reflector, as we would expect. At $z = 0$, i.e. at the beam waist, the mode has a planar wave front as well as the smallest spot size. (3) The third exponential factor expresses the longitudinal phase shift in the z direction. The phase shift is important in determining the resonance frequency.

Note that the electric field distribution of any TEM_{lm} mode is independent of the size of the reflector. The Gaussian field description already included the diffraction effect without explicitly invoking Kirchhoff's formula. Only the diffraction loss is dependent on the reflector size.

Since U^*U is the intensity, the amplitude variation of the field is the dominant concern in conventional optics. We emphasize again that, in laser optics, the quadratic phase variation is equally important. For example, (1) high coupling efficiency between a specific laser mode and the mode of another optical component requires good phase matching as well as amplitude matching of the two modes; (2) phase variations are important in analyzing the diffraction; and (3) as the laser light encounters a lens, the quadratic phase variation will affect the focusing of the laser radiation.

2.1.7 Example illustrating the properties of confocal cavity modes

Consider a confocal cavity with end reflectors separated by 30 cm and $a = 0.5$ mm. The medium between the mirrors is air, i.e. $n = 1$. The wavelength is 1 μm. The reflectors are 99% reflection and 1% transmission in intensity. The confocal resonator modes will have a beam waist size on the mirror of $\sqrt{b\lambda/\pi} = 0.3$ mm, which is independent of and much smaller than the mirror size. The mode pattern in the x and y directions will not be dependent on the mirror size.

The mode pattern will depend only on the mode order, l and m, and $b\lambda$. According to Eq. (2.11), the radiation field of the mode assumes its far field pattern when $4z^2/b^2 \gg 1$. The beam divergence angle at the far field is given by Eq. (2.13) as

$\sqrt{2\lambda/\pi b}$, which is 1.5×10^{-3} radians and is independent of the mode order. Note that the condition for far field is different from the far field condition for Kirchhoff's diffraction given in Eq. (1.18) that depends on λ and which is much harder to satisfy.

The diffraction loss per pass will depend on the mode order, l and m. For this cavity, $a^2/b\lambda = 0.83$. From Fig. 2.3, the diffraction loss for the TEM_{00} mode is approximately 10^{-3} per pass. The diffraction loss per pass for the TEM_{01} or TEM_{10} mode jumps to 2×10^{-2}, while the loss per pass for the TEM_{11} mode is 5×10^{-2}. The mirror size, a, is much larger than the spot size. The mode patterns in the x and y variations are the same. According to Eq. (2.8), the diffraction loss per pass will be independent of whether the mirrors are square or round in cross-section as long as the area of the mirror is approximately the same. Since the transmission is 1%, the total loss per pass is 1.1×10^{-2} for the TEM_{00} mode, 3×10^{-2} for the TEM_{01} or TEM_{10} mode, and 6×10^{-2} for the TEM_{11} mode. Notice the sensitivity of the diffraction loss per pass to changes in $a^2/b\lambda$. In order to obtain a much larger loss per pass for the TEM_{01}, TEM_{10} or TEM_{11} mode, it is necessary to reduce the mirror size a. Conversely, at $a = 0.525$ mm, the total losses for these modes are: 1.02×10^{-2}, 1.5×10^{-2} and 2×10^{-2}, respectively. The increase of total loss per pass for the higher order modes is much less significant for the larger mirrors. A favorite practical trick to increase the differential losses of the higher order modes is to put an aperture in front of the mirror to reduce a. In other cases, the effective a of the cavity may be limited by other considerations, such as the size of the laser tube.

2.2 Modes in non-confocal cavities

In this section we will find the modes of non-confocal cavities with arbitrary spherical end reflectors at a given distance of separation by identifying them with the modes of a virtual equivalent confocal cavity as follows.

(1) We will first show that the reflectors of any given confocal resonator can be replaced by other reflectors at various locations and with specific radius of curvature. Such a replacement will not change the resonant mode pattern. We call this technique the formation of a new cavity for known modes of confocal resonators.

(2) We will then solve the inverse problem: how to find the virtual equivalent confocal resonator for a given pair of non-identical spherical mirrors at a given distance of separation.

(3) Once we have found the virtual equivalent confocal resonator, we will obtain the properties of the modes of the original resonator, such as the field pattern, diffraction loss, resonance frequencies, etc., from the modes of the virtual resonator.

(4) We will illustrate how to find the modes in non-confocal cavities via an example.

2.2.1 *Formation of a new cavity for known modes of confocal resonators*

Let us first examine closely the consequence of the confocal resonator modes found in Section 2.1. Equation (2.14) implies that there is a constant phase surface for any resonator mode whenever the x, y and z satisfy the condition

$$z + \frac{r^2}{2R(z)} = \text{constant}.$$

It is clear that if a reflector with curvature $R(z)$ is placed at this z position to replace one of the confocal mirrors at $z = \pm d/2$, we will have the same Gaussian transverse mode as in the case of the original confocal cavity. The frequency at which resonance will occur will be shifted because η is a function of z, and the round trip distance of propagation will be different from that of the original confocal resonator. However, the transverse mode variation will be the same. The spot size on this mirror at z is given by the ω in Eq. (2.14). The diffraction loss per pass will depend on the size of the reflectors.

In other words, a new optical cavity can be formed with mirrors at z_1 and z_2, provided that

$$R_1 = z_1 + \frac{z_0^2}{z_1},$$
$$R_2 = z_2 + \frac{z_0^2}{z_2}. \qquad (2.15)$$

The transverse lm modes of the original confocal resonator are also modes of this new cavity. The resonant modes of the new cavity will have the same transverse field variation as the modes of the original cavity. The diffraction loss of the modes will be the same in the original cavity and in the new cavity when the mirror size varies proportionally to $\omega(z)$. Figure 2.4 illustrates the surfaces of constant phase at two z positions. Note that one of z_1 and z_2 can have negative values, producing a negative R. Negative R means we have a curved mirror at $z < 0$ bending toward $z = 0$. As $|z_2|$ or $|z_1|$ becomes very large, $|R_1|$ and $|R_2|$ become approximately the same as $|z_1|$ or $|z_2|$, i.e. the surface of constant phase is approximately the same as a spherical wave originating from $z = 0$. As $|z_1|$ or $|z_2|$ becomes very small, $|R_1|$ or $|R_2|$ becomes very much larger than $|z_1|$ or $|z_2|$. At $z = 0$, the surface of constant phase is a plane.

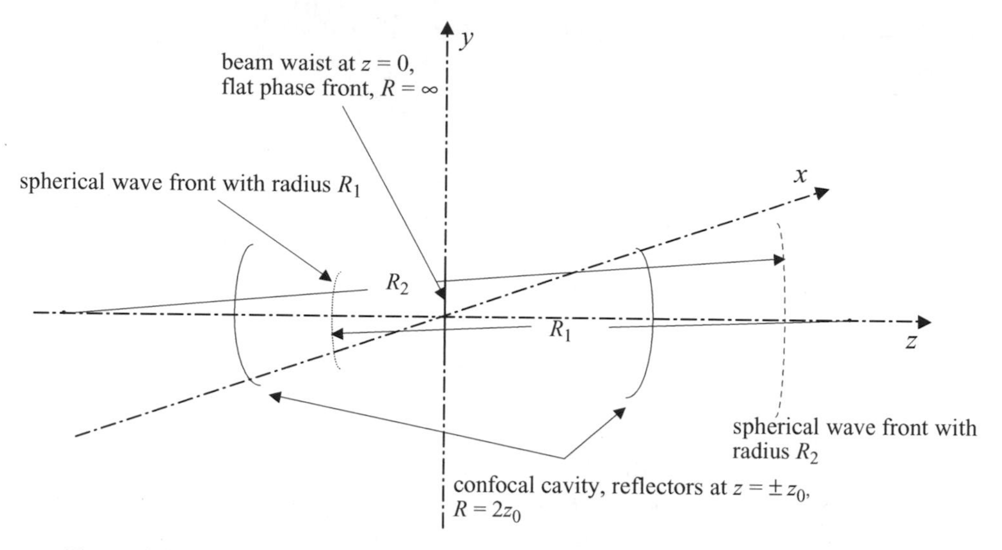

Figure 2.4. Illustration of constant phase fronts of the modes of confocal resonators. The confocal cavity is shown as two spherical reflectors at $z = \pm z_0$. The radius of these confocal spherical reflectors is $2z_0$. The modes of the confocal cavity have a spherical wave front inside and outside the cavity. Outside, a spherical wave front (dashed line) is shown to have a radius of curvature R_2. Inside, a spherical wave front (dotted line) is shown to have a radius of curvature R_1. The beam waist (solid line) is at $z = 0$; the modes have a flat wave front at this position.

2.2.2 *Finding the virtual equivalent confocal resonator for a given set of reflectors*

If there are two mirrors with curvatures R_1 and R_2, separated by a distance D, we can find z_1 and z_2 to fit R_1 and R_2 according to Eqs. (2.15) as follows:

$$
\begin{aligned}
z_1 &= \frac{R_1}{2} \pm \frac{1}{2}\sqrt{R_1^2 - 4z_0^2}, \\
z_2 &= \frac{R_2}{2} \pm \frac{1}{2}\sqrt{R_2^2 - 4z_0^2}.
\end{aligned}
\tag{2.16}
$$

Here, $\pm\, z_0$ are the positions of the mirrors for the virtual equivalent confocal resonator that will have the same transverse modes. However, we still need to determine z_0.

In order to find z_0, we shall first observe some important conditions for z_0. Assuming $z_2 > z_1$, we obtain

$$
D = z_2 - z_1 = \frac{R_2}{2} - \frac{R_1}{2} \pm \frac{1}{2}\sqrt{R_2^2 - 4z_0^2} \mp \frac{1}{2}\sqrt{R_1^2 - 4z_0^2}.
$$

Rearranging terms and squaring both sides to eliminate the square root, we obtain

$$z_0^2 = \frac{D(-R_1 - D)(R_2 - D)(R_2 - R_1 - D)}{(R_2 - R_1 - 2D)^2}. \quad (2.17)$$

Clearly, z_0 must be a positive quantity in order to obtain real values of the equivalent confocal resonator position.

Equation (2.17) allows us to calculate z_0 with a real value only when the right hand side is positive. The requirement for the right hand side to be positive also imposes certain conditions on R_1, R_2 and D as follows. Let us assume that R_1 is negative at negative z_1. Then, we must have

$$D(|R_1| - D)(|R_2| - D)(|R_1| + |R_2| - D) > 0.$$

There are only two ways to satisfy this condition: (1) $0 < D < |R_1|$ or $|R_2|$, whichever is smaller; (2) $|R_1| + |R_2| > D > |R_1|$ or $|R_2|$, whichever is larger. Condition (1) can be expressed as $0 < (1 - D/|R_1|)(1 - D/|R_2|)$. Condition (2) can be expressed as $(1 - D/|R_1|)(1 - D/|R_2|) < 1$. Hence the criterion for the existence of a resonator mode, equivalent to a confocal resonator mode with z_0 given in Eqs. (2.15), is

$$0 < \left(1 - \frac{D}{|R_1|}\right)\left(1 - \frac{D}{|R_2|}\right) < 1. \quad (2.18)$$

If we plot this in a rectangular coordinate system with the two axes as $D/|R_1|$ and $D/|R_2|$, then the lower limit of Eq. (2.18), where the product of the two quantities in the brackets is zero, is represented by two straight lines, $D/|R_1| = 1$ and $D/|R_2| = 1$. On the other hand, the upper limit,where the product of the two brackets is unity, is represented by a hyperbola. Figure 2.5 shows this plot. The shaded regions show the combinations of R_1, R_2 and D that satisfy the inequality in Eq. (2.18). Resonators with these combinations are called stable resonators. The regions outside the shaded regions are called the unstable resonator regions. The confocal resonator configuration has $D = |R_1| = |R_2|$. Thus, the confocal resonator can easily be pushed into the unstable region by a slight misalignment of the cavity. In reality, the assumptions used in our diffraction loss calculation break down near the boundaries of stable and unstable regions. More precise calculations show the diffraction loss increases rapidly from the stable to the unstable configuration. There is no sudden change in diffraction loss from the stable to the unstable configuration. Unstable resonator modes not only exist, they are often used in very high power lasers so that optical energy is not concentrated in a small physical region to avoid material damage by the high electric field [7].

In summary, when the given R_1, R_2 and D satisfy the stability criterion in Eq. (2.18), z_0, z_1 and z_2 are determined from Eqs. (2.16) and (2.17); z_0 provides us with the specifications of the virtual equivalent confocal resonator. Note that

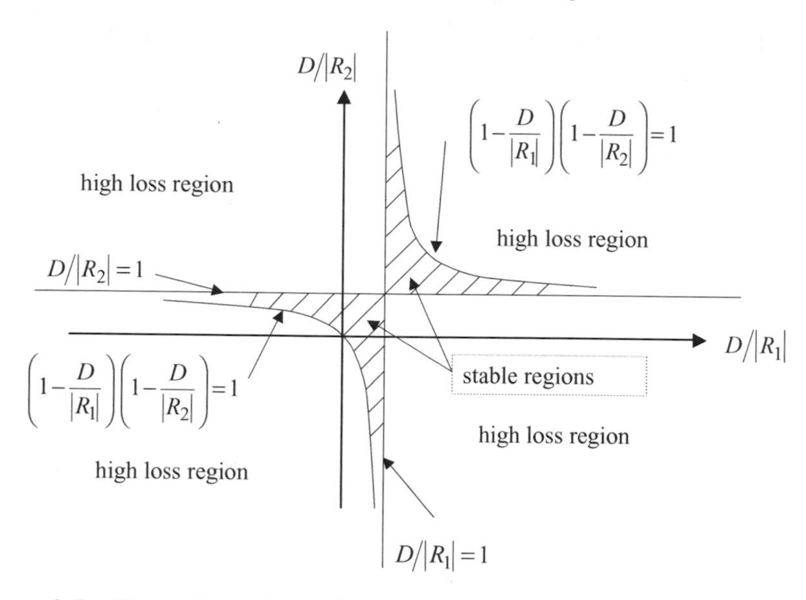

Figure 2.5. The stable and unstable regions of laser cavities. The straight lines are the plots of the lower limit of Eq. (2.18), and the hyperbola is the plot of the upper limit. The shaded region (i.e. the stable region) shows the $D/|R_1|$ and $D/|R_2|$ values that satisfy Eq. (2.18). In this region, modes have a low or moderate diffraction loss per pass. Cavities in the high loss region do not have $D/|R_1|$ and $D/|R_2|$ values that satisfy Eq. (2.18). It is called the unstable region.

the $\pm$ sign in Eqs. (2.16) gives us two answers for z_1 and for z_2. The correct choice is the one that gives the correct D.

2.2.3 *Formal procedure to find the resonant modes in non-confocal cavities*

A formal procedure can now be set up to find the resonant modes in non-confocal cavities according to the analysis presented in Section 2.2.2. We will first test the stability of the given R_1, R_2 and D according to Eq. (2.18). For stable cavities, we will find the field pattern, the diffraction loss and the resonant frequency of their resonant modes by the following seven steps.

(1) Calculate z_0, z_1 and z_2 from Eqs. (2.16) and (2.17); z_1 and z_2 determine the center position (i.e. the $z = 0$ plane) of the equivalent virtual confocal cavity and z_0 determines the separation and the radius of curvature of the equivalent virtual confocal cavity.
(2) The minimum spot size of all modes at $z = 0$ is $\omega_0 = \sqrt{\lambda z_0/\pi}$.
(3) The spot sizes on the two reflectors are

$$\begin{aligned}\omega_{s1} &= \omega_0\sqrt{1+(z_1/z_0)^2},\\ \omega_{s2} &= \omega_0\sqrt{1+(z_2/z_0)^2}.\end{aligned} \tag{2.19}$$

(4) Let the sizes of the two mirrors be a_1 and a_2. In order to calculate the diffraction loss of a non-confocal resonator, we first find the equivalent sizes of the virtual confocal mirrors, $a_{eq,1}$ and $a_{eq,2}$, that will be proportional to a_1 and a_2. The proportionality is the ratio of the spot sizes on the actual mirrors to the spot sizes on the equivalent confocal resonator. The confocal resonator with $a_{eq,1}$ and $a_{eq,2}$ will have the same diffraction loss as the actual cavity with a_1 and a_2. Using Eqs. (2.19) as a guide, we obtain

$$a_{eq,1} = (\sqrt{2}\omega_0)a_1/\omega_{s1},$$
$$a_{eq,2} = (\sqrt{2}\omega_0)a_2/\omega_{s2}.$$

(5) For a symmetrical cavity, the diffraction loss per pass is calculated directly from the confocal resonator with the size a_{eq}. For asymmetrical cavities, the diffraction loss per pass is the average of the diffraction losses. The averaged diffraction loss per pass for the cavity is one-half the sum of the diffraction loss for the two different virtual confocal cavities, one with mirror size $a_{eq,1}$ and one with mirror size $a_{eq,2}$.

(6) In general, the resonance wavelength, λ_{lmq}, is determined by

$$(2\pi D/\lambda_{lmq}) = q\pi + (l+m+1)\left[\tan^{-1}(z_2/z_0) - \tan^{-1}(z_1/z_0)\right].$$

The differences in resonance frequency for different longitudinal order q and transverse order l and m are

$$f_{l,m,q+1} - f_{l,m,q} = c/2D,$$
$$f_{l',m',q} - f_{l,m,q} = \frac{c}{2\pi D}\left(\frac{\pi}{2} - \tan^{-1}\frac{z_2}{z_0} - \tan^{-1}\frac{z_1}{z_0}\right)(l' - l + m' - m).$$

Note again that the difference in resonance frequency for two adjacent longitudinal orders is just $1/T$, where T is the round trip propagation time inside the cavity, $T = 2D/c$. If the cavity is filled with a dielectric that has an index of refraction n, $T = 2nD/c$. The transverse modes are still degenerate. All modes which have the same $l+m$ orders will have the same resonance frequency.

(7) Practical resonators do not use end mirrors with square cross-sections. It is clear from the previous discussions that the mode patterns (i.e. the Hermite polynomials and the Gaussian envelope) will be affected only by the curvature and the position of the reflector, not by the shape of the cross-section, e.g. whether it is square or round. Thus the modes derived for the square mirrors are equally applicable to the round mirrors. From Eq. (2.8), it is clear that the diffraction loss per pass depends primarily on the area Ω of the mirror. Round or square mirrors with the same size are likely to have the same diffraction loss per pass. Thus, Fig. 2.3 may also be used to estimate the diffraction loss for the round mirrors.

2.2.4 Example of resonant modes in a non-confocal cavity

Let us consider a semi-spherical cavity which has one flat reflector with $a_1 = 2$ mm and one spherical reflector with radius of curvature $R_2 = 0.7$ m and $a_2 = 0.6$ mm, separated by a distance $D = 30$ cm. The wavelength is 1 μm. The medium between the reflectors is air. Clearly, the stability criterion in Eq. (2.18) is satisfied so that we can find the modes and their diffraction losses by means of the virtual equivalent confocal cavity. Following the process outlined in (3) above, we obtain the following results.

For the equivalent virtual confocal cavity, $z_0 = [D(R_2 - D)]^{1/2} = 0.346\,410$ m, $z_1 = 0$, $z_2 = 0.3$ m. Notice that there are two solutions for z_2 given in Eqs. (2.16). The correct solution is the one which yields $z_2 - z_1 = D$.

The spot sizes are $\omega_0 = 0.332\,063$ mm, $\omega_{s1} = 0.332\,063$ mm and $\omega_{s2} = 1.322\,88 \times 0.332\,063 = 0.439\,28$ mm.

The appropriate sizes of the equivalent confocal reflectors for the calculation of diffraction loss are $a_{eq,1} = 2.828\,43$ mm and $a_{eq,2} = 0.641\,427$ mm.

For reflector 1, $a^2/2z_0\lambda$ is 11.5. For reflector 2, $a^2/2z_0\lambda$ is 0.59. Therefore, the diffraction loss per pass of the TEM_{00} mode obtained from Fig. 2.3 for the flat mirror is negligible, while the diffraction loss per pass for the second mirror is 5×10^{-3}. The averaged diffraction loss per pass for the cavity is 2.5×10^{-3}. The averaged diffraction loss per pass for the TEM_{01} mode will be approximately 5%.

2.3 Gaussian beam solution of the vector wave equation

We have learned that the Gaussian beam is a solution of the integral equation of the fields in cavities with spherical reflectors. It is based on the scalar wave equation. It is instructional to learn that Gaussian beams are also natural mathematical solutions of Maxwell's equations without the restrictions of the scalar wave equation and without the existence of a cavity [6].

We have seen many special forms of the solutions of Maxwell's equations such as plane waves, cylindrical waves, spherical waves, etc. We will learn in this section how a Gaussian beam is just another form of these types of solutions. Knowing that the Gaussian beam is a natural solution of Maxwell's equations, we feel more comfortable in approximating various radiation fields from components that are not lasers, e.g. output from a single-mode fiber, by Gaussian beams. The advantage of the use of Gaussian beam approximation is that the wave equation is satisfied without the use of Kirchhoff's diffraction formula. Fields propagating through reasonably large apertures retain the same functional variation, except for a reduction in its amplitude.

Consider Maxwell's equations,

$$\begin{aligned} \nabla \times \underline{h} &= \varepsilon(\partial \underline{e}/\partial t), \\ \nabla \times \underline{e} &= -\mu(\partial \underline{h}/\partial t), \\ \nabla \cdot (\varepsilon \underline{e}) &= 0. \end{aligned}$$

In the most general case, ε can be a function of (x, y, z). Using $\nabla \times \nabla \times \underline{e}$, we obtain

$$\nabla^2 \underline{e} - \varepsilon\mu \frac{\partial^2 \underline{e}}{\partial t^2} = -\nabla \left(\frac{1}{\varepsilon} \underline{e} \cdot \nabla\varepsilon \right).$$

If $\nabla\varepsilon \perp \underline{e}$ (such as the ε variation in an optical fiber) or if $\nabla\varepsilon$ is small, we can then replace the right hand side with zero. If we further assume the time variation to be $\exp(j\omega t)$, then the equation for the electric field is again

$$\nabla^2 \underline{e} + k^2(\underline{r})\,\underline{e} = 0, \tag{2.20}$$

where

$$k^2(\underline{r}) = \omega^2/\mu\varepsilon(\underline{r}).$$

When the medium is homogeneous, k is a constant. The significance of Eq. (2.20) is that an equation similar to the scalar wave equation can be obtained for e under more general situations than for TEM waves.

In the case of plane waves in classical textbooks, we have assumed that the electric and the magnetic fields do not vary in the lateral directions. The plane waves are solutions of Maxwell's equations in that format. We will now show that Gaussian waves are cylindrically symmetric solutions of Eq. (2.20).

Let E be a linearly polarized field and

$$E(x, y, z) = \psi(x, y, z)\, e^{-jkz}. \tag{2.21}$$

We will now show that, in a homogeneous medium, a circular symmetric ψ has a functional form identical to that of a Gaussian beam. We do this in five mathematical steps.

Step 1. Substituting Eq. (2.21) into Eq. (2.20) in cylindrical coordinates with $\partial\psi/\partial\theta = 0$, we obtain

$$\nabla_t^2 \psi - 2jk \frac{\partial \psi}{\partial z} = 0,$$

where

$$\nabla^2 = \nabla_t^2 + \frac{\partial^2}{\partial z^2} = \frac{\partial^2}{\partial r^2} + \frac{1}{r}\frac{\partial}{\partial r} + \frac{\partial^2}{\partial z^2}.$$

Step 2. Let

$$\psi = \exp\left\{ -j \left[p(z) + \frac{k}{2q(z)} r^2 \right] \right\}.$$

Substituting this functional form into the equation, we obtain

$$-\left(\frac{k}{q}\right)^2 r^2 - 2j\left(\frac{k}{q}\right) - k^2r^2\frac{\partial}{\partial z}\left(\frac{1}{q}\right) - 2k\frac{\partial p}{\partial z} = 0.$$

This equation must hold for all values of r. Thus the terms involving different powers of r must vanish simultaneously, i.e.

$$\frac{1}{q^2} + \frac{\partial}{\partial z}\left(\frac{1}{q}\right) = 0,$$
$$\frac{\partial}{\partial z}p = \frac{-j}{q}.$$

Step 3. Let $1/q = (dS/dz)/S$, then the equation for S is $d^2S/dz^2 = 0$. The solution for S is obviously

$$S = az + b,$$
$$q = S/(dS/dz) = z + b/a = z + q_0. \tag{2.22}$$

Substituting this solution into the equation for $p(z)$, we obtain

$$\frac{\partial p}{\partial z} = -\frac{j}{z+q_0},$$
$$p(z) = -j\ln\left(1 + \frac{z}{q_0}\right).$$

Step 4. The objective of finding the solutions for p and q is to show that ψ has the functional form of a TEM_{00} Gaussian beam. We can accomplish this by replacing q_0 by a new constant $q_0 = j\pi\omega_0^2/\lambda$. After such a substitution, we obtain

$$e^{-jp(z)} = e^{-\ln\left(1-\frac{j\lambda z}{\pi\omega_0^2}\right)} = \frac{1}{\sqrt{1 + (\lambda^2 z^2/\pi^2\omega_0^4)}} e^{\left[j\tan^{-1}\left(\frac{\lambda z}{\pi\omega_0^2}\right)\right]}$$

$$e^{\frac{-jkr^2}{2(z+q_0)}} = \exp\left\{\frac{-r^2}{\omega_0^2\left[1 + \left(\frac{\lambda z}{\pi\omega_0^2}\right)^2\right]}\right\}\exp\left\{\frac{-jkr^2}{2z\left(1 + \left(\frac{\pi\omega_0^2}{\lambda z}\right)^2\right)}\right\}.$$

Step 5. Substituting the above results into the expression for ψ, we obtain an expression for the E in Eq. (2.21) identical to the TEM_{00} mode in Eq. (2.14),

$$E = \frac{1}{\sqrt{1 + (\lambda^2 z^2/\pi^2 \omega_0^4)}} \exp\left\{ \frac{-r^2}{\omega_0^2 \left[1 + \left[\frac{\lambda z}{\pi \omega_0^2} \right]^2 \right]} \right\}$$
$$\times \exp\left\{ \frac{-jkr^2}{2z \left[1 + \left(\frac{\pi \omega_0^2}{\lambda z} \right)^2 \right]} \right\} e^{-jkz} e^{\left[j \tan^{-1}\left(\frac{\lambda z}{\pi \omega_0^2} \right) \right]}.$$

In summary, the Gaussian beam is a natural solution of Maxwell's vector wave equations with $\nabla\varepsilon \perp \underline{e}$ or $\nabla\varepsilon \approx 0$. We have only derived the Gaussian mode for a homogeneous medium. Yariv [6] shows that when $k^2(r) = k^2 - kk_2 r^2$ in an inhomogeneous graded index medium, the solution of Eq. (2.20) for a circular symmetric mode is still a Gaussian beam.

2.4 Propagation and transformation of Gaussian beams (the *ABCD* matrix)

2.4.1 Physical meaning of the terms in the Gaussian beam expression

We note that for a given Gaussian beam, as a solution of cavity resonance, we can describe its functional variation at various values of z by

$$E = A(x, y) e^{-jkz} e^{-jp(z)} e^{-jk\frac{r^2}{2q(z)}}, \tag{2.23}$$
$$A(x, y) = E_0 H_l \left[\frac{\sqrt{2}x}{\omega(z)} \right] H_m \left[\frac{\sqrt{2}y}{\omega(z)} \right],$$

where the coordinate z starts at the beam waist where the spot size is ω_0. The labeling of the parameters by p and q is inspired by the discussion in the preceding section. Please keep in mind that the E given here is taken from Eq. (2.14), not from the solution of ψ. However, it has been shown in Section 2.3 that E is also a natural solution of Maxwell's equations when $l = m = 0$.

The first factor, A, describes the x and y variation (i.e. the field pattern) of E. At two different z positions, z_1 and z_2, the A function will be the same. A is different for different l and m order of the mode.

The second factor, $\exp(-jkz)$, and the third factor, $\exp(-jpz)$, are simple functions of z. They specify the phase of the beam as the beam propagates from one z position to another. They are independent of x and y. Parameter p is dependent on the mode order, l and m, and

$$e^{-jp(z)} = \frac{\omega_0}{\omega(z)} e^{j(l+m+1)\tan^{-1}\left(\frac{\lambda z}{\pi\omega_0^2}\right)}. \tag{2.24}$$

Thus, $p + kz$ determines the phase of E at different z.

The $1/q$ factor carries the most important physical meaning of the Gaussian beam. This factor has a real part which specifies the curvature of the phase front and an imaginary part which specifies the Gaussian variation of the amplitude at any z. To be more specific,

$$\frac{1}{q} = \frac{1}{R} - \frac{j2}{k\omega^2};$$

q is independent of the mode order, l and m.

Also, q will be different at different z positions,

$$\frac{1}{q} = \frac{1}{z + q_0}.$$

From Eq. (2.22), the q values at two z values are related to each other by

$$q(z_2) - q(z_1) = z_2 - z_1. \tag{2.25}$$

2.4.2 Description of Gaussian beam propagation by matrix transformation

It is important to note that as a Gaussian beam propagates the E is always given by Eq. (2.23). The relationship between q at z_1, call it q_1, and q at z_2, call it q_2, is a linear relationship. Instead of writing the Gaussian beam as a function of coordinates x, y and z, we may write the relation between q_1 and q_2 in the formal form of a linear transformation,

$$q_2 = \frac{Aq_1 + B}{Cq_1 + D}, \tag{2.26}$$

where $A = 1,\ B = z_2 - z_1,\ C = 0,\ D = 1,\ q_1 = q(z_1)$ and $q_2 = q(z_2)$. In other words, q_2 is transformed from q_1 by a linear transformation with the above $ABCD$ coefficients.

A linear transformation relationship also exists between q values for Gaussian beams transmitting or reflecting from various optical components. When a Gaussian beam is incident on an ideal thin lens, we learned in Eq. (1.27) that the transmitted field immediately after the lens, E_t, is related to the incident field, E_{inc}, by the transmission function of the lens, which is a quadratic phase shift; i.e.

$$E_t = E_{inc}\, e^{j\frac{\pi}{\lambda f}(x^2+y^2)} = Ae^{-jkz}e^{-jp(z)}e^{-j\frac{\pi}{\lambda}\left(\frac{1}{q}-\frac{1}{f}\right)r^2}.$$

Therefore, the transmitted beam will have the same form as given in Eq. (2.23). Let q_1 be the q parameter before the lens and let q_2 be the q parameter after the lens. q_2 is related to q_1 of the incident beam by

$$\frac{1}{q_2} = \frac{1}{q_1} - \frac{1}{f}. \tag{2.27}$$

When we separate the imaginary and real parts of Eq. (2.27), we obtain

$$\frac{1}{R_2} = \frac{1}{R_1} - \frac{1}{f}, \qquad \omega_2 = \omega_1. \tag{2.28}$$

This implies that the spot size is not changed by transmission through a thin lens. However, the radius of curvature of the phase front is changed according to Eq. (2.28). We conclude that q_2 and q_1 are again related by Eq. (2.26) with $A = 1$, $B = 0$, $C = -1/f$ and $D = 1$. The p does not change when the beam propagates through a thin lens.

If the lens is set in a finite aperture, the transmitted Gaussian beam will have the same functional variation as for an infinite aperture. However, the amplitude will be reduced. The reduction in amplitude will be identical to the amplitude reduction calculated from the diffraction loss per pass caused by the same aperture.

The *ABCD* transformation method is applicable to the propagation of Gaussian modes through many other optical elements. The *ABCD* transformation coefficients of various optical elements are given in many textbooks [6]; see Table 2.1. The diffraction loss is not shown in Table 2.1; it may be calculated using the procedure outlined in Section 2.2.3.

If a Gaussian beam propagates through more than one optical element, the q parameters at various positions can be determined by *ABCD* transformations in succession. In other words, for two successive transformations:

$$q_3 = \frac{A_2 q_2 + B_2}{C_2 q_2 + D_2}, \qquad q_2 = \frac{A_1 q_1 + B_1}{C_1 q_1 + D_1},$$

and thus

$$q_3 = \frac{(A_1 A_2 + B_2 C_1) q_1 + (A_2 B_1 + B_2 D_1)}{(A_1 C_2 + C_1 D_2) q_1 + (B_1 C_2 + D_1 D_2)} = \frac{A q_1 + B}{C q_1 + D}.$$

The *ABCD* coefficients for q_3 in terms of q_1 in the above equation are simply the coefficients obtained by multiplying matrix $A_1 B_1 C_1 D_1$ by matrix $A_2 B_2 C_2 D_2$, as follows:

$$\begin{Vmatrix} A & B \\ C & D \end{Vmatrix} = \begin{Vmatrix} A_2 & B_2 \\ C_2 & D_2 \end{Vmatrix} \times \begin{Vmatrix} A_1 & B_1 \\ C_1 & D_1 \end{Vmatrix}. \tag{2.29}$$

Table 2.1. *The ABCD transformation matrix for some common optical elements and media.*

Transformation description	Figure	Matrix
Homogeneous medium Length d	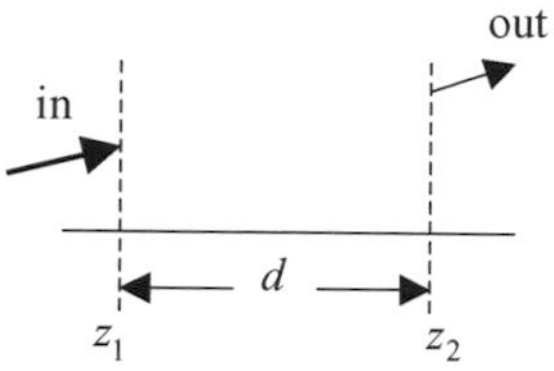 in, out, d, z_1, z_2	$\begin{bmatrix} 1 & d \\ 0 & 1 \end{bmatrix}$
Thin lens Focal length f ($f > 0$, converging; $f < 0$, diverging)	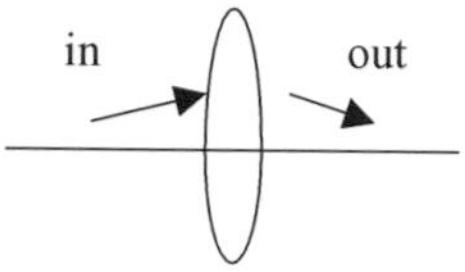 in, out	$\begin{bmatrix} 1 & 0 \\ \dfrac{-1}{f} & 1 \end{bmatrix}$
Dielectric interface Refractive indices n_1, n_2	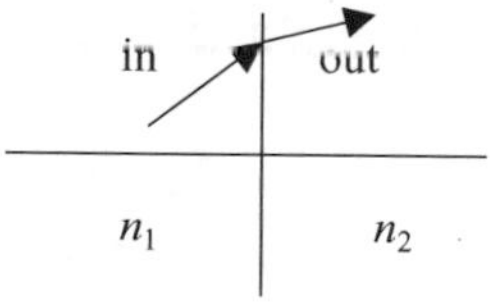 in, out, n_1, n_2	$\begin{bmatrix} 1 & 0 \\ 0 & \dfrac{n_1}{n_2} \end{bmatrix}$
Spherical dielectric interface Radius R	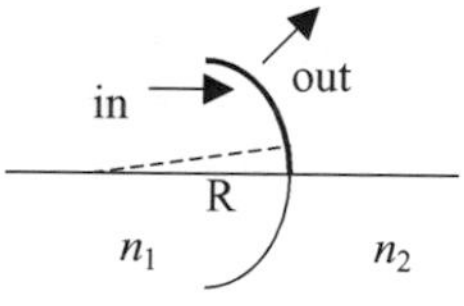 in, out, R, n_1, n_2	$\begin{bmatrix} 1 & 0 \\ \dfrac{n_2 - n_1}{n_2 \cdot R} & \dfrac{n_1}{n_2} \end{bmatrix}$
Spherical mirror Radius of curvature R	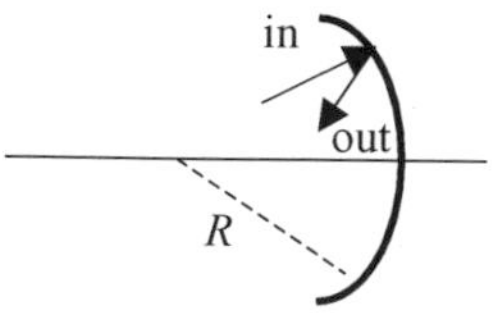 in, out, R	$\begin{bmatrix} 1 & 0 \\ \dfrac{-2}{R} & 1 \end{bmatrix}$

After the Gaussian beam has propagated through many optical elements, this matrix multiplication process can be repeated many times to obtain the *ABCD* coefficients for the total transformation matrix. Thus the *ABCD* coefficients are called the *ABCD* transformation matrices. It can be shown that any *ABCD* matrix is a unitary matrix, i.e. $AD - BC = 1$. It is important to keep in mind that the order of multiplication must follow the order in which the Gaussian beam is propagating through various elements. It cannot be taken for granted that permutation of the order of matrix multiplication will give the same result.

The p changes only when the z position changes. Therefore, when the TEM_{lm} mode passes through any element which has zero thickness, such as a thin lens, p does not change. After the mode propagates through many elements and distances, the new p is obtained by using the total distance of propagation as z. $A(x, y)$ does not change.

2.4.3 Example of a Gaussian beam passing through a lens

As the first example, consider a Gaussian beam at $\lambda = 1\ \mu\mathrm{m}$ with $\omega_0 = 0.4$ mm at $z = 0$. It propagates through a thin lens with $f = 2$ mm at $z = 0.1$ m. Let us find the field pattern at $z = 0.1$ m after the lens.

There are two ways to find the answer. (1) We can find the answer using Eq. (2.14) for the Gaussian beam. The given Gaussian beam has $z_0 = \pi\omega_0^2/\lambda = 0.502\,665$ m. From Eq. (2.14), we know the field pattern for any TEM_{lm} mode incident on the lens at $z = 0.1$ m. It has a Gaussian amplitude variation with $\omega = 0.407\,839$ mm, a radius of curvature for the phase front $R = 2.626\,62$ m, and a phase shift given by $\eta = 0.1964$ radian. According to Eq. (2.28), the radiation field emerging from the thin lens will have the same phase and amplitude variation. However, the radius of curvature for the phase front will now be $Rf/(R - f)$, which is $2.001\,52 \times 10^{-3}$ m. We would intuitively expect such an answer because the lens should create a focused spot near its focal plane.

(2) The answer could also be obtained very quickly from the $ABCD$ matrix transformation as follows:

$$\begin{Vmatrix} A & B \\ C & D \end{Vmatrix} = \begin{Vmatrix} 1 & 0 \\ \dfrac{-1}{0.002} & 1 \end{Vmatrix} \times \begin{Vmatrix} 1 & 0.1 \\ 0 & 1 \end{Vmatrix} = \begin{Vmatrix} 1 & 0.1 \\ -500 & -49 \end{Vmatrix}.$$

At $z = 0$, the q is $jk\omega_0^2/2$, which is $j0.502\,655$. Therefore at the exit plane of the lens,

$$\frac{1}{q} = \frac{Cq_1 + D}{Aq_1 + B} = \frac{-500\,(j0.502\,655) - 49}{j0.502\,655 + 0.1}$$
$$= \frac{-131.231 - j0.502\,645}{0.262\,662}$$
$$= -499.619 - j1.913\,66.$$

The real part of the $1/q$ is $1/R$, and the imaginary part of the $1/q$ is $-\lambda/\pi\omega^2$. Note that the complete expression for the field is given in Eqs. (2.23) and (2.24) with this q value.

2.4.4 Example of a Gaussian beam passing through a spatial filter

As the second example, let us re-consider example (6) of Section 1.5 when the incident beam is a Gaussian beam. We will show that the *ABCD* transformation matrix method lets us find the main propagation characteristics of the incident beam without any integration. We will need to perform integration only when we want to know the diffraction loss.

Figures 1.11 and 1.12 have already illustrated the geometrical configuration of this spatial filtering setup. Let the incident beam be a TEM_{00} Gaussian beam incident on the film at $z = 0$. The beam waist is at $z = 0$ with spot size ω_0, where $\omega_0 \ll d$. Notice now the effective beam size is controlled by ω_0 and not by d:

$$E = E_0 e^{-jkz} e^{-jp(z)} e^{-jk\frac{r^2}{2q_0}}.$$

Therefore,

$$\frac{1}{q_0} = -\frac{j2}{k\omega_0^2}.$$

For $d \gg \omega_0$, the aperture size d does not change the functional form of the Gaussian beam. It may introduce a reduction of the amplitude because of the diffraction loss caused by the aperture. Immediately after the film with transmission function t, at $z = 0$,

$$\begin{aligned} E &= \frac{1}{2} E_0 e^{-jkz} e^{-jp(z)} e^{-jk\frac{r^2}{2q_0}} \\ &+ \frac{1}{4} E_0 e^{j2\pi Hx} e^{-jkz} e^{-jp(z)} e^{-jk\frac{r^2}{2q_0}} \\ &+ \frac{1}{4} E_0 e^{-j2\pi Hx} e^{-jkz} e^{-jp(z)} e^{-jk\frac{r^2}{2q_0}}. \end{aligned} \tag{2.30}$$

Thus, the *ABCD* transformation matrix in Table 2.1 does not directly apply. However, each of the three terms in Eq. (2.30) is still a Gaussian beam. The first term is the same as the incident Gaussian beam with one-half the amplitude. For $\lambda H \ll 1$, $\exp(j2\pi Hx)\exp(-jkz)$ is a propagating beam in the xz plane at an angle $-\theta$ with respect to the z axis where $\sin\theta = \lambda H$. Similarly, the third term is a propagating beam in the xz plane at angle θ with respect to the z axis. For small θ, the three beams are still approximately Gaussian in their three respective directions of propagation, i.e. the z axis, the $+\theta$ axis and the $-\theta$ axis. Therefore, we will treat them as three separate Gaussian beams along those directions.

After the lens at $z = 0$, we have

$$E = \frac{1}{2} E_0 e^{-jkz} e^{-jp(z)} e^{-jk\frac{r^2}{2q_1}} + \frac{1}{4} E_0 e^{-jkz} e^{-jp(z)} e^{-j2\pi Hx} e^{-jk\frac{r^2}{2q_1}} + \frac{1}{4} E_0 e^{-jkz} e^{-jp(z)} e^{+j2\pi Hx} e^{-jk\frac{r^2}{2q_1}},$$

where

$$\frac{1}{q_1} = \frac{1}{q_0} - \frac{1}{f},$$

$$q_1 = \frac{-\left(k\omega_0^2\right)^2 f + j2k\omega_0^2 f^2}{\left(k\omega_0^2\right)^2 + (2f)^2}.$$

In front of the screen at $z = f$, the three beams are

$$E = \frac{1}{2} E_0 e^{-jkf} e^{-jp(z=f)} e^{-jk\frac{r^2}{2q_2}} + \frac{1}{4} E_0 e^{-jkf} e^{-j2\pi Hx} e^{-jp(z=f)} e^{-jk\frac{r^2}{2q_2}} + \frac{1}{4} E_0 e^{-jkf} e^{+j2\pi Hx} e^{-jp(z=f)} e^{-jk\frac{r^2}{2q_2}}, \tag{2.31}$$

where

$$q_2 = q_1 + f = \frac{2jf^2}{k\omega_0^2 + 2jf},$$

$$\frac{1}{q_2} = +\frac{1}{f} - j\frac{k\omega_0^2}{2f^2} = \frac{1}{R_2} - \frac{j2}{k\omega_2^2}.$$

R_2 is the curvature of the Gaussian beam and ω_2 is the spot size at $z = f$. Therefore,

$$R_2 = f \quad \text{and} \quad \omega_2 = \frac{\lambda f}{\pi \omega_0},$$

implying that the curvature of the beam is f and that the spot size is proportional to f/ω_0. This result agrees with our intuition since we expect an ideal lens to focus a plane wave into a spherical wave with focused spot size proportional to the focal length and inversely proportional to the incident beam size. (For small θ, we have approximated the distance along the respective directions of propagation by z in this calculation.)

The centers of the three beams are at $z = 0$ and $z \approx \pm\,\theta f \approx \pm\lambda HF$. The beam centered at $z = 0$ is always blocked by the screen. In order for the two beams in the $\pm\theta$ direction to pass, we need

$$\lambda Hf + \frac{f\lambda}{\pi\omega_0} < l < 2\lambda Hf - \frac{2f\lambda}{\pi\omega_0}. \tag{2.32}$$

This is the same result obtained in example (6) in Section 1.5.

When the two transmitted beams traveled to $z = 2f$, in front of the second lens, the q parameter of the Gaussian beams is q_3, where

$$q_3 = q_2 + f,$$

$$q_3 = \frac{fk\omega_0^2}{k\omega_0^2 - 2jf},$$

$$\frac{1}{q_3} = \frac{1}{f} - \frac{2j}{k\omega_0^2}.$$

After the lens, the parameter q_4 is

$$\frac{1}{q_4} = \frac{1}{q_3} - \frac{1}{f} = -\frac{2j}{k\omega_0^2}.$$

Therefore we get back two original Gaussian beams, now propagating in the $\pm\theta$ directions with the same spot size. There will be some diffraction loss associated with the aperture and the screen.

2.4.5 Example of a Gaussian beam passing through a prism

The objective of this example is to describe analytically the output beam refracted by a prism. A thin prism is illustrated in Fig. 2.6. Let the prism be made of material with refractive index n at wavelength λ. Let the prism axis be the x axis and let the base of the prism be parallel to the y axis. The prism has a wedge angle α. The vertex of the prism is placed at $x = h$ and $z = 0$. Let a Gaussian beam,

$$E_{\text{inc}} = A(x, y)\, e^{-jkz} e^{-jp(z)} e^{-jk\frac{(x^2+y^2)}{2q(z)}},$$

be incident on the prism. The symbols in the expression for the incident E have already been defined and explained in Eq. (2.23).

Similarly to the thin lens discussed in Section 1.4.3, there is a phase change for any beam propagating through a thin prism. Diffraction can be neglected. For the geometry shown in Fig. 2.6, the phase change from any incident beam to the outgoing beam can be derived from phase changes of small optical rays passing through the prism at different x positions. The transfer function t for any beam passing through a prism can be written as

$$t = e^{-jk(n-1)\alpha(h-x)}. \tag{2.33}$$

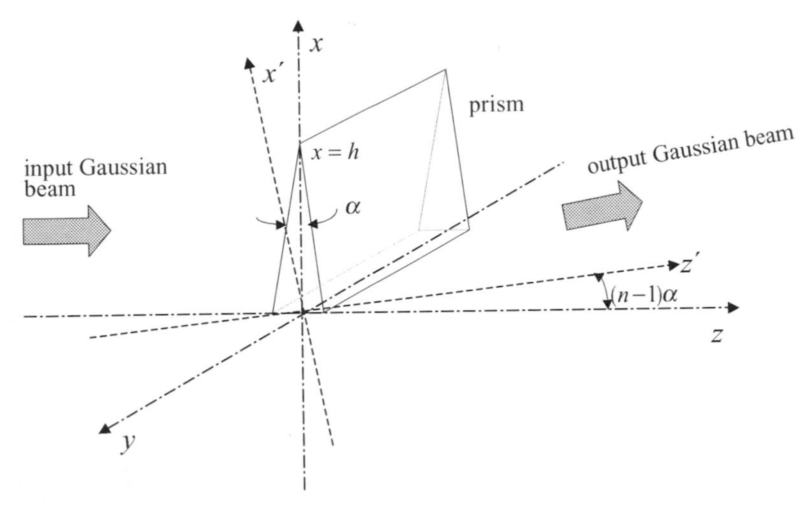

Figure 2.6. Illustration of a Gaussian beam propagating through a prism. The phase shift of the optical wave propagating through a thin prism can be represented as a phase shift equivalent to tilting the wave from the z direction to the z' direction of propagation. The tilt angle is $(n-1)\alpha$, where n is the index of the prism material at that wavelength and α is the vertex angle of the prism.

Here we have assumed that the beams are located well below $x = h$ so that the diffraction from the prism vertex at $x = h$ can be neglected. α is small so that $\sin\alpha \approx \alpha$. Therefore, the output beam will be

$$E_{\text{out}} = A\,(x, y)\, e^{-jk(n-1)\alpha h} e^{-jkz} e^{-jp(z)} e^{jk(n-1)\alpha x} e^{-jk\frac{(x^2+y^2)}{2q(z)}}.$$

If we define a new set of coordinates, x' and z', such that they are rotated from x and z by the angle $(n-1)\alpha$, as shown in Fig. 2.6, where

$$\begin{aligned} x' &= x\cos[(n-1)\alpha] - z\sin[(n-1)\alpha] \approx x - z\,(n-1)\,\alpha, \\ z' &= x\sin[(n-1)\alpha] + z\cos[(n-1)\alpha] \approx x\,(n-1)\,\alpha + z, \end{aligned} \tag{2.34}$$

then we can rewrite E approximately as

$$E_{\text{out}} = A e^{-jk(n-1)\alpha k} e^{-jkz'} e^{-jp(z')} e^{-jk\frac{(x'^2+y^2)}{2q(z')}}. \tag{2.35}$$

Here, we have neglected terms involving α^2, and we have made the approximations $p(z) \approx p(z')$ and $q(z) \approx q(z')$; $e^{-jk(n-1)\alpha k}$ is just a constant phase factor. Therefore E_{out} describes approximately a Gaussian beam propagating in the new z' direction without any change of Gaussian beam parameters. Since n is wavelength dependent, the direction of the output beam will be wavelength dependent, as expected for chromatic dispersion.

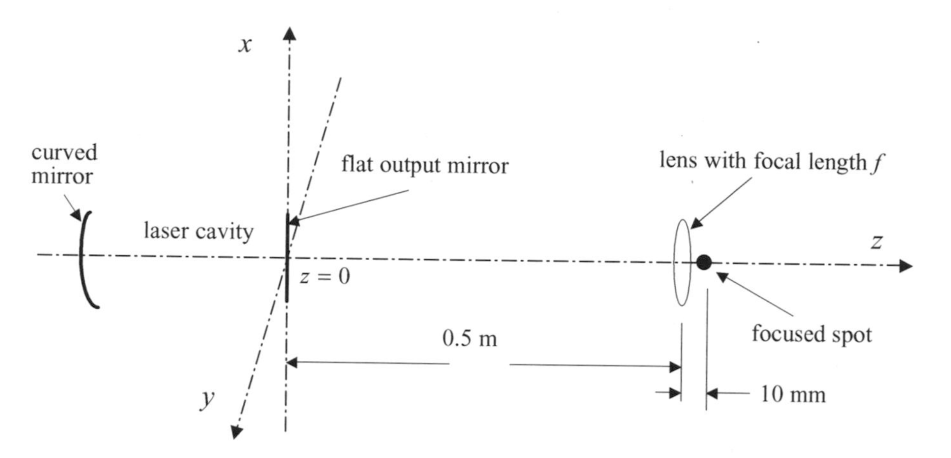

Figure 2.7. Illustration of a Gaussian beam focused by a lens. The laser is oscillating in the TEM mode of the laser cavity. The laser radiation is focused to a specific spot 10 mm beyond the lens. The Gaussian beam transformation technique is used to find the optimal focal length of the lens.

2.4.6 Example of focusing a Gaussian beam

Intuitively, we know that in order to focus a beam to a distance d away from a lens or mirror, we will use a lens or a mirror with a focal length of d. The smaller the d, the smaller is the focused spot. However, we would always wonder whether the focusing will be affected by the nature of the Gaussian beam or by the location of the lens. It is also instructive to see how we could analyze the focusing of a Gaussian beam by the *ABCD* transformation method. This example shows us how to calculate the value of f that will yield the smallest focused spot at a given distance away and how to determine the size of the focused spot.

Figure 2.7 shows a laser oscillating in the TEM_{00} mode and a lens focusing the laser mode. The ω_0 of the TEM_{00} oscillating mode is 1 mm on the flat mirror located at $z = 0$. Let the wavelength be 1 μm. A lens of focal length f is used to focus the laser beam to a distance 10 mm beyond the lens.

For the semi-spherical laser cavity, the beam waist of the resonant mode is on the flat mirror. The Gaussian beam parameter, q_1, of this oscillating mode at $z = 0$ is

$$\frac{1}{q_1} = -j\frac{1}{\pi} \quad 1/\text{meter}.$$

At $z = 0.5$ m away, the Gaussian beam parameter q_2 is

$$q_2 = q_1 + 0.5 = j\pi + 0.5,$$

$$\frac{1}{q_2} = \frac{0.5}{\pi^2 + (0.5)^2} - j\frac{\pi}{\pi^2 + (0.5)^2}.$$

Immediately after the lens, q_3 is

$$\frac{1}{q_3} = \left[\frac{0.5}{\pi^2 + (0.5)^2} - \frac{1}{f}\right] - j\frac{\pi}{\pi^2 + (0.5)^2}.$$

We still have a Gaussian beam beyond the lens. At the intended focusing position,

$$\frac{1}{q_4} = \frac{1}{q_3 + 0.01}.$$

We would obtain the smallest focused spot if the Gaussian beam waist were located at that position. Therefore, the correct f for us to use is the f value that will yield a zero for the real part of $1/q_4$. In other words, q_4 must be imaginary. Or, the real part of q_3 should be -0.01. Numerical solution of that condition yields $f = 0.009\,995\,16$ m. In order to obtain the spot size at the focus, we need to find the imaginary part of $1/q_4$. Note that $\text{Im}\,[q_4] = \text{Im}\,[q_3]$. Substitution of the f value into $1/q_3$ yields a spot size of 9.88 μm at the focus. Clearly, a change of the position of the lens or a change of the Gaussian parameter q_1 will change very slightly the optimum f value. On the other hand, if we reduce the distance between the focused spot and the lens, we obtain a smaller focused spot size.

2.4.7 Example of Gaussian mode matching

Let there be a Gaussian beam with parameter q_a at location A. Let there be an optical instrument that requires a Gaussian beam with parameter q_b at location B, as illustrated in Fig. 2.8. A lens with focal length f can be placed at a specific distance d from A to match the Gaussian beam with q_a at A to a Gaussian beam with q_b at B. We can find the values of f and d by the *ABCD* transformation method as follows.

We know q_b is related to q_a:

$$q_b = \frac{(q_a + d)\,f}{f - (q_a + d)} + (L - d)\,. \tag{2.36}$$

q_a and q_b have two differences: the difference in their real parts (i.e. the curvature of the Gaussian beam wave front), and the difference in their imaginary parts (i.e. the Gaussian spot size). We have two algebraic equations for f and d that can be easily obtained from Eq. (2.36) to match the two differences in q_a and q_b. Spurr and Dunn [8] have shown that high school geometry can be used to solve these algebraic problems arising from Gaussian beam optics.

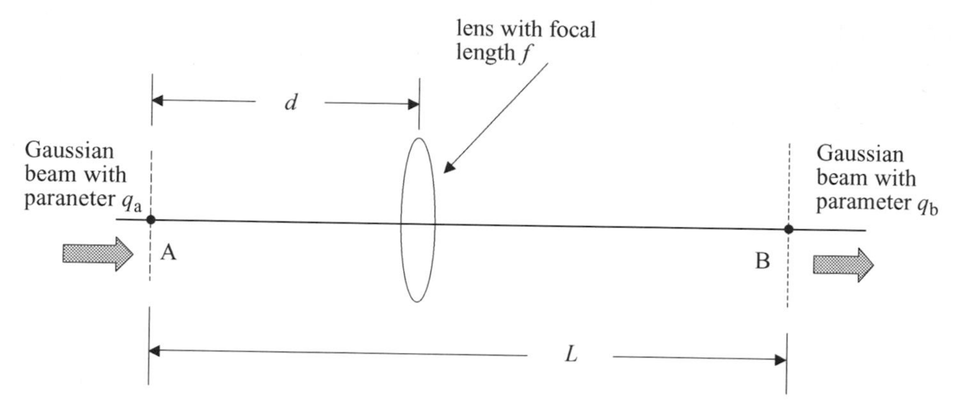

Figure 2.8. Matching a Gaussian beam at A to a Gaussian beam at B. A lens can be used to match a Gaussian beam at A to a different Gaussian beam at B. The Gaussian beam transformation technique could be used to determine the position and the proper focal length of the lens.

2.5 Modes in complex cavities

When there are many optical elements in a cavity, the q parameter of the Gaussian beam at different positions in such a cavity can be found by considering the transformation of q after a round trip in the cavity. Let the q parameter at any point in the cavity be q_s. The final q parameter after a round trip is $(Aq_s + B)/(Cq_s + D)$. For a stable mode in the cavity, it must also be the original q_s. Thus, the equation for q_s is

$$q_s = \frac{Aq_s + B}{Cq_s + D}. \tag{2.37}$$

This is a quadratic algebraic equation for $1/q_s$. The solution is

$$\begin{aligned} \frac{1}{q_s} &= \frac{D - A}{2B} \pm \frac{j\sqrt{1 - [(D + A)/2]^2}}{B} \\ &= \frac{D - A}{2B} \pm \frac{j \sin\theta}{B}, \end{aligned} \tag{2.38}$$

where $\cos\theta = (D + A)/2$. We have learned earlier that

$$\frac{1}{q_s} = \frac{1}{R} - j\frac{\lambda}{\pi\omega^2}.$$

For a stable resonator, R is the radius of curvature of the spherical phase front and ω is the spot size. Therefore, the magnitude of $\cos\theta$ must be less than unity. Or,

$$\left|\frac{D + A}{2}\right| < 1. \tag{2.39}$$

For simple cavities, Eq. (2.39) is identical to Eq. (2.18). $|(D + A)/2 | = 1$ is represented also by the boundary between stable and unstable regions shown in Fig. 2.5.

Once the q at various positions in the cavity is known, we can find the position at which q is purely imaginary. This is the position of the origin of the z axis, i.e. $z = 0$, for the virtual equivalent confocal resonator. At this position, the beam waist is ω_0. The lmth mode of the equivalent virtual confocal resonator is given by Eqs. (2.23) and (2.24) in terms of this coordinate and the complex q values. The phase shift for the round trip propagation depends on the mode order, l and m, and the total distance of propagation from $z = 0$. The resonance frequency is determined by the wavelength at which the round trip phase shift is 2π. The diffraction loss per pass of each optical element encountered in the round trip path can be calculated by the same procedure as we have used for reflectors in non-confocal resonators at the end of Section 2.2.

2.5.1 Example of the resonance mode in a ring cavity

A ring cavity is illustrated in Fig. 2.9. There are three flat mirrors at A, B and C, separated by distance d between A and B and $2d$ between B and C as well as A and C. A lens with focal length 1 m is placed midway between mirrors A and B. The recirculating resonance mode is the mode that starts with Gaussian parameter q_1 at mirror A, is transmitted through the lens, reflected by mirrors B and C, and propagates back to mirror A. Let $d = 1$ m and $\lambda = 1$ μm. We can find the recirculating resonant modes and the diffraction loss per pass from the *ABCD* transformation matrix method.

The transformation matrix M from q_2 at mirror B to q_1 through q_3, q_4 and q_5, in the counterclockwise direction in Fig. 2.9 is:

$$M = \begin{Vmatrix} 1 & \dfrac{d}{2} \\ 0 & 1 \end{Vmatrix} \begin{Vmatrix} 1 & 0 \\ -\dfrac{1}{f} & 1 \end{Vmatrix} \begin{Vmatrix} 1 & \dfrac{9d}{2} \\ 0 & 1 \end{Vmatrix}.$$

For $d = 1$ and $f = 1$,

$$M = \begin{Vmatrix} +\dfrac{1}{2} & \dfrac{11}{4} \\ -1 & -\dfrac{7}{2} \end{Vmatrix}.$$

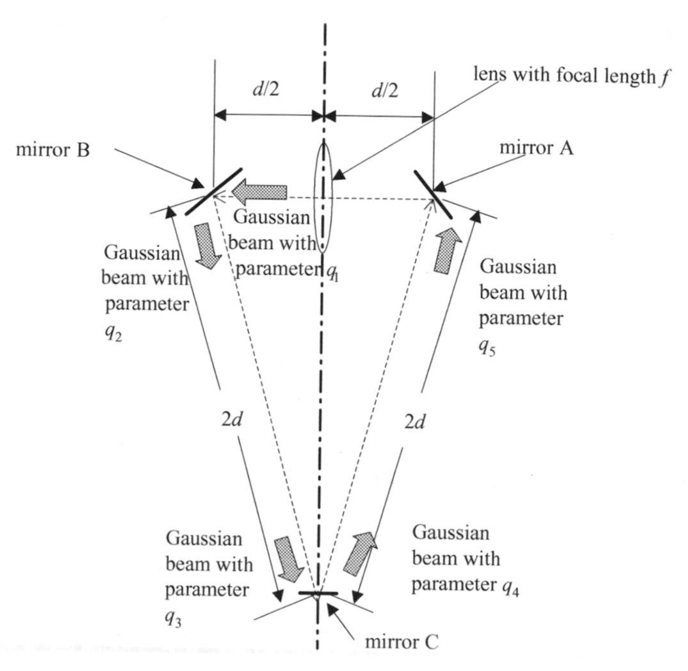

Figure 2.9. The optical elements and their positions in a ring cavity. In a ring cavity, the resonant mode is the recirculating mode that reproduces the field pattern with integer multiples of 2π total phase shift after multiple diffraction. The optical path of the recirculating mode is shown by the block arrows. The Gaussian beam parameter values of q before and after each reflector are also shown.

If we require that $q_1 = q_2$ in a round trip, we have

$$q_1 = \frac{\dfrac{q_1}{2} + \dfrac{11}{4}}{-q_1 - \dfrac{7}{2}}.$$

Therefore,

$$\frac{1}{q_1} = -\frac{8}{11} \pm j\frac{\sqrt{20}}{11}.$$

If we examine the mode starting from mirror C in a similar manner, we obtain

$$\frac{1}{q_3} = \frac{-\left(\frac{3}{2}\right)q - \frac{5}{4}}{-q_3 - \frac{3}{2}}, \quad \text{or} \quad q_3 = \pm j\frac{2}{\sqrt{5}}.$$

The values of $1/q$ at each mirror tell us the curvature and the spot size of the Gaussian beam at that mirror. We can obtain the diffraction loss per pass of each mirror from the mirror size and the spot size. In particular, q_3 is imaginary. Thus we know that the beam waist of the recirculating resonant mode is at mirror C. The size of the beam waist, ω_0, at mirror C is determined by the value of q_3. From ω_0 we obtain the z_0 of the equivalent confocal resonator mode.

References

1 G. D. Boyd and J. P. Gordon, "Confocal Multimode Resonator for Millimeter through Optical Wavelength Masers," *Bell System Technical Journal*, **40**, 1961, 489
2 P. M. Morse and H. Feshback, *Methods of Theoretical Physics*, Chapter 8, New York, McGraw-Hill, 1953
3 D. Slepian and H. O. Pollak, "Prolate Spheroidal Wave Functions," *Bell System Technical Journal*, **40**, 1961, 43
4 C. Flammer, *Spheroidal Wave Functions*, Stanford, CA, Stanford University Press, 1957
5 K. T. Hecht, *Quantum Mechanics*, Section F, Chapter 4, New York, Springer-Verlag, 2000
6 A. Yariv, *Quantum Electronics*, Chapter 6, New York, John Wiley and Sons, 1989
7 A. E. Siegman, *Lasers*, Chapters 21–23, Sausalito, CA, University Science Books, 1986
8 M. Spurr and M. Dunn, "Euclidean Light: High-School Geometry to Solve Problems in Gaussian Beam Optics," *Optics and Photonic News*, **13**, 2002, 40

3

Guided wave modes and their propagation

In Chapters 1 and 2 we discussed the propagation of laser radiation and the cavity modes as TEM waves. The amplitude and phase variations of these waves are very slow in the transverse directions. However, in applications involving single-mode optical fibers and optical waveguides, the assumption of slow variation in the transverse directions is no longer valid. Therefore, for electromagnetic analysis of such structures, we must go back to Maxwell's vector equations. Fortunately, the transverse dimensions of the components in these applications are now comparable to or smaller than the optical wavelength; solving Maxwell's equations is no longer a monumental task.

Many of the theoretical methods used in the analysis of optical guided waves are very similar to those used in microwave analysis. For example, modal analysis is again a powerful mathematical tool for analyzing many devices and systems. However, there are also important differences between optical and microwave waveguides. In microwaves, we usually analyze closed waveguides inside metallic boundaries. Metals are considered as perfect conductors at most microwave frequencies. In these closed structures, we have only a discrete set of waveguide modes that have an electric field terminating at the metallic boundary. We must avoid the use of metallic boundaries at the optical wavelength because of their strong absorption of radiation. Thus, we use open dielectric waveguides and fibers in optics, with boundaries extending theoretically to infinity. These are open waveguides. There are three important differences between optical and microwave waveguide modes and their utilization.

(1) In open dielectric waveguides, the discrete optical modes have an evanescent field outside the core region (the core is sometimes called vaguely the optical waveguide). There may be a significant amount of energy carried in the evanescent tail. The evanescent field may be used to achieve mutual interaction with other adjacent waveguides or structures. The evanescent field interaction is very important

in device applications, such as the dielectric grating filter, the distributed feedback laser and the directional coupler.

(2) The analysis is mathematically more complex for open than for closed waveguides. In fact, there exists no analytical solution of three-dimensional open channel waveguide modes (except the modes of the round step-index fiber) in the closed form. One must use either numerical analysis or approximations in order to find the field distribution of channel waveguide modes.

(3) In addition to the set of guided modes that have discrete eigen values, there is an infinite set of continuous modes. Only the sum of the discrete and continuous modes constitutes a complete set of orthogonal functions. Continuous modes are called radiation modes because they are propagating waves outside the waveguide. At any dielectric discontinuity, the boundary conditions of the electric and magnetic fields are satisfied by the summation of both the guided wave modes and the continuous modes on both sides of the boundary. Continuous modes are excited at any discontinuity. Energy in the continuous modes is radiated away from the discontinuity.

Most classical optics books do not discuss the guided wave modes and their propagation because single-mode optical fibers and waveguides cannot be excited efficiently by incoherent light.

In this chapter, we will discuss first the rigorous mathematical analysis of simple planar waveguides. Through such an analysis, concepts such as evanescent field, TE versus TM modes, guided versus radiation modes, discrete versus continuous modes and how to match the boundary conditions can be more easily understood. We will also discuss some of the planar waveguide devices in fiber optical communications. The geometry of channel waveguides is normally too complex for us to solve Maxwell's equations in closed form. The exception is the solution of the modes in step-index round fibers. We will discuss briefly the discrete (i.e. the guided) modes of single-mode optical fibers in Section 3.6. However, we will discuss first the modes of open channel waveguides with a rectangular cross-section of the core. Instead of rigorous analysis, an approximate method based on the planar guided wave modes, called the effective index method, will be presented to analyze the guided wave modes of three-dimensional channel waveguides. It is accurate only for well guided modes, i.e. modes with a short evanescent tail. However, understanding of the effective index analysis will enable us to understand the basic properties of channel guided wave modes. For example, we will recognize the similarity between the guided modes of the optical fiber and the channel waveguide. Finally, we will discuss the excitation of discrete modes by modal analysis. Such an analysis is extremely important in practical applications such as the coupling of fibers to lasers and modulators.

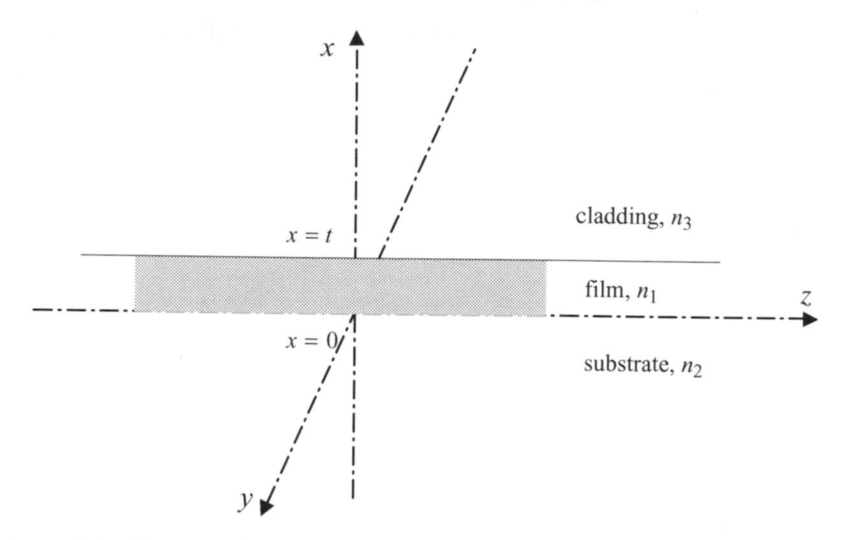

Figure 3.1. Planar optical waveguide. A symmetric planar waveguide consists of a film with a higher refractive index n_1 sandwiched between a substrate and a cladding with lower refractive indices, n_2 and n_3. The effective index and the cut-off condition of the modes are determined by the film thickness t and the refractive indices.

3.1 Asymmetric planar waveguides

A typical uniform dielectric thin film planar waveguide is shown in Fig. 3.1, where the film, the cladding and the substrate are all uniform and infinitely wide in the y and z directions. The film typically has a thickness of the order of a wavelength or less, supported by a substrate, and a cladding many wavelengths (or infinitely) thick. The index of the film (i.e. the waveguide core), n_1, is higher than the indices of the surrounding layers.

From a mathematical point of view, all modes are unique eigen solutions of Maxwell's equations satisfying the boundary conditions of the continuity of the tangential fields at the boundaries. Since the structure is identical in any direction in the yz plane, we will temporarily choose the $+z$ axis as the direction of propagation in our mathematical analysis of the planar waveguide in Sections 3.1 to 3.3. No generality is lost in this choice. For planar waveguide modes, we further assume $\partial/\partial y \equiv 0$. This assumption is similar to the assumption made for plane waves in a homogeneous medium in many textbooks, except this assumption covers only the y variation in this book.

If we analyze optical plane waves propagating in multi-layered media such as that shown in Fig. 3.1, we find that there are three typical situations.

(1) In the first situation, a plane wave is obliquely incident on the structure from either $x \ll 0$ or $x \gg t$, propagating in a direction in the xz plane that makes an angle θ_j with respect to the x axis. The angle, θ_j, will be different in different layers, where

j designates the layer with index n_j. For example, plane waves in the film with index n_1 will have a functional form, $\exp(\pm jn_1k\sin\theta_1 z)\ \exp(\pm jn_1k\cos\theta_1 x)$ $\exp(j\omega t)$. There will be reflected and diffracted waves at the top and bottom boundaries of the film. The continuity of the tangential electric field requires that $n_1k\sin\theta_1 = n_2k\sin\theta_2 = n_3k\sin\theta_3$ at the boundaries. There is a continuous range of real values of θ_j that will satisfy Maxwell's equations and the boundary conditions in all the layers. Plane waves with real values of θ_j represent radiation waves for $x < 0$ and for $x > t$ because they propagate in the x direction. In the language of modal analysis, the multiple reflected and refracted waves constitute the radiation modes with continuous eigen values in the x direction, i.e. $k_x = n_jk\cos\theta_j$, and k_x is always real.

(2) In the second situation, there are plane waves trapped in the film by total internal reflections from the top and the bottom boundaries of the film at $x = 0$ and $x = t$. In this case, the plane waves in the film will still have the functional variation of $\exp(\pm jn_1k\sin\theta_1 z)\ \exp(\pm jn_1k\cos\theta_1 x)\ \exp(j\omega t)$ with real values of θ_1. When θ_1 is sufficiently large, total internal reflection occurs at the boundaries. For total internal reflection, the factor $n_1k\sin\theta_1$ is larger than n_jk of the surrounding media, and θ_j (for $j \neq 1$) becomes imaginary because of the boundary conditions. Now the fields in the regions $x < 0$ and $x > t$ decay exponentially away from the boundaries. Since the trapped waves are bounced back and forth between the two boundaries, they will cancel each other and yield zero total field except at specific values of θ_1. As we shall show later, the non-zero waves trapped in the film at these specific θ_1 constitute the guided waves. In other words, trapped waves can only have discrete eigen values of propagation constant k_x (i.e. $n_1k\cos\theta_1$) in the film and discrete imaginary θ_j values outside the film.

(3) Let us assume that the index of the substrate is higher than the index of the cladding. In the third situation, there are radiation modes (i.e. propagating plane waves) in two regions of x, usually in the substrate, and in the high-index film region with index n_1. The value of θ_1 is sufficiently large that plane waves are totally internally reflected at the boundary between the film and the cladding. Only the field in the cladding region now decays exponentially away from the film boundary.

In Sections 3.2 to 3.4, we will solve rigorously Maxwell's equations and obtain all the modes for the asymmetric planar waveguide shown in Fig. 3.1. We will identify the solutions thus obtained with the three types of waves discussed above.

3.1.1 TE and TM modes in planar waveguides

The variation of the refractive index in the transverse direction is independent of z in Fig. 3.1. From discussions of electromagnetic theory in classical electrical engineering textbooks, we know that modes for structures that have constant transverse cross-section in the direction of propagation can often be divided into TE (transverse electric) type and TM (transverse magnetic) type. TE means no electric field component in the direction of propagation. TM means no magnetic field component in the direction of propagation.

For the planar waveguide case, if we substitute $\partial/\partial y = 0$ into the $\nabla \times \underline{e}$ and $\nabla \times \underline{h}$ Maxwell equations, we obtain two separate groups of equations:

$$\left.\begin{aligned} \frac{\partial E_y}{\partial z} &= \mu \partial H_x/\partial t, \\ \frac{\partial E_y}{\partial x} &= -\mu \partial H_z/\partial t, \\ \frac{\partial H_z}{\partial x} - \frac{\partial H_x}{\partial z} &= -\varepsilon \partial E_y/\partial t; \end{aligned}\right\} \tag{3.1a}$$

and

$$\left.\begin{aligned} \frac{\partial H_y}{\partial z} &= -\varepsilon \partial E_x/\partial t, \\ \frac{\partial H_y}{\partial x} &= \varepsilon \partial E_z/\partial t, \\ \frac{\partial E_z}{\partial x} - \frac{\partial E_x}{\partial z} &= \mu \partial H_y/\partial t. \end{aligned}\right\} \tag{3.1b}$$

Clearly, E_y, H_x and H_z are related only to each other, and H_y, E_x and E_z are related only to each other. Since the direction of propagation is z, the solutions of the first group of equations (3.1a) are the TE modes. The solutions of the second group of equations (3.1b) are the TM modes. Thus we have shown the separation of all planar waveguide modes into TE and TM types.

Since ε is only a function of x, the z variation of the solution must be the same in all layers. This is a consequence of the requirement for the continuity of E_y or H_y for all z. For propagating waves in the $+z$ direction, we will have the $\exp(-j\beta z)$ variation, whereas the waves in the $-z$ direction will have the $\exp(j\beta z)$ variation. The TE wave equations for E_y, $E_y(x, z) = E_y(x)E_y(z)$, can now be written as

$$\left[\frac{\partial^2}{\partial x^2} + (\omega^2 \mu \varepsilon(x) - \beta^2)\right] E_y(x)E_y(z) = 0, \tag{3.2a}$$

$$\left[\frac{\partial^2}{\partial z^2} + \beta^2\right] E_y(z) = 0 \tag{3.2b}$$

and

$$\left[\frac{\partial^2}{\partial x^2} + (\omega^2 \mu \varepsilon(x) - \beta^2)\right] E_y(x) = 0. \tag{3.2c}$$

Similar equations exist for TM modes.

3.2 TE planar waveguide modes

The TE planar waveguide modes are the eigen solutions of the equation

$$\left[\frac{\partial^2}{\partial x^2}+\frac{\partial^2}{\partial z^2}+\omega^2\mu\varepsilon(x)\right]E_y(x,z)=0,$$

where

$$\varepsilon(x)=n_3^2\varepsilon_0,\quad x\geq t,$$
$$\varepsilon(x)=n_1^2\varepsilon_0,\quad t>x>0,$$
$$\varepsilon(x)=n_2^2\varepsilon_0,\quad 0\geq x,$$
$$H_x=-\frac{j}{\omega\mu}\frac{\partial E_y}{\partial z}\quad\text{and}\quad H_z=\frac{j}{\omega\mu}\frac{\partial E_y}{\partial x}.$$

Here, ε_0 is the free space electric permittivity, all layers have the same magnetic permeability μ, and the time variation is $\exp(j\omega t)$. Note that when E_y is known, H_x and H_y can be calculated directly from E_y. The boundary conditions are the continuity of the tangential electric and magnetic fields at $x=0$ and $x=t$. As we shall see in the following subsections, the TE modes can be further classified into three groups. One group, the guided waves, is characterized as plane waves trapped inside the film, and the other two groups are two different kinds of combination of radiating plane waves known as substrate modes and air modes.

3.2.1 TE planar guided wave modes

Mathematically, Eqs. (3.2) plus the boundary conditions (the continuity of the tangential electric and magnetic field at $x=0$ and $x=t$) have unique solutions. The significance of a unique solution is that we may choose whatever functional form we like for $E_y(x)$ and $E_y(z)$. As long as they satisfy the differential equation plus the boundary conditions, it is the correct solution. It follows that, instead of solving Eqs. (3.2), we will choose a solution and demonstrate that they satisfy the differential equation and all the boundary conditions.

As in the second situation described in Section 3.1, we look for solutions with sinusoidal variations for $t>x>0$ and with decaying exponential variations for $x>t$ and $x<0$. Since we have chosen the time variation as $\exp(+j\omega t)$, we choose here the $\exp(-j\beta z)$ variation for $E_y(z)$ to represent a forward propagating wave in the $+z$ direction. In short, we will assume the following functional form for $E_y(x,z)$:

$$E_m(x,z)=A_m\sin(h_m t+\phi_m)\,e^{-p_m(x-t)}e^{-j\beta_m z},\qquad x\geq t,\tag{3.3a}$$
$$E_m(x,z)=A_m\sin(h_m x+\phi_m)\,e^{-j\beta_m z},\qquad t>x>0,\tag{3.3b}$$
$$E_m(x,z)=A_m\sin\phi_m\,e^{q_m x}e^{-j\beta_m z},\qquad 0\geq x,\tag{3.3c}$$

where, in order to satisfy Eq. (3.2c),

$$(\beta_m/k)^2 - (p_m/k)^2 = n_3^2,$$
$$(\beta_m/k)^2 + (h_m/k)^2 = n_1^2,$$
$$(\beta_m/k)^2 - (q_m/k)^2 = n_2^2.$$

The subscript m stands for the mth-order solution of Eq. (3.2c), which is clearly satisfied in all the individual regions. We have also chosen the functional form so that the continuity of E_y is automatically satisfied at $x = 0$ and $x = t$. In order to satisfy the magnetic boundary conditions at $x = 0$ and $x = t$, h_m, q_m and p_m must be the mth set of the root of the transcendental equations, which are also called the characteristic equations,

$$\tan[(h_m/k)kt + \phi_m] = -h_m/p_m,$$
$$\tan\phi_m = h_m/q_m. \tag{3.4}$$

For a given normalized thickness kt, there is only a finite number of roots of the characteristic equations yielding a discrete set of real values for h, p and q. For this reason, the guided wave modes are also called the discrete modes. They are labeled by the subscript m ($m = 0, 1, 2, \ldots$). The lowest order mode with $m = 0$ has the largest β value, $\beta_0 > \beta_1 > \beta_2 > \beta_3 \cdots$ and $h_0 < h_1 < h_2 \cdots$. Moreover, one can show that the number of times in which $\sin(h_m x + \phi_m)$ is zero is m.

3.2.2 TE planar guided wave modes in a symmetrical waveguide

In order to visualize why there should be only a finite number of modes, let us consider the example of a symmetrical waveguide. In that case, $n_2 = n_3 = n$ and $p_m = q_m$. The quadratic equations for h_m and β_m and the transcendental equation now become

$$\left(\frac{h_m}{k}\right)^2 + \left(\frac{p_m}{k}\right)^2 = n_1^2 - n^2$$

and

$$\tan\left[\left(\frac{h_m}{k}\right)kt\right] = \frac{-2(h_m/p_m)}{1 - (h_m^2/p_m^2)}.$$

Since

$$\tan\left[2\left(\frac{h_m}{k}\right)\frac{kt}{2}\right] = \frac{2\tan\left[\left(\frac{h_m}{k}\right)\frac{kt}{2}\right]}{1 - \tan^2\left[\left(\frac{h_m}{k}\right)\frac{kt}{2}\right]},$$

the above equation is equivalent to two equations,

$$\tan\left[\left(\frac{h_m}{k}\right)\frac{kt}{2}\right] = \frac{p_m/k}{h_m/k}$$

or

$$\frac{h_m}{k}\tan\left[\left(\frac{h_m}{k}\right)\frac{kt}{2}\right] = \frac{p_m}{k}$$

and

$$\tan\left[\left(\frac{h_m}{k}\right)\frac{kt}{2}\right] = -\frac{h_m/k}{p_m/k}$$

or

$$-\frac{h_m}{k}\cot\left[\left(\frac{h_m}{k}\right)\frac{kt}{2}\right] = \frac{p_m}{k}.$$

In the coordinate system of p_m/k and h_m/k, the solutions of the above equations are given by the intersections of the two curves representing the quadratic equation $(h_m/k)^2 + (p_m/k)^2 = n_1^2 - n^2$ and one of the two equivalent tangent equations. In short, there are two sets of equations. The solutions for the first tangent equation and the quadratic equation are known as the even modes because they lead to field distributions close to a cosine variation in the film. They are symmetric with respect to $x = t/2$. The solutions from the second tangent equation and the quadratic equation are called the odd modes because the fields in the film have distributions close to sine variations. They are anti-symmetric with respect to $x = t/2$.

Let us examine the even modes in detail. If we plot the quadratic equation of h_m/k and p_m/k, it is a circle with radius $(n_1^2 - n^2)^{1/2}$. The curve describing the first tangent equation will be obtained from those values of h_m/k and p_m/k whenever the left hand side (LHS) equals the right hand side (RHS) of the tangent equation. The RHS is just p_m/k. The LHS has a tangent which is a multi-valued function. It starts from zero whenever $(h_m/k)kt/2$ is $0, \pi$, or $m\pi$. It approaches + or − infinity when $(h_m/k)kt/2$ approaches $+\pi/2$ or $-\pi/2$, or $(m\pi + \pi/2)$ or $(m\pi - \pi/2)$, where m is an integer. The curves representing these two equations are illustrated in Fig. 3.2. Clearly, there is always a solution as long as $n_1 > n$, i.e. there is an intersection of the two curves, no matter how large (or how small) is the circle (i.e. the n_1 value). This is the fundamental mode, labeled $m = 0$. However, whether there will be a $m \geq 1$ solution depends on whether the radius is larger than $2\pi/kt$. There will be $m = j$ solutions when the radius is larger than $2j\pi/kt$. Notice that $h_0 < h_1 < h_2 < \cdots$ and $\beta_0 > \beta_1 > \beta_2 > \cdots$. When the radius of the circle is equal to $2j\pi/kt$, the value for p/k is zero. This is the cut-off point for the jth $(j > 1)$ mode.

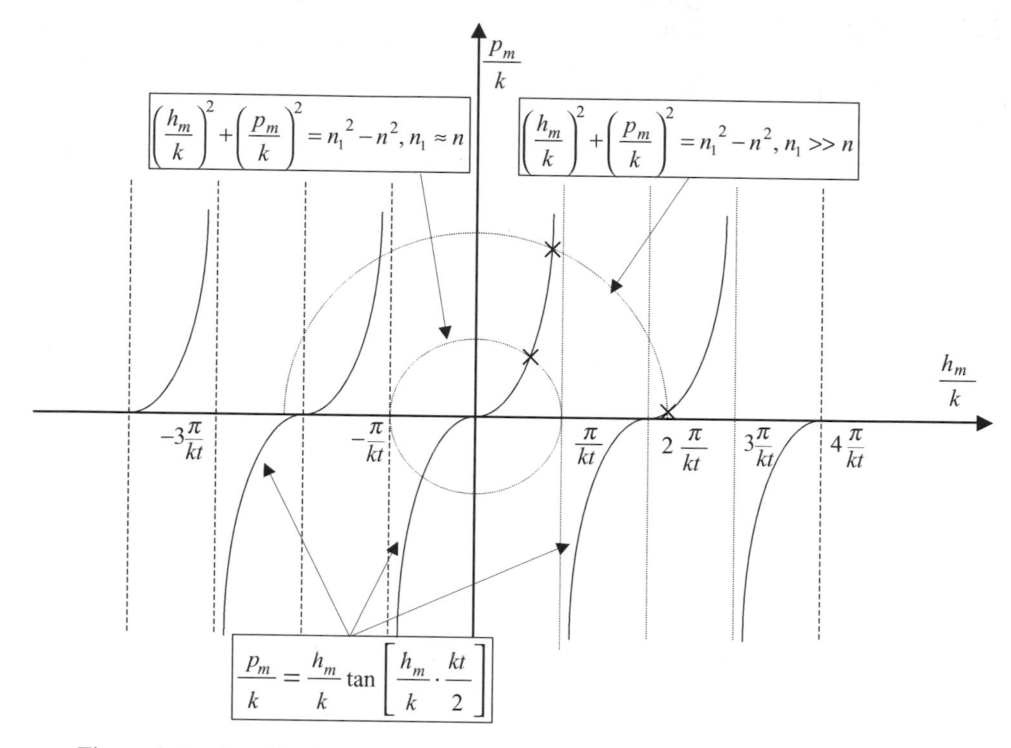

Figure 3.2. Graphical solution for h_m and p_m, for even TE guided wave modes, in a symmetrical planar waveguide. The intersections of the two equations, marked by $\times$, are h_m/k and p_m/k, solutions of the mth-order mode. The larger the $n_1^2 - n^2$, the larger the circle and the larger the h_m/k solution. The cut-off of the mth-order mode occurs when the radius of the circle is $2m\pi/kt$. There is no cut-off for the $m = 0$ mode.

3.2.3 Cut-off condition for TE planar guided wave modes

There are conditions imposed on the refractive indices without which there is no guided wave mode solution for the asymmetric waveguides. The first condition is

$$n_1 > n_2 \text{ and } n_3.$$

Without any lost of generality, let $n_1 > n_2 \geq n_3$. In addition, there is a minimum thickness t_m, called the cut-off thickness, which will permit the mth solution to Eq. (3.2c) to exist. However, other than the symmetric waveguide, for which there is always an $m = 0$ even mode, there is a cut-off condition for even the $m = 0$ mode in asymmetric waveguides. At the cut-off of the mth mode, $q_m = 0$, $\beta_m/k = n_2$, $p_m/k = (n_2^2 - n_3^2)^{1/2}$, $\phi_m = \pm(m + 1/2)\pi$ and $h_m = k(n_1^2 - n_2^2)^{1/2}$. Thus the cut-off

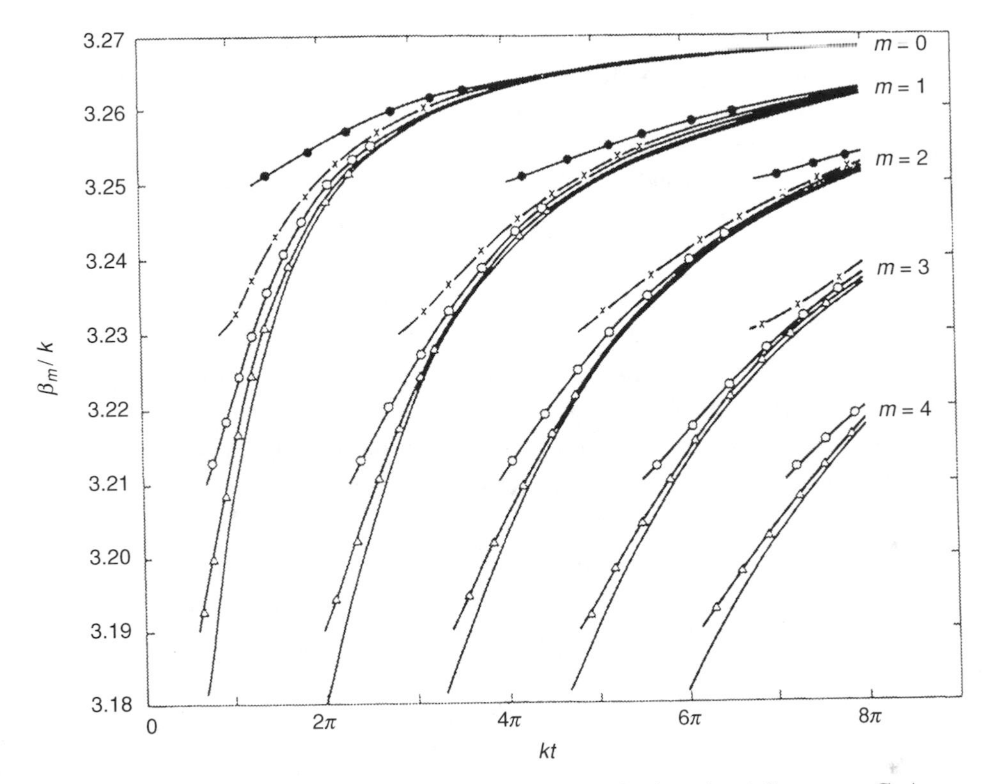

Figure 3.3. Propagation wave number of TE_m modes in epitaxially grown GaAs waveguides. Solid line, $\Delta n = 0.10$; $\triangle$, $\Delta n = 0.08$; $\circ$, $\Delta n = 0.06$; $\times$, $\Delta n = 0.04$; $\bullet$, $\Delta n = 0.02$. The copyright figure is taken from ref. [1] with permission from Elsevier.

thickness can be calculated from Eqs. (3.4) to be:

$$kt_m = \left\{\left(m + \frac{1}{2}\right)\pi - \tan^{-1}\left[\left(n_1^2 - n_2^2\right)/\left(n_2^2 - n_3^2\right)\right]^{1/2}\right\}\left(n_1^2 - n_2^2\right)^{-1/2}. \quad (3.5)$$

The thicker the film, the larger the number of guided wave modes the film can support. For all guided wave modes above cut-off, $n_1 \geq |\beta_m/k| > n_2$.

3.2.4 Properties of TE planar guided wave modes

Figure 3.3 shows the propagation wave number, β_m/k, of TE planar guided wave modes in epitaxially grown GaAs waveguides with air as the cladding, where $n_3 = 1$ [1]. The Δn, i.e. $n_1 - n_2$, depends on the alloy composition of the epitaxially grown thin film. Notice that we have only real eigen values for β, h, p and q. Since β is real, these modes propagate in the z direction without attenuation. The fields of these modes are evanescent in the air and in the substrate. This is the most important characteristic of guided waves.

Figure 3.3 demonstrates clearly that, at a given thickness t, the higher order modes have lower β/k values. Thus, the evanescent decay of the higher order modes will be slower in the n_2 and n_3 layers. When there is scattering or absorption loss on the bottom surface of the n_3 layer or above the n_1 layer, it will not affect the mode pattern significantly. It will cause attenuation as the mode propagates in the z direction. In Chapter 4, a mathematical method to calculate the attenuation will be discussed. The slower the evanescent decay, the larger the attenuation rate. For this reason, higher order guided wave modes often have a larger attenuation rate.

If we had allowed imaginary values for p_m or q_m, there would have been be many more solutions. However, those solutions will not be the total internally reflected field in the film. They will radiate away and they represent the radiation modes discussed in Sections 3.2.5 and 3.2.6.

Physically, the electric field of the mth TE guided wave mode inside the film is just a plane wave in the n_1 layer (with the electric field polarized in the y direction), totally internally reflected back and forth from the two boundaries at $x = 0$ and $x = t$. Its propagation direction in the xz plane makes an angle θ_m with respect to the x axis:

$$\begin{aligned} \beta_m &= n_1 k \sin\theta_m, \\ h_m &= n_1 k \cos\theta_m. \end{aligned} \tag{3.6}$$

When β_m and h_m are given by the mth solution of Eqs. (3.4), the total round trip phase shift of such a plane wave after reflection from both the air and the substrate boundary is $2m\pi$. In some technical papers and relevant books, instead of solving Maxwell's equations directly, as we did in Eqs. (3.3) and (3.4), the guided wave modes are found by requiring the round trip phase shift of totally internally reflected plane wave to be $2m\pi$. This is the condition that the total field for all the plane waves reflected back and forth is non-zero.

The TE guided wave modes are orthogonal to each other and to any other TE or TM modes of the same waveguide. It is customary to normalize the constant A so that a unit amount of power (1 W) per unit length in the y direction is carried out by a normalized mode. Thus,

$$\frac{1}{2}\,\mathrm{Re}\left[\int_{-\infty}^{+\infty} E_{yn} H_{xm}^{*}\, dx\right] = (\beta_m/2\omega\mu)\int E_n E_m^{*}\, dx = \delta_{nm}, \tag{3.7}$$

$$A_m^2 = \frac{4\omega\mu}{\beta_m}\left[\frac{1}{p_m} + \frac{1}{q_m} + t\right]^{-1}.$$

3.2.5 TE planar substrate modes

In the range $n_2 > |\beta/k| > n_3$, the electric field has an exponential variation for $x > t$ and sinusoidal variation in the film and in the substrate. In view of the typical situation (3) discussed in Section 3.1, these are called substrate modes. In that case, we have

$$
\begin{aligned}
E^{(s)}(x, z; \beta) &= A^{(s)} \sin(ht + \phi) \exp[-p(x - t)] \exp(-j\beta z), & x &\geq t, \\
E^{(s)}(x, z; \beta) &= A^{(s)} \sin(hx + \phi) \exp(-j\beta z), & t &> x > 0, \\
E^{(s)}(x, z; \beta) &= \left[C^{(s)} \exp(-j\rho x) + C^{(s)*} \exp(+j\rho x)\right] \exp(-j\beta z), & 0 &\geq x,
\end{aligned}
$$

$$
\left.
\begin{aligned}
(h/k)^2 + (\beta/k)^2 &= n_1^2, \\
(\beta/k)^2 - (p/k)^2 &= n_3^2, \\
(\rho/k)^2 + (\beta/k)^2 &= n_2^2,
\end{aligned}
\right\} \tag{3.8}
$$

$$\tan[(h/t)kt + \phi] = -h/p,$$

$$C^{(s)} = A^{(s)}[\sin\phi + j(h \cos\phi/\rho)]/2.$$

$C^{(s)}$ and $A^{(s)}$ are normalized so that

$$(\beta/2\omega\mu) \int_{-\infty}^{\infty} E^{(s)}(x, z; \beta) E^{(s)*}(x, z; \beta')\, dx = \delta(\rho - \rho'), \tag{3.9}$$

which requires that

$$C^{(s)} C^{(s)*} = \frac{\omega\mu}{\beta\pi}.$$

Unlike guided wave modes, which have $n_1 > |\beta_m/k| > n_2$ and n_3, β, p, h, ρ and ϕ of the substrate modes have a continuous range of values which satisfy the above equations within the range $n_2 > |\beta/k| > n_3$. Thus these modes are called continuous modes. The field in the air region has an evanescent variation. However, the field in the substrate region has the form of two propagating planc waves with propagation constant ρ, one in the $+x$ direction and the other in the $-x$ direction. Thus they are also called the substrate radiation modes.

In the plane wave description of the substrate modes, β/n_1k, h/n_1k, β/n_2k and ρ/n_2k are direction cosines of the plane waves with respect to the z axis and the x axis in the film region, and in the substrate region, respectively. The plane waves in the film are totally internally reflected only at the boundary $x = t$.

3.2.6 TE planar air modes

As discussed in the typical situation (1) in Section 3.1, a third class of solutions of Eq. (3.2c) can be represented in terms of a plane wave with its accompanying reflected and refracted beams at each boundary, without total internal reflection at either boundary. It is well known that for each set of angles of incidence, reflection and refraction, there are always two independent plane wave solutions. One is a wave incident on the film from the air side plus its accompanying reflected and refracted waves, and the other is a wave incident from the substrate side plus its accompanying reflected and refracted waves. They all have the same z variation.

Mathematically, there are always two independent solutions of Maxwell's equations for a given set of propagation constants. By linearly combining the two independent solutions, one can always obtain two orthogonal independent modes for each set of propagation constants. These orthogonal modes are called air modes because they propagate in both media with indices n_2 and n_3, and because the cladding medium with n_3 is often the air.

If the structure were symmetrical, these two orthogonal modes would represent odd and even variations with respect to $x = t/2$ inside the film. For asymmetrical structures, such as the one shown in Fig. 3.1, the x variations are more complex. Nevertheless, there are still two modes for each set of propagation constants, and these two modes differ from each other by a $\pi/2$ phase shift of the sinusoidal variations in the x direction in the film which has index n_1. The mathematical expressions for E_y of the air modes are as follows:

$$\begin{aligned} E'(x, z; \beta) &= \{D' \exp[-j\sigma(x-t)] \\ &\quad + D'^* \exp[+j\sigma(x-t)]\} \exp(-j\beta z), && x \geq t, \\ E'(x, z; \beta) &= A' \sin(hx + \phi) \exp(-j\beta z), && t > x > 0, \\ E'(x, z; \beta) &= [C' \exp(-j\rho x) + C'^* \exp(+j\rho x)] \exp(-j\beta z), && 0 \geq x, \end{aligned} \tag{3.10a}$$

for the first set, and for the second set

$$\begin{aligned} E''(x, z; \beta) &= \{D'' \exp[-j\sigma(x-t)] \\ &\quad + D''^* \exp[+j\sigma(x-t)]\} \exp(-i\beta z), && x \geq t, \\ E''(x, z; \beta) &= A'' \sin\left(hx + \phi + \frac{\pi}{2}\right) \exp(-j\beta z), && t > x > 0, \\ E''(x, z; \beta) &= [C'' \exp(-j\rho x) + C''^*(+j\rho x)] \exp(-j\beta z), && 0 \geq x, \end{aligned} \tag{3.10b}$$

with

$$\begin{aligned} (\beta/k)^2 + (\sigma/k)^2 &= n_3^2, \\ (\beta/k)^2 + (h/k)^2 &= n_1^2, \\ (\beta/k)^2 + (\rho/k)^2 &= n_2^2. \end{aligned}$$

Imposing the boundary conditions at $x = 0$ and $x = t$, we obtain

$$C' = A'[\sin\phi + j(h\cos\phi/\rho)]/2, \tag{3.11a}$$

$$D' = A'\left[\sin(ht+\phi) + j\frac{h}{\sigma}\cos(ht+\phi)\right]/2. \tag{3.11b}$$

A″, C″ and D″ are obtained when ϕ is replaced by $\phi + \pi/2$ in Eqs. (3.11a) and (3.11b). All modes form an orthogonal normalized set, as defined in Eqs. (3.7) and (3.9). For both sets of modes, a continuous range of solutions of ρ, σ, β and h exist, where $n_3 \geq |\beta/k| \geq 0$.

3.3 TM planar waveguide modes

The TM modes are eigen solutions of the wave equation (with $\partial/\partial y = 0$ and $\exp(j\omega t)$ time variation):

$$\left[\frac{\partial^2}{\partial x^2} + \frac{\partial^2}{\partial z^2} + \omega^2\varepsilon(x)\mu\right] H_y(x,z) = 0, \tag{3.12a}$$

$$E_x = \frac{j}{\omega\varepsilon(x)}\frac{\partial H_y}{\partial z},$$

$$E_z = \frac{-j}{\omega\varepsilon(x)}\frac{\partial H_y}{\partial x},$$

where $\varepsilon(x)$ is the same as that in Section 3.2. Or, in a manner similar to Eq. (3.2c), we can write

$$\left[\frac{\partial^2}{\partial x^2} + (\omega^2\mu\varepsilon(x) - \beta^2)\right] H_y(x,z) = 0. \tag{3.12b}$$

3.3.1 *TM planar guided wave modes*

Like the TE modes, the y component of the magnetic field for the nth TM guided wave mode is

$$\begin{aligned}
H_n(x,z) &= B_n \sin(h_n t + \phi_n)\exp[-p_n(x-t)]\exp(-j\beta_n z), && x \geq t,\\
H_n(x,z) &= B_n \sin(h_n x + \phi_n)\exp(-j\beta_n z), && t > x > 0,\\
H_n(x,z) &= B_n \sin\phi_n \exp[q_n x]\exp(-j\beta_n z), && 0 \geq x,
\end{aligned} \tag{3.13}$$

with

$$\begin{aligned}
(\beta_n/k)^2 - (p_n/k)^2 &= n_3^2,\\
(\beta_n/k)^2 + (h_n/k)^2 &= n_1^2,\\
(\beta_n/k)^2 - (q_n/k)^2 &= n_2^2.
\end{aligned}$$

Continuity of the tangential electric field requires that h_n, q_n and β_n also satisfy the transcendental equation

$$\tan[(h_n/k)kt + \phi_n] = -\frac{n_3^2 h_n}{n_1^2 p_n},$$
$$\tan \phi_n = \left(\frac{n_2}{n_1}\right)^2 \frac{h_n}{q_n}. \tag{3.14}$$

Note that, unlike TE guided wave modes, TM guided wave modes have an electric field perpendicular to the interface boundary of the cladding and the film.

3.3.2 TM planar guided wave modes in a symmetrical waveguide

It is instructive to see what happens to the TM modes in a symmetrical waveguide, i.e. $n_2 = n_3 = n$. The solution obtained in this example will also be used directly in the effective index method to find the TE modes in channel waveguides. In this case, $p_n = q_n$. The quadratic equation for h_n and β_n and the transcendental equation now becomes

$$\left(\frac{h_n}{k}\right)^2 + \left(\frac{p_n}{k}\right)^2 = n_1^2 - n^2,$$
$$\tan\left[\left(\frac{h_n}{k}\right)kt\right] = -\frac{2(n^2 h_n/n_1^2 p_n)}{1 - \left(\frac{n^2 h_n}{n_1^2 p_n}\right)^2}.$$

As we have seen in the case of TE guided wave modes in symmetrical waveguide structures, the above tangent equation is equivalent to two equations,

$$\tan\left[\left(\frac{h_n}{k}\right)\frac{kt}{2}\right] = -\frac{n^2 h_n/k}{n_1^2 p_n/k},$$

and

$$\tan\left[\left(\frac{h_n}{k}\right)\frac{kt}{2}\right] = \frac{n_1^2 p_n/k}{n^2 h_n/k},$$

or

$$-\frac{n^2}{n_1^2}\left(\frac{h_n}{k}\right)\cot\left[\left(\frac{h_n}{k}\right)\frac{kt}{2}\right] = \frac{p_n}{k},$$

and

$$\frac{n^2}{n_1^2}\left(\frac{h_n}{k}\right)\tan\left[\left(\frac{h_n}{k}\right)\frac{kt}{2}\right] = \frac{p_n}{k}.$$

These equations again point to the existence of two orthogonal sets of modes, modes symmetric and anti-symmetric with respect to $t/2$. The $n = 0$ symmetric TM mode has no cut-off thickness t. These equations are very similar to the equations for the TE modes, except for the ratio $(n/n_1)^2$, which is always smaller than unity. Therefore, for the same order (i.e. $m = n$), the p_n values of the TM modes are slightly smaller than the p_m values of the TE modes for the same thickness t and indices.

3.3.3 Cut-off condition for TM planar guided wave modes

Again, for a given normalized thickness kt, there is only a finite number of discrete modes, labeled by the subscript n ($n = 0, 1, 2, \ldots$), where $h_0 < h_1 < h_2 < \cdots$ and $n_1 > \beta_0 > \beta_1 > \beta_2 > \cdots > n_2$. The cut-off thickness for the nth TM mode is given by $q = 0$ and by

$$kt_n = \left\{ n\pi + \tan^{-1}\left[\frac{n_1^2}{n_3^2}\sqrt{\frac{n_2^2 - n_3^2}{n_1^2 - n_2^2}} \right] \right\} \left(n_1^2 - n_2^2\right)^{-1/2}. \tag{3.15}$$

Note that the cut-off thickness t_n for TM modes is always larger than the cut-off thickness t_m for TE modes of the same order. Thus it is possible to design the waveguide with appropriate n_1, n_2 and t so that only the lowest order TE mode can exist.

3.3.4 Properties of TM planar guided wave modes

Figure 3.4 shows the propagation wave number β_n/k of TM planar guided wave modes in epitaxially grown GaAs waveguides where the cladding is the air with $n_3 = 1$ [1]. As in Fig. 3.3, the Δn, i.e. $n_1 - n_2$, depends on the composition of the epitaxially grown thin film. Because of the dependence on $(n_2/n_1)^2$ and $(n_3/n_1)^2$, which are always smaller than unity, the β/k of the TM modes are usually slightly smaller than those of the corresponding TE modes. The most important difference between the TM and TE modes is, of course, the polarization of the optical electric field. On many occasions, metallic electrodes are fabricated on top of the n_3 layer to allow the application of a DC or RF electric field. The difference in the polarization of the optical electric field may make a difference to the attenuation of the guided wave mode in the z direction caused by the metal. When there is metallic absorption, the TM modes have higher attenuation. On other occasions, such as the coupling of the radiation field into a planar waveguide, the coupling efficiency is dependent critically on the matching of the polarization of the incident radiation field with the polarization of the guided wave mode.

Similarly to TE guided wave modes, TM planar guided wave modes inside the film with index n_1 can also be described by a plane wave that has a magnetic field polarized in the y direction. It is totally internally reflected back and forth from the two boundaries, in a propagation direction in the xz plane making an angle θ_n with

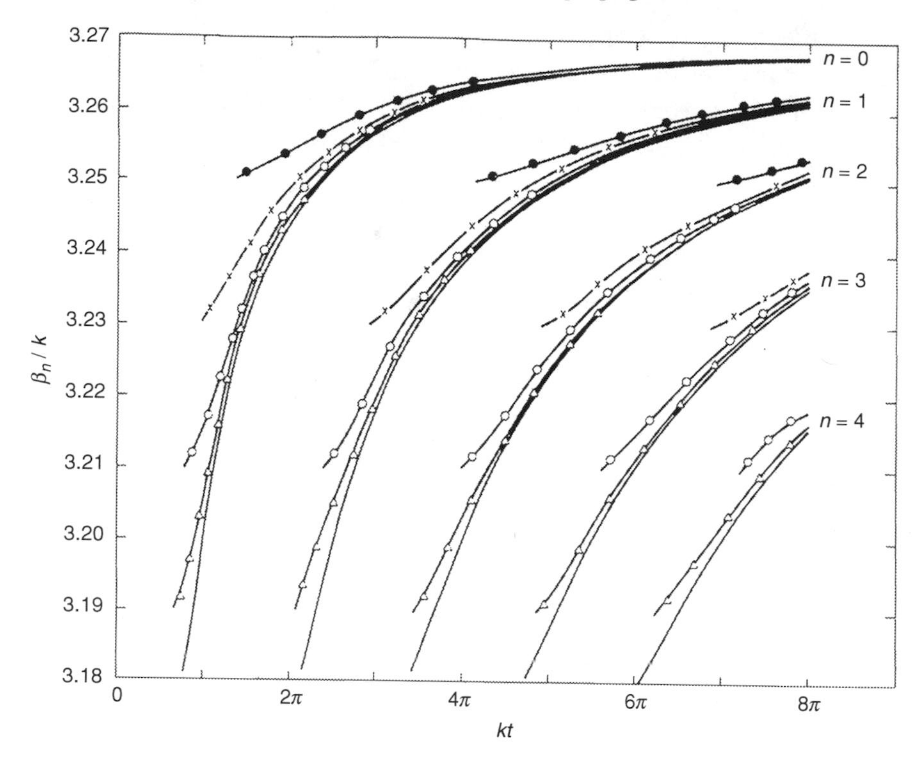

Figure 3.4. Propagation wave number of TM_n modes in epitaxially grown GaAs waveguides. Solid line, $\Delta n = 0.10$; $\triangle$, $\Delta n = 0.08$; $\circ$, $\Delta n = 0.06$; $\times$, $\Delta n = 0.04$; $\bullet$, $\Delta n = 0.02$. The copyright figure is taken from ref. [1] with permission from Elsevier.

respect to the x axis. The nth TM guided wave modes can be found by requiring the round trip phase shift to be $2n\pi$.

The TM planar guided wave modes are orthogonal to each other and to the TE modes. When TM modes are normalized,

$$\frac{1}{2}\,\mathrm{Re}\left[\int_{-\infty}^{+\infty} H_{yn} E_{xm}^{*}\, dx\right] = \frac{\beta_n}{2\omega}\int_{-\infty}^{+\infty} H_n H_m^{*}\frac{1}{\varepsilon(x)}\, dx = \delta_{nm} \tag{3.16}$$

and

$$B_n^2 = \frac{4\omega\varepsilon_0}{\beta_n}\left[\left(\frac{n_1^2}{n_3^2 p_n}\right)\left[\frac{p_n^2 + h_n^2}{h_n^2 + \left(\frac{n_1^2}{n_3^2}\right)^2 p_n^2}\right] + \left(\frac{n_1^2}{n_2^2 q_n}\right)\left[\frac{q_n^2 + h_n^2}{h_n^2 + \left(\frac{n_1^2}{n_3^2}\right) q_n^2}\right] + t\right]^{-1}. \tag{3.17}$$

3.3.5 TM planar substrate modes

For the substrate TM modes, the y component of the magnetic field is

$$\left.\begin{aligned} H^{(s)}(x,z;\beta) &= B^{(s)}\sin(ht+\phi)e^{-p(x-t)}e^{-j\beta z}, && x \geq t, \\ H^{(s)}(x,z;\beta) &= B^{(s)}\sin(hx+\phi)e^{-j\beta z}, && t > x > 0, \\ H^{(s)}(x,z;\beta) &= \left[D^{(s)}e^{-j\rho x} + D^{(s)*}e^{+j\rho x}\right]e^{-j\beta z}, && 0 \geq x, \end{aligned}\right\} \tag{3.18}$$

$$D^{(s)} = \left(\frac{B^{(s)}}{2}\right)\left[\sin\phi + j\left(\frac{n_2^2 h\cos\phi}{n_1^2\rho}\right)\right],$$

$$\tan[(h/k)kt+\phi] = -\frac{n_3^2 h}{n_1^2 p}.$$

D and B are obtained from the orthogonalization and normalization conditions,

$$\frac{\beta}{2\omega}\int_{-\infty}^{+\infty} H^{(s)}(\beta)H^{(s)*}(\beta')/\varepsilon(x)\,dx = \delta(\rho-\rho'), \tag{3.19}$$

$$D^{(s)}D^{(s)*} = \frac{\omega\varepsilon_0 n_2^2}{\beta\pi}.$$

β, p, h, ρ and ϕ have a continuous range of solutions within the range, $n_2 > |\beta/k| > n_3$.

3.3.6 TM planar air modes

There are again two orthogonal TM air modes for each set of propagation constants. For the first set of modes,

$$\begin{aligned} H'(x,z;\beta) &= \left\{E'e^{-j\sigma(x-t)} + E'^{*}e^{j\sigma(x-t)}\right\}e^{-j\beta z}, && x \geq t, \\ H'(x,z;\beta) &= B'\sin(hx+\phi)e^{-j\beta z}, && t > x > 0, \\ H'(x,z;\beta) &= [F'e^{-j\rho x} + F'^{*}e^{j\rho x}]e^{-j\beta z}, && 0 \geq x, \end{aligned} \tag{3.20a}$$

and, for the second set of modes,

$$\begin{aligned} H''(x,z;\beta) &= \left\{E''e^{-j\sigma(x-t)} + E''^{*}e^{j\sigma(x-t)}\right\}e^{-j\beta z}, && x \geq t, \\ H''(x,z;\beta) &= B''\sin\left(hx+\phi+\frac{\pi}{2}\right)e^{-j\beta z}, && t > x > 0, \\ H''(x,z;\beta) &= [F''e^{-j\rho x} + F''^{*}e^{j\rho x}]e^{-j\beta z}, && 0 \geq x. \end{aligned} \tag{3.20b}$$

For both sets of orthogonal modes, a continuous range of solutions of ρ, σ, β and h exist, where $n_3 \geq |\beta/k| \geq 0$. For the first set of modes, the continuity of the electric and magnetic fields at $x = 0$ and $x = t$ requires

$$E' = \frac{1}{2}B'\left\{\sin(ht + \phi) + j\frac{hn_3^2\cos(ht + \phi)}{\sigma\, n_1^2}\right\}, \tag{3.21a}$$

$$F' = \frac{1}{2}B'\left\{\sin\phi + j\frac{hn_2^2\,\cos\phi}{\rho\, n_1^2}\right\}. \tag{3.21b}$$

For the second set of modes, ϕ is replaced by $\phi + \pi/2$ in Eqs. (3.21a) and (3.21b).

3.4 Generalized properties of guided wave modes in planar waveguides and applications

The most important characteristics of guided wave modes are the exponential decay of their evanescent tails, the distinct polarization associated with each mode and the excitation of continuous modes at any defect or dielectric discontinuity that causes diffraction loss of the guided wave mode. The evanescent tail ensures that there is only minor perturbation of the mode pattern for structure changes several decay lengths away from the surface of the high-index layer.

Since propagation loss of the guided wave modes is usually caused by scattering or absorption, the attenuation rate of the guided mode will be very low as long as there is very little absorption or scattering loss in or near the high-index layer. The most common causes for absorption loss are either the placement of a metallic electrode nearby or the use of semiconductor cladding or substrate that has conduction due to electrical carriers. Besides absorption, the propagation losses are caused most commonly by volume scattering in the layers or by surface scattering at the dielectric interfaces. Volume scattering is created in the materials as they are grown or deposited. Surface scattering is usually created through fabrication processes.

On the other hand, the evanescent tail also enables us to interact purposely with the guided wave mode by placing perturbations close to the surface of the high-index layer. For example, in Chapter 4, we will discuss the directional coupler formed by two adjacent waveguides or a grating filter fabricated on top of a waveguide.

The exponential decay rate of any guided wave mode is determined only by the index of the layer (either at $x > t$ or at $x < 0$) and the β/k value of the mode. The β/k value is called the effective index, n_{eff}, of the mode. The effective index times the velocity of light in free space is the phase velocity of the guided wave mode. For the same polarization, lower order modes will have larger effective index and faster exponential decay. For the same $\Delta\varepsilon$ of defects or interface roughness, modes that have a smaller effective index will be scattered more strongly into radiation modes, i.e. substrate and air modes. Therefore, higher order modes usually have larger attenuation.

In order to excite effectively a specific guided wave mode, the incident radiation must have a polarization close to the polarization of that mode. For incident radiation with polarization between the TE and TM polarizations, both TE and TM modes will be excited. Since TM and TE modes have different effective indices, they have different phase velocities. When both TE_0 and TM_0 modes are excited by a given incident radiation, the total polarization of the two modes will rotate as they propagate, due to the difference in phase velocities.

3.4.1 Planar guided waves propagating in other directions in the yz plane

In Section 3.1, we presented the analysis of the planar modes when they propagate in the direction of the z axis. In reality, planar guided wave modes for a waveguide structure as shown in Fig. 3.1 can propagate in any direction in the yz plane with the same x functional variation as given in Eqs. (3.3) and (3.13). For a planar guided wave mode propagating in a direction θ with respect to the z axis, it will have a z variation of $\exp(-jn_{\text{eff}}k(\cos\theta)z)$ and a y variation of $\exp(-jn_{\text{eff}}k(\sin\theta)y)$. For such a planar guided wave, there is no amplitude variation in the direction perpendicular to the direction of propagation.

There can be superposition of TE_m modes propagating in different θ directions to form diverging or focusing waves in the yz plane with identical x variation. Similarly, there can be superposition of TM_n modes propagating in different θ directions to form diverging or focusing waves in the yz plane that have the same x variation.

3.4.2 Helmholtz equation for the generalized guided wave modes in planar waveguides

In short, there may be a number of planar guided waves with the same $E_m(x)$ or $H_n(x)$ simultaneously propagating in different θ directions in the yz plane. These modes all have the same x variation. Superposition of such planar guided waves can give very complex y and z variations.

We will now consider any generalized TE_m guided wave mode to be a product $E_m(x)E_{m,t}(y, z)$:

$$E_m(x, y, z) = E_m(x)E_{m,t}(y, z).$$

$E_m(x)$ is the mth solution of Eq. (3.2c) which has the eigen value β_m, or $n_{m,\text{eff}}k$. ($n_{m,\text{eff}}$ is known as the effective index of the mth TE planar guided wave mode.) $E_{m,t}(y, z)$ is a function of y and z satisfying the two-dimensional scalar wave equation

$$\left(\frac{\partial^2}{\partial y^2}+\frac{\partial^2}{\partial z^2}+n_{m,\mathrm{eff}}^2k^2\right)E_{m,t}(y,z)=0, \tag{3.22a}$$

$$\left[\frac{\partial^2}{\partial x^2}+\left(\omega^2\mu\varepsilon(x)-n_{m,\mathrm{eff}}^2k^2\right)\right]E_m(x)=0, \tag{3.22b}$$

$$\left.\begin{aligned} E_m(x) &= A_m\sin(h_mt+\phi_m)\exp[-p_m(x-t)], && x\geq t,\\ E_m(x) &= A_m\sin(h_mx+\phi_m), && t>x>0,\\ E_m(x) &= A_m\sin\phi_m\exp(q_mx), && x<0. \end{aligned}\right\} \tag{3.22c}$$

The $\exp(-jn_{m,\mathrm{eff}}k\sin\theta\ y)\exp(-jn_{m,\mathrm{eff}}k\cos\theta\ z)$ solution for Eqs. (3.22c) is just the plane wave (i.e. plane wave in the yz plane in the θ direction and guided wave variation in the x direction) solution for $E_{m,y}(y,z)$. There are many other possible solutions. There is a strong similarity between the equation for $E_{m,t}(y,z)$ in Eqs. (3.22c) and the Helmholtz equation, Eq. (1.4). All the techniques used to solve the scalar wave equation, Eq. (1.4), can be applied here to the $E_{m,t}(y,z)$, as long as (1) the polarization of the electric field is dominantly in the transverse direction, (2) the x variation is in the form of $E_m(x)$ of the TE_m mode and (3) the transverse variation in the yz plane is slow within a distance comparable to λ. The major difference is that Eq. (1.4) is a scalar wave equation in three dimensions whereas Eq. (3.22a) is a scalar wave equation in two dimensions. The mathematical details of how to solve scalar wave equations in two dimensions and in three dimensions are very different [2].

Similar comments can be made for TM_n guided wave modes for the magnetic field.

3.4.3 Applications of generalized guided waves in planar waveguides

In order to appreciate the importance of the more complex yz variation and the generalized guided wave mode in planar waveguides, we will consider four applications using planar TE_m waveguides.

(1) Radiation from a line source in the yz plane

Let there be a single TE mode planar waveguide (i.e. the index and the thickness combination allows only the TE_0 mode to exist). A line source of guided wave TE_0 mode is placed at the origin of the yz plane. A line source is represented mathematically as a unit impulse function δ in the yz plane. The solution for such an

$E_{0,t}(y, z)$ in Eq. (3.22a) is a cylindrical wave at distances far away from the origin,

$$E_{0,t}(y, z) = \frac{A}{\sqrt{\rho}} e^{-jn_{0,\mathrm{eff}}k\rho}, \tag{3.23}$$

where

$$\rho = \sqrt{y^2 + z^2}. \tag{3.24}$$

Note that E approaches infinity as ρ approaches zero. This solution is similar to the spherical wave shown in Chapter 1 except for the $1/\sqrt{\rho}$ variation instead of the $1/R$. This modification is necessary if we consider the power P radiated by such a cylindrical wave in the yz plane in the form of a TE guided wave:

$$\begin{aligned} P &= \left\{ \frac{n_{\mathrm{eff},0}k}{2\omega\mu} \int_{-\infty}^{\infty} |E_0(x)|^2\, dx \right\} \int_{-\pi}^{\pi} |E_{0,t}(y, z)|^2 \rho \sin\theta\, d\theta \\ &= 2\pi\, A^2. \end{aligned} \tag{3.25}$$

In evaluating P, we already know from Eq. (3.7) that the result of the integration in x within the curly brackets is unity. Therefore the P becomes proportional to A^2. In other words, the square root dependence in ρ is necessary for power conservation, i.e. for P to be independent of ρ. Notice also that the $E_{0,t}$ in Eq. (3.23) satisfies Eq. (3.22a) only for large ρ when higher orders of $1/\rho$ can be neglected.

(2) Diffraction and collimation of guided waves in the yz plane

Let a single TE mode planar waveguide be terminated abruptly at $z = 0$, i.e. the waveguide exists only for $z > 0$. Its end surface is the $z = 0$ plane. When the radiation from a laser with electric field polarized in the y direction is focused perpendicularly on this end surface, it will excite only the TE_0 guided wave mode and the TE substrate and air modes. No TM mode is excited. We will discuss later, in Section 3.7, how to calculate this excitation. We observe here that at any significant distance away from $z = 0$, i.e. $z \gg \lambda$, the substrate modes and air modes would have been radiated away. Therefore, in order to find the propagation of the TE_0 beam in the yz plane for this excitation, we can ignore the radiation modes.

Let us approximate the excitation on the $z = 0$ plane by a TE_0 planar guided wave mode $E_0(x)$ which has uniform intensity A for $|y| < l_y$ and 0 for $|y| \geq l_y$. In other words, we have a one-dimensional slit in the y direction applied to a TE_0 planar waveguide. We will now use the Green's function technique to solve the scalar wave equation shown in Eq. (3.22a) with this boundary condition.

However, in order to apply the Green's function technique as we did in Chapter 1, we now need a Green's function for a two-dimensional Helmholtz equation. Such a Green's function is given in classical electromagnetic theory books, for example ref. [3], by

$$\left.\begin{aligned} 4\pi\, G_\beta(y, z; y_0, z_0) = j\pi H_0^{(2)}(\beta\rho) &\to -2\ln(\beta\rho) \quad \text{as} \quad \beta\rho \to 0 \\ &\to \sqrt{\frac{2\pi}{j\beta\rho}} e^{-j\beta\rho} \quad \text{as} \quad \beta\rho \to \infty. \end{aligned}\right\} \tag{3.26}$$

Here the source is at (y, z) and the observation point is at (y_0, z_0), $\beta = n_{\text{eff},0}k$, $\rho = \sqrt{(y - y_0)^2 + (z - z_0)^2}$, and $H_0^{(2)}$ is the Hankel function of the zeroth order and the second kind. This G corresponds to G_1 in Chapter 1. The method of images can be used again to yield a Green's function such that $G = 0$ at $z = 0$,

$$G_\rho(y, z; y_0, z_0) = \frac{j}{4}\left[H_0^{(2)}(\beta\rho) - H_0^{(2)}(\beta\rho_i)\right].$$

Here, $\rho_i = \sqrt{(y_i - y)^2 + (z_i - z)^2}$, and y_i and z_i are images of y_0 and z_0 across the $z = 0$ axis. It is instructive to see that the mathematics of scalar wave equations in two dimensions becomes very complicated as compared to wave equations in three dimensions. Fortunately, for large $\beta\rho$, the Hankel function has a simple approximation, as shown in Eq. (3.26). Therefore, for the far field, where higher order terms in the binomial expansion can be neglected,

$$E_{0,t}(y, z) = \int_{-l_y}^{+l_y} A\, h(y, z = 0; y_0, z_0)\, dy, \tag{3.27}$$

$$h(y, z; y_0, z_0) \approx \sqrt{\frac{n_{\text{eff},0}k}{j2\pi(z_0 - z)}}\, e^{-jn_{\text{eff},0}k\rho},$$

$$-jn_{\text{eff},0}k\rho \approx -jn_{\text{eff},0}k(z_0 - z) - jn_{\text{eff},0}k\frac{y_0^2}{2(z_0 - z)} + jn_{\text{eff},0}k\frac{yy_0}{z_0 - z} + \cdots.$$

The integration over y at the far field is a Fourier transform of the excitation field. It yields

$$E_{0,t} = 2l\sqrt{\frac{n_{\text{eff},0}k}{j2\pi(z_0 - z)}}\, e^{-jn_{\text{eff},0}k(z_0 - z)} e^{-jn_{\text{eff},0}k\frac{y_0^2}{2(z_0 - z)}} \operatorname{sinc}\left(\frac{2n_{\text{eff},0}l_y y_0}{\lambda(z_0 - z)}\right). \tag{3.28}$$

$E_{0,t}$ is the Fraunhofer diffraction pattern in the yz plane. For our example, $z = 0$. Note the divergent cylindrical wave front in the far field. The result agrees with the $E_{0,y}$ in Eq. (3.23). The sinc function shows just the amplitude distribution pattern of the diffracted field as the observation angle $\theta = y_0/z_0$ is varied. The diffracted field has a main lobe and side lobes.

(3) Cylindrical guided wave lens

Similarly to the three-dimensional case, an ideal guided wave cylindrical lens at $z = 0$, placed parallel to the xy plane with the axis of the cylinder along the x axis, can be represented by a quadratic phase shift, $\exp[j(\pi n_{m,\text{eff}}/\lambda f)y^2]$. It will convert a planar guided wave normally incident on the lens into a convergent cylindrical guided wave focused at $z = f$. It will also collimate a divergent guided wave into a collimated guided wave. Similarly to the three-dimensional case, including the quadratic phase modification into the diffraction integral is sufficient to represent the diffraction effect of the lens in the yz plane.

In practice, it is difficult to obtain waveguide structures such that the effective index of the guided wave mode within the lens is much larger than the effective index of the guided wave outside the lens. This is similar to the problems involved in making a three-dimensional lens out of a material that has an index not much larger than the index of air. Such lenses will be very weak. For these reasons, Fresnel lenses and geodesic lenses are usually used.

(4) Star coupler

As the fourth example, we will consider a device called a planar waveguide star coupler, as shown in Fig. 3.5 [4]. It is used in wavelength division multiplexed (WDM) fiber optical systems. It consists of two arrays of N uniformly spaced identical channel waveguides. Each waveguide has width a. The ends of channel waveguides in each array are located on a circular arc with radius R. There are two circular arcs facing each other. The center of the circle of the array on the left is at O', which is also the middle of the circular arc for the array on the right. Vice versa, the center of the circle on the right is at O, which is also the middle of the circular arc for the array on the left. The center position of the kth waveguide on the left arc is given by $R\theta_{0.k}$, and the center position of the jth waveguide on the right arc is given by $R\theta'_{0.j}$. The region between the two arrays is a single-mode planar waveguide. The power entering the single-mode planar waveguide region (from any one of the $2N$ waveguides) will be diffracted and propagated in the yz plane as the generalized guided wave of the planar waveguide. Waveguides on the opposite circular arc are excited by the radiation carried by this generalized guided wave.

The objective of the star coupler is to maximize the power transfer between any one of the channel waveguides in the left array and any one of the waveguides in the right array. Ideally, there is no power loss, and the input power from any waveguide

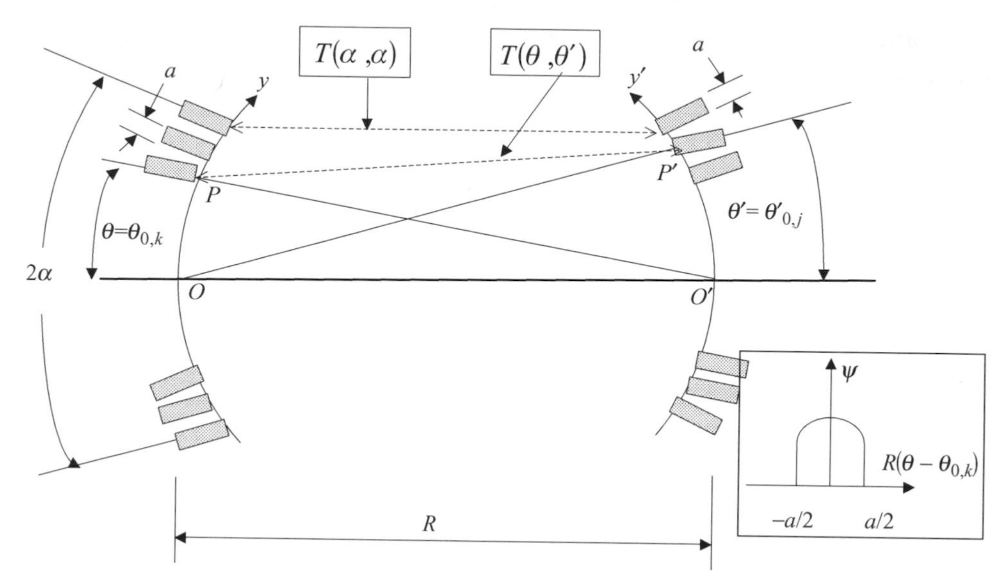

Figure 3.5. The star coupler. Illustration of the planar arrangement of two circular arrays of channel waveguides (shaded) with fields related by a Fourier transformation. The channel waveguides are arranged in two confocal circular arcs that face each other, connected by a planar waveguide. The objective of the coupler is to provide a uniform transmission of optical power $T(\theta, \theta')$ from any channel waveguide at location P (at $\theta_{0,k}$) to channel waveguides at any P' (at $\theta'_{0,j}$) location. The electric field pattern within each channel waveguide is shown as ψ in the inset. The figure is taken from ref. [4] with copyright permission from the IEEE.

is divided uniformly into the N output channels. In that case, the transfer efficiency will be $1/N$. However, this is impossible to achieve in practice. In this example we will analyze the star coupler using the generalized planar TE_0 guided wave mode. In particular, we will calculate the field at the output array produced by the radiation from a given channel waveguide in the input array. We will calculate the excitation of the mode of the channel waveguide in the output array by this field, thereby determining the power transfer from the input channel to the output channel.

We have not yet discussed the fields of a channel waveguide mode. Let us assume that the E_y of the guided wave mode for all input and output channels in the yz plane is $\psi(y)$ or $\psi(y')$, where y (or y') is the coordinate along the left (or right) circular arc, as shown in Fig. 3.5. Transmission between two elements (i.e. channel waveguides), i.e. the P channel waveguide on the left circular arc centered about $\theta_{0,k}$ and the P' channel waveguide on the right circular arc centered about $\theta'_{0,j}$, is determined by (1) calculating the generalized planar guided wave field at $y' = R\theta'$ diffracted from P, as we have done in the preceding example, and (2) calculating the coupling of that field into P'.

In order to calculate the field radiated from P to $R\theta'$, we note that the distance between y and y' in the first-order approximation of the binomial expansion is

$$\begin{aligned}\rho &= \sqrt{[R\cos\theta' - (R - R\cos\theta)]^2 + (R\sin\theta' - R\sin\theta)^2} \\ &\approx R - R\sin\theta'\sin\theta \\ &\approx R - R\theta\theta'.\end{aligned}$$

Thus, for large $\beta\rho$, the field produced by P at P' is

$$E_y(R\theta') \approx \sqrt{\frac{n_{\text{eff}}k}{j2\pi R}}\, e^{-jn_{\text{eff}}kR} \int_{\theta_{0,k}-\frac{a}{2R}}^{\theta_{0,k}+\frac{a}{2R}} \psi(R\theta)\, e^{+j2\pi\left(\frac{n_{\text{eff}}}{\lambda}\theta'\right)R\theta}\, R\, d\theta,$$

where we have assumed that the field for the kth channel waveguide is confined approximately within the waveguide, as shown in the inset of Fig. 3.5. Note that the phase factor, $-jn_{\text{eff}}kR$, is now a constant on the circular arc on the right. Thus the circular arcs serve a function similar to the spherical reflectors in a confocal resonator in three dimensions.

Using a change of variable, $u = (2R/a)(\theta - \theta_{0,k})$, we obtain the following:

$$E_y(R\theta') \approx a\sqrt{\frac{n_{\text{eff}}}{j\lambda R}} e^{-jn_{\text{eff}}kR} e^{+j2\pi\left(\frac{n_{\text{eff}}R\theta_{0,k}\theta'}{\lambda}\right)} \phi(R\theta'),$$

where

$$\phi(R\theta') = \frac{1}{2}\int_{-1}^{+1} \psi\left(\frac{au}{2}\right) e^{+j2\pi\left(\frac{n_{\text{eff}}a\theta'}{2\lambda}\right)u}\, du.$$

Since $\psi(au/2)$ is identical for all the waveguides, the ϕ factor is independent of $\theta_{0,k}$. The E_y is only dependent on the center position $R\theta_{0,k}$ of the input channel through the factor $\exp(j2\pi \frac{n_{\text{eff}}R\theta_{0,k}\theta'}{\lambda})$. Let the total E_y at $R\theta'$ be expressed as a summation of the fields of the channel guides, $\psi_i(R\theta')$, on the right circular arc array plus the stray guided wave fields in the gaps between channel guides, $\zeta(R\theta')$. Let us assume, as an approximation, that there is negligible overlap among all the ψ_i and the ζ. Then,

$$E_y(R\theta') = \sum_i b_i\psi_i(R\theta') + \zeta(R\theta').$$

Here, $\psi_i(R\theta')$ is the ψ centered about $\theta_{0,i}$. Multiplying both sides by $\psi_j^*(R\theta')$ and integrating with respect to $R\theta'$ from $-\infty$ to $+\infty$, we obtain

$$\int_{\theta_{0,j}-\frac{a}{2R}}^{\theta_{0,j}+\frac{a}{2R}} E_y(R\theta')\psi(R\theta')R\, d\theta' \approx b_j \int_{\theta_{0,j}-\frac{a}{2R}}^{\theta_{0,j}+\frac{a}{2R}} |\psi(R\theta')|^2 R\, d\theta'.$$

Utilizing once more the change of variable $u' = (2R/a)(\theta' - \theta_0')$, we obtain

$$|b_j|^2 \left[\frac{a}{2} \int_{-1}^{+1} \left| \psi\left(\frac{a}{2}u' + R\theta_{0,j}\right) \right|^2 du' \right]^2 = \left[\frac{n_{\text{eff}} a^4}{\lambda R} |\phi(R\theta_{0,k})|^2 |\phi(R\theta_{0,j})|^2 \right],$$

or

$$|b_j|^2 = \frac{4 n_{\text{eff}} a^2}{\lambda R} \frac{|\phi(R\theta_{0,k})|^2 |\phi(R\theta_{0,j})|^2}{\left[\int_{-1}^{+1} \left| \psi\left(\frac{a}{2}u + R\theta_{0,k}\right) \right|^2 du \right]^2}. \tag{3.29}$$

Since the power contained in the total E_y is proportional to $\int |E_y|^2 \, R \, d\theta$, which is approximately equal to $\sum_i |b_i|^2 \int |\psi|^2 R \, d\theta$, $|b_j|^2$ is the power transfer from the channel waveguide centered at $\theta_{0,k}$ to the channel waveguide centered at $\theta_{0,j}$.

In an actual star coupler, R, N and a are designed to optimize the power transfer. Dragone [4] optimized the design which gives $0.34(1/N)$ to $0.55(1/N)$ of the input power to any one of the output channels.

3.5 Rectangular channel waveguides and effective index analysis

Rectangular waveguides are important in many practical applications because the rectangular cross-section is an idealized cross-section of the actual waveguides fabricated by most micro-fabrication processes such as etching. Figure 3.6 illustrates the index profiles of two rectangular channel waveguides. In either case, the center portion, at $W/2 \geq y$, consists of a ridge with a finite width W. Because of the complexity of the geometry of the dielectric boundaries, there is no analytical solution of the modes of such a structure. There are only approximate solutions such as those given in ref. [5] and computer programs that can calculate numerically the guided wave modes. These computer programs use numerical methods such as the beam propagation method or the finite element method for simulation. However, the guided wave modes could be obtained easily by an approximate method called the effective index method, which will be presented here. This method is reasonably accurate for strongly guided modes (i.e. modes well above cut-off). It is based on the solutions of the planar guided wave modes discussed in Sections 3.1 to 3.3. The effective index analysis might also provide us with insight into the properties of channel guided wave modes.

Let us consider the rectangular channel waveguides in Fig. 3.6 where there are a rectangular core region, $y \leq |W/2|$, and a cladding region, $y \geq |W/2|$. If W is large, then we would have approximately a planar waveguide in the core. The propagation

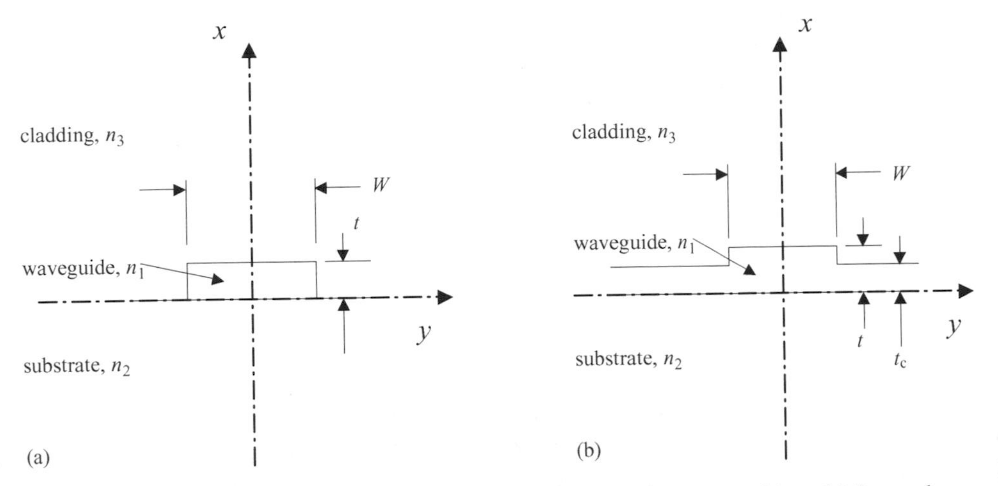

Figure 3.6. Index profiles of two examples of channel waveguides. (a) Lateral cross-section of a waveguide where the high-index core is etched down to the substrate outside the waveguide. (b) Lateral cross-section of a ridged waveguide where the high-index core outside the ridge is etched partially down to a thickness t_c.

of the $m = 0$ planar TE guided wave mode in the core along its longitudinal direction z is given by $\exp(\pm j\beta_0 z)$ where β_0/k is its effective index, n_{e1}. In Fig. 3.6(b), there is also a different planar waveguide mode in the cladding region when we ignore the ridge. Let the effective index of the $m = 0$ TE planar guided wave mode of the structure in the cladding region be n_{e2}. Since the high-index layer is thicker for $y \leq |W/2|$, $n_{e1} > n_{e2}$. In Fig. 3.6(a), there is no guided wave mode in the cladding region; there are only continuous substrate and air modes for $y \geq |W/2|$. These continuous modes will have $n_{e1} > n_2 > \beta/k > n_3$.

Let us consider first the channel waveguide in Fig. 3.6(b). The core planar guided wave mode in the $y \leq |W/2|$ region can propagate in any direction in the yz plane. Let us consider a core planar guided wave propagating in a direction making a very small angle δ with respect to the z axis. Let δ be so small that $n_{e1} \cos\delta > n_{e2}$. When this core planar guided wave is incident on the vertical boundary at $y = |W/2|$, it excites the cladding planar waveguide mode at $y > |W/2|$ plus continuous modes. However, in order to match the boundary condition at $y = |W/2|$ as a function of z, the cladding planar guided wave mode cannot have a real propagating wave number in the y direction. It must have an exponentially decaying y variation. In other words, the core planar guided wave is now totally internally reflected back and forth between the two boundaries at $y = \pm W/2$. The sum of all the reflected core planar guided waves yields a non-zero solution when the round trip phase shift of the total internal reflection at specific values of δ is a multiple of 2π. These special sets of totally internally reflected core planar waveguide modes constitute the channel guided wave modes.

Let us now consider the mathematical details of the approach discussed in the preceding paragraph. At the $y = |W/2|$ boundaries, the electric field E_y of the core planar guided wave mode is no longer the field transverse to the boundaries. E_y is now approximately perpendicular to the boundaries, which are the $y = \pm W/2$ planes. The tangential field of the core guided wave is the magnetic field that has two components, the H_x and the quasi H_z fields. The dominant component is H_x. Therefore, at the $y = |W/2|$ boundary, we will match the H_x of the core and cladding modes.

The transverse field in the cladding that is the closest match to the x variation of H_x at the $y = |W/2|$ boundary is the H_x of the cladding TE planar guided wave of the same order. In order to satisfy the boundary condition for all z values, the z variation of this cladding guided wave mode must be equal to $\exp(-jn_{e1}\cos\delta)$. If we let the y variation of the cladding guided wave be $\exp(-j\gamma y)$, γ must satisfy the equation

$$\gamma^2 = n_{e2}^2 - n_{e1}^2 \cos^2 \delta. \tag{3.30}$$

This relationship is a consequence of Eq. (3.22a). Thus, γ is imaginary when $n_{e1} \cos\delta > n_{e2}$. An imaginary γ represents an exponentially decaying cladding guided wave in the y direction, not a propagating cladding guided wave. In other words, the core guided wave is totally internally reflected at the $y = |W/2|$ boundaries. The channel guided wave mode is obtained by requiring the total round trip phase shift (with total internal reflection at the $y = \pm W/2$ boundaries) of the core planar guided wave (at angle δ) to be $2n\pi$.

In short, the mathematics of analyzing the total internal reflection of the core planar guided wave in the y direction is equivalent to analyzing the total reflection of the equivalent plane wave propagating in the yz plane at angle δ with approximately the E_y and H_x polarization. The equivalent material refractive index is n_{e1} and n_{e2} and the magnetic field is the transverse field. In other words, we can use the TM planar guided wave mode equation for a symmetric waveguide, i.e. Eqs. (3.13), with y replacing x and letting n_{e1} be the core index and n_{e2} be the index of the substrate and top cladding. The solutions of that equation are the channel guided wave modes that we are looking for. This is the effective index method.

It is important to use the TM equation because the field tangential to the $y = |W/2|$ plane is the magnetic field. The most important quantity to be obtained is the effective index, i.e. the β_n/k or $n_{e1}\cos\delta$, of the channel waveguide in the z direction. Knowing this effective index, we know both the δ in the core and the exponential decay constant, γ, in the cladding. Since δ is very small, the channel guided wave mode obtained from the TE core planar guided mode is still approximately a y polarized TE mode propagating in the z direction. Naturally, the x variation is approximately the same as the core planar guided wave for $y < |W/2|$ and the same as the cladding planar guided wave for $y > |W/2|$.

Note that the boundary conditions at $y = \pm W/2$ are not satisfied exactly by just the core and the cladding guided waves. In order to satisfy the boundary conditions accurately, many other modes, especially the substrate and air modes, must be involved. Therefore, the effective index is only reasonably accurate for well guided modes. Notice also that we no longer have purely TE or TM modes. We have basically TE-like modes with a small E component in the z direction. Similarly, we have TM-like modes with a small H component in the z direction. These modes are called hybrid modes.

For the waveguide shown in Fig. 3.6(a), the x variation of the tangential field of the core guided wave propagating at angle δ is matched by the summation of the continuous cladding modes at $y = |W/2|$. For $n_{e1} \cos\delta > n_3$ and n_2, in order to satisfy the boundary condition as a function of z, all continuous modes will decay exponentially away from the $y = |W/2|$ boundary. Thus the core guided wave mode is again totally internally reflected back and forth. The sum of all the reflected core planar guided waves yields a non-zero solution when the round trip phase shift of total internal reflection at specific values of δ is a multiple of 2π. These special sets of totally internally reflected core planar waveguide modes constitute the channel guided wave modes.

The effective index method can also be used to obtain approximately modes of channel waveguide structures such as that shown in Fig. 3.6(a). In this case, we know the n_{e1} of the core TE planar guided wave mode, but we do not know n_{e2}. Since a combination of substrates and air modes is used to match the x variation of the core guided wave at $y = \pm|W/2|$, the value of n_{e2} is somewhere between n_3 and the substrate index n_2. The effective index n_{e2} to be used for the cladding region in the TM equation in y will depend on the profile of the core TE mode. For high-index waveguides with deep sided walls, we will most likely use n_3 for the cladding. For a core guided wave with a long evanescent tail in the x direction in the substrate, we may use the substrate index. Fortunately, for well guided channel modes in the core, the solution of n_{eff} and the y variation is not very sensitive to the value of the effective index used for the cladding. Clearly, the approximation of the effective index method may not be very good for such a structure. It is also difficult to say anything about the x variation of the field in the cladding. The best we can do is to estimate the γ in the cladding region and to assume that for $|y| - |W/2| \ll \gamma$ the x variation is similar to the core guided wave mode.

Similarly, a channel guided wave mode with approximately TM polarization can be obtained from a TM planar guided wave mode in the core and in the cladding region. In that case the equivalent TE guided wave equation will be used to find the effective index of the channel waveguide mode and the y variation.

3.5.1 Example for the effective index method

Consider first a GaAs planar waveguide with $n_1 = 3.3$, $n_2 = 3.188$ and $t = 0.9$ μm in the core region operating at $\lambda = 1.5$ μm. This waveguide is coated with a dielectric film with $n_3 = 1.68$. The dielectric film has been etched away at $y \geq |W/2|$, where $W = 3$ μm. In the cladding region, $t = 0.6$ μm. We would like to find the effective index and the field of the lowest order TE-like channel waveguide mode.

The first step of our calculation is to find the effective index of the TE_0 planar guided wave in the core region at $W/2 \geq |y|$ and in the cladding region at $|y| > W/2$. From Eqs. (3.4), we find the TE planar guided wave modes and $n_{e1} = 3.257$ and $n_{e2} = 3.247$. Using the example shown in Section 3.3.2, we will solve the following equations to obtain the y variation of the lowest order channel waveguide mode (i.e. $n = 0$):

$$\tan\left[(h'_n/k)\frac{kW}{2}\right] = \frac{n_{e1}^2 p'_n/k}{n_{e2}^2 h'_n/k}, \tag{3.31a}$$

$$\left(\frac{h'_n}{k}\right)^2 + \left(\frac{p'_n}{k}\right)^2 = n_{e1}^2 - n_{e2}^2. \tag{3.31b}$$

The solution is $(h'_0/k) = 0.096\,51$, which gives $n_{\mathrm{eff},0} = 3.2556$ and $p'_0/k = 0.236$. The field distributions are

$$\begin{aligned}
E_y &= A\sin(h_0x + \phi_0)\sin(h'_0y + \phi'_0)e^{-jn_{\mathrm{eff},0}kz}, && 0 < x < t,\ y \leq |W/2|, \\
E_y &= A\sin\phi_0 e^{q_0x}\sin(h'_0y + \phi'_0)e^{-jn_{\mathrm{eff},0}kz}, && x \leq 0,\ y \leq |W/2|, \\
E_y &= A\sin(h_0t + \phi_0)^{-p_0^{\mathrm{c}}(x-t)}\sin(h'_0y + \phi'_0) && \\
&\quad \times e^{-jn_{\mathrm{eff},0}kz}, && x \geq t,\ y \leq |W/2|, \\
E_y &= A\sin\left(h_0^{\mathrm{c}}x + \phi_0^{\mathrm{c}}\right)\sin\left(\frac{h'_0W}{2} + \phi'_0\right) && \\
&\quad \times e^{-p'_0(y-\frac{W}{2})}e^{-jn_{\mathrm{eff},0}kz}, && 0 < x < t,\ y > W/2, \\
E_y &= A\sin\left(h_0^{\mathrm{c}}x + \phi_0^{\mathrm{c}}\right)\sin\left(-\frac{h'_0W}{2} + \phi'_0\right) && \\
&\quad \times e^{+p'_0(y+\frac{W}{2})}e^{-jn_{\mathrm{eff},0}kz}, && 0 < x < t,\ y < -W/2.
\end{aligned} \tag{3.32}$$

Here, $\tan\phi_0^c = h_0^c/p_0^c$, where h_0^c and ϕ_0^c are parameters from the planar guided wave TE_0 mode in the cladding region (given in Eqs. (3.3) and (3.4) with $\beta = 3.247k$). h_m, ϕ_0, q_0, h_0 and p_0 are parameters of the planar guided wave TE_0 mode in the core (given in Eqs. (3.3) and (3.4) with $\beta_m = 3.257k$). We cannot find the field distributions accurately in the regions ($x > t$, $|y| > W/2$) and ($x < 0$, $|y| > W/2$) from the effective index method. We may estimate that the fields will decay exponentially in the x and y directions with decay constants q_0, p_0 and p_0', respectively.

3.5.2 Properties of channel guided wave modes

Channel waveguides are used mostly in guided wave devices such as the directional coupler, the Y-branch splitter, the index-guided laser, the guided wave modulator, the waveguide photodetector, the waveguide demultiplexer and the waveguide filter. Therefore the properties of the channel-guided wave mode most important to these applications are $n_{\mathrm{eff},m}$ or $n_{\mathrm{eff},n}$, the attenuation rate, the polarization of the mode and the evanescent tails described by p_m, q_m and γ. Most active channel waveguide devices are a few centimeters or less in length. Thus, unlike for optical fibers, any reasonable attenuation rate, such as 1 dB/cm or less, may be acceptable in practical applications. Active channel waveguide devices mostly involve one guided wave mode interacting with another guided wave mode, and these will be discussed in detail in Chapter 4. On the other hand, passive waveguide splitters, combiners and demultiplexers could be much longer. The reduction of scattering loss is an important engineering issue for these devices. The excitation efficiency of the channel waveguide mode by end excitation from another component, such as a laser or an optical fiber, will depend critically on the matching of the polarization and the field pattern of the waveguide mode with the mode of the fiber or the laser. The excitation of the channel waveguide will be discussed in Section 3.7.

3.5.3 Phased array channel waveguide demultiplexer in WDM systems

Let us consider an application of channel waveguides in a component called a PHASAR demultiplexer in wavelength division multiplexed (WDM) optical fiber systems [6]. Consider two star couplers, as discussed in Section 3.4.3, interconnected by an array of identical channel waveguides, each of length L_j, as shown in Fig. 3.7. On the input side of the first star coupler, there is the transmitting waveguide. The field distribution at the input is given by $E_y = \psi_k$ of the input channel and zero elsewhere. In terms of the star coupler discussed in Section 3.4.3, the

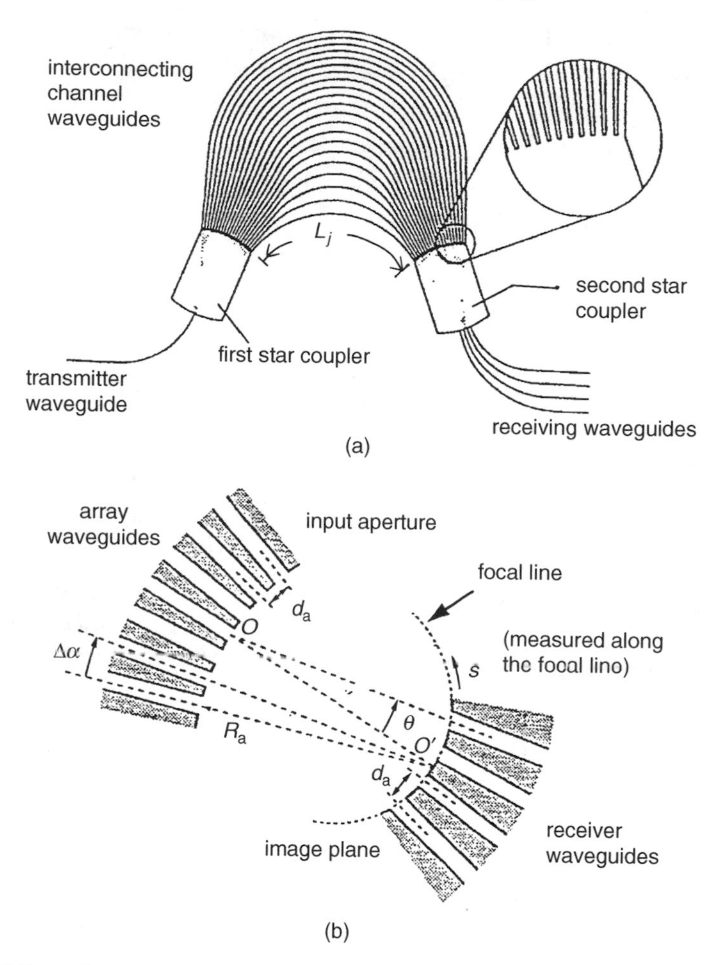

Figure 3.7. (a) Layout of the PHASAR demultiplexer. (b) Geometry of the receiver side. The two star couplers are connected by an array of interconnecting channel waveguides that have different lengths. Optical radiation from the input waveguide is transmitted to the interconnecting waveguides by the input star coupler. The input radiation to the output coupler will have phase shifts controlled by the wavelength as well as by the length increments of the interconnecting waveguides. The objective is to create an appropriate phase shift so that radiation at a different wavelength is transmitted to a different receiving waveguide. The geometry of the output coupler is shown in (b), where the array waveguides are the interconnecting waveguides. This copyright figure is taken from ref. [6] with permission from the IEEE.

input consists of a circular array of channel waveguides on the input side within which only the kth waveguide (i.e. the transmitting waveguide) is excited. All other waveguides have zero power. The transmitting channel waveguide at the kth position will create a field distribution $E_y(R\theta')$ on all the output channels in the first star coupler. If all the interconnecting waveguides have equal length, and if the stray field ζ in the gap between channel guides is small, a field distribution identical to $E_y(R\theta)$ in the first coupler will be created on the input side of the second star coupler. By reciprocity, this field distribution on the input side of the second star coupler will create a field distribution on the output side which is zero except for the ψ_k at the position of the kth output waveguide. In other words, the power in the transmitting waveguide of the first coupler will now be transmitted exclusively to the kth output channel of the second star coupler. The situation does not change if the lengths of the interconnecting waveguides between the two star couplers differ from each other such that the phase shift between adjacent waveguides is 2π, i.e.

$$\frac{2\pi n_{\text{eff.c}}}{\lambda}(L_j - L_{j-1}) = \frac{2\pi n_{\text{eff.c}}}{\lambda}\Delta L = 2\pi,$$

where $n_{\text{eff,c}}$ is the effective index of the channel waveguide.

Let the spacing between adjacent channel waveguides be d_{a} ($d_{\text{a}} = R\Delta\alpha$). Then, according to Section 3.4.3, the E_y in the first star coupler, created by the transmitting waveguide at the kth channel position, has a phase

$$\exp\left(j2\pi\frac{n_{\text{eff}}R}{\lambda}(k\Delta\alpha)(m\Delta\alpha)\right)$$

at the center of the mth waveguide in the output array, where $kR\Delta\alpha$ and $mR\Delta\alpha$ are the center angular position of the kth and mth channel waveguides in the input and output arrays of the star coupler, respectively, as shown in Fig. 3.5, and k and m are integers, ranging from $-(N-1)/2$ to $(N-1)/2$; n_{eff} is the effective index of the planar waveguide in the star coupler. The difference in E_y caused by excitation from the kth waveguide or the $(k+1)$th waveguide is just a phase difference, $m\Delta\phi = 2\pi(R\Delta\alpha)(n_{\text{eff}}/\lambda)(m\Delta\alpha)$, at the center of the mth waveguide. Conversely, when the radiation in the array of input waveguides in the second star coupler has a total E_y field that contains this extra phase factor $m\Delta\phi$ for the mth waveguide, $m = -(N-1)/2$ to $(N-1)/2$, the total radiation will be coupled to the $(k+1)$th output waveguide instead of the kth output waveguide.

The central idea of the demultiplexer is that, when the kth waveguide is the output guide at λ_1, and when the apppropriate phase shift $m\Delta\phi$ is obtained as the wavelength is shifted from λ_1 to λ_2, we would have shifted the output from the kth waveguide to the $(k+1)$th waveguide.

Let the difference in length of the adjacent interconnecting waveguides be ΔL. The mth interconnecting waveguide then has a length $m\Delta L$ longer than the waveguide at the origin. Now consider in detail the second star coupler at two different wavelengths, λ_1 and λ_2. Let the output channel be the kth waveguide at λ_1. This extra phase factor $m\Delta\phi$ (which is needed to shift the output to the $(k+1)$th waveguide) will be obtained at λ_2 when

$$m\Delta\phi = \frac{2\pi}{c} n_{\text{eff,c}}(\Delta f) m \Delta L$$

or

$$\frac{R\Delta\alpha}{\Delta f} = \frac{d_{\text{a}}}{\Delta f} = \left(\frac{n_{\text{eff,c}}}{n_{\text{eff}}}\right)\left(\frac{\Delta L}{\Delta\alpha}\right)\frac{1}{f_2}.$$

Here, $f_1 = c/\lambda_1, f_2 = c/\lambda_2$ and $\Delta f = f_1 - f_2$. The ratio $d_{\text{a}}/\Delta f$ is called the dispersion of the interconnecting waveguides. In practice, there may be optical carriers at a number of closely equal spaced wavelengths, $\lambda_1, \lambda_2, \lambda_3, \ldots$ (i.e. $\Delta f =$ constant) in the transmitting channel. When the above dispersion relationship is satisfied, optical carriers at different wavelengths are transmitted to a different output waveguide. This device is called a PHASAR wavelength demultiplexer in WDM fiber systems [6]. The properties of the channel waveguides important to this application are $n_{\text{eff,c}}$, the uniformity of $n_{\text{eff,c}}$ in different channels and the attenuation of the waveguides.

3.6 Guided wave modes in single-mode round optical fibers

There are many books that discuss the modes of various optical fibers (see, for example, ref. [7]). We will not repeat those discussions here. However, guided wave modes in round step-index optical fibers are important and will be presented here because they are the only analytical solutions of channel waveguides. These analytical solutions allow us to show the similarities and the differences between modes of round optical fibers and the rectangular channel waveguides. Step-index fibers are not used in practical applications.

The cross-section of a step-index optical fiber with infinitely thick cladding is shown in Fig. 3.8. The core index n_1 is larger than the cladding index n_2. In contrast to channel waveguides with rectangular cross-sections, there are now analytical solutions of guided wave modes in single-mode step-index fibers because of the cylindrical symmetry. Although the field distribution and the effective index (especially the dispersion) of modern graded index fibers used in communication systems are different from those of the step-index fibers, step-index fiber modes are used here simply to demonstrate many properties of the modes of round fibers. We propose to demonstrate the following points. (1) There is a big difference in mathematical complexity between cylindrical and rectangular geometry. (2) There are, in general, only hybrid, not TE or TM, modes. (3) Unlike modes of rectangular channel waveguides, modes of round fibers are degenerate. In order to obtain linearly polarized transverse modes in fibers, we depend on the degeneracy in weakly

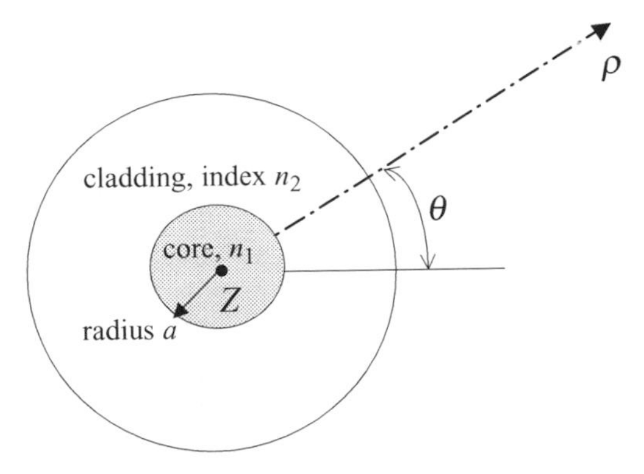

Figure 3.8. Cross-section of a step-index cladded core fiber and the cylindrical coordinates used for the analysis. The variation of the refractive index in the lateral direction is shown in cylindrical coordinates. The longitudinal direction Z is perpendicular to the figure, and is indicated by the dot at the center.

guiding fibers and linear combinations of these degenerate modes. (4) Properties of the modes of step-index fibers are important for understanding the dispersion properties of fibers and the random polarization rotation of the propagating radiation in fibers.

3.6.1 Guided wave solutions of Maxwell's equations

The vector wave equations obtained from Maxwell's equations in a homogeneous medium with refractive index n are [7]:

$$\begin{aligned}(\nabla^2 + n^2k^2)\underline{E} &= 0,\\ (\nabla^2 + n^2k^2)\underline{H} &= 0.\end{aligned} \tag{3.33}$$

In addition, we have the curl equations relating $\underline{E}$ and $\underline{H}$. If we assume that guided wave modes have the $\exp(-j\beta z)$ variation along the z direction, which is also the fiber axis, then in cylindrical coordinates we can write

$$\left(\nabla_{\rm t}^2 + k_{\rm t}^2\right) E_z = 0, \tag{3.34a}$$

$$\left(\nabla_{\rm t}^2 + k_{\rm t}^2\right) H_z = 0, \tag{3.34b}$$

$$\nabla^2 = [\nabla_{\rm t}^2] + \frac{\partial^2}{\partial z^2} = \left[\frac{1}{\rho}\frac{\partial}{\partial \rho}\left(\rho\frac{\partial}{\partial \rho}\right) + \frac{1}{\rho^2}\frac{\partial^2}{\partial \theta^2}\right] + \frac{\partial^2}{\partial z^2},$$

$$k_{\rm t}^2 = n^2k^2 - \beta^2.$$

The remaining transverse components of the fields are related to E_z and H_z as follows:

$$\begin{aligned}
E_\rho &= -\frac{j}{k_t^2}\left[\beta\frac{\partial E_z}{\partial \rho} + \frac{\omega\mu}{\rho}\frac{\partial H_z}{\partial \theta}\right], \\
E_\theta &= -\frac{j}{k_t^2}\left[\frac{\beta}{\rho}\frac{\partial E_z}{\partial \theta} - \omega\mu\frac{\partial H_z}{\partial \rho}\right], \\
H_\rho &= \frac{j}{k_t^2}\left[\frac{\omega\varepsilon_0 n^2}{\rho}\frac{\partial E_z}{\partial \theta} - \beta\frac{\partial H_z}{\partial \rho}\right], \\
H_\theta &= -\frac{j}{k_t^2}\left[\omega\varepsilon_0 n^2\frac{\partial E_z}{\partial \rho} + \frac{\beta}{\rho}\frac{\partial H_z}{\partial \theta}\right].
\end{aligned} \tag{3.35}$$

The solutions of Eqs. (3.34a) and (3.34b) are:

$$E_z = A J_m(k_{t1}\rho)\cos(m\theta), \tag{3.36a}$$

$$H_z = B J_m(k_{t1}\rho)\sin(m\theta), \tag{3.36b}$$

respectively, where

$$k_{t1} = \sqrt{n_1^2 k^2 - \beta^2}$$

for $a \geq \rho$ and

$$E_z = C H_m^{(2)}(jk_{t2}\rho)\cos(m\theta), \tag{3.37a}$$

$$H_z = D H_m^{(2)}(jk_{t2}\rho)\sin(m\theta), \tag{3.37b}$$

$$jk_{t2} = \sqrt{\beta^2 - n_2^2},$$

for $\rho > a$. There is a second set of solutions in which E_z has the $\sin(m\theta)$ variation and H_z has the $\cos(m\theta)$ variation. J_m is the Bessel function of the first kind and order m; $H_m^{(2)}$ is the Hankel function of the second kind of order m; m is an integer. Similarly to the guided waves in planar and channel waveguides, the Hankel function gives an exponential decay as $\rho \to \infty$ in the cladding. E_ρ, E_θ, H_ρ and H_θ are obtained from E_z and H_z from Eqs. (3.35). Continuity of E_z, H_z, E_θ and H_θ at $\rho = a$ yields the relationship among the A, B, C and D coefficients and the characteristic equation which determines the discrete values of β of the mode. The effective index, n_{eff}, of the mode is β/k. Similarly to the channel waveguide modes, each mode has a cut-off condition. The higher the order of the mode, the larger the value of $ka\sqrt{n_1^2 - n_2^2}$ for cut-off.

3.6.2 Properties of the guided wave modes

It is interesting to note that the axially symmetric modes have $m = 0$. In that case, we again have TE (with non-zero H_z, E_θ and H_ρ, called H_{0p} modes) and TM (with non-zero E_z, H_θ and E_ρ, called E_{0p} modes) modes. However, the lowest order mode that has the largest decay constant in the cladding, jk_{t2}, is not an axially symmetric mode. For $m \neq 0$, only a superposition of E_z and H_z solutions can satisfy all boundary conditions. The modes lose their transverse character, and are known as hybrid modes. The lowest order mode is the HE_{11} mode, which has $m = 1$ and the lowest order radial solution of the characteristic equations. There is no cut-off for the HE_{11} mode. In HE modes the longitudinal electric field is larger than the longitudinal magnetic field. There are also EH modes, in which the longitudinal magnetic field is dominant. The TM (i.e. E_{0p}) modes are the axially symmetric members of the HE family of modes. The H_{0p} modes are the axially symmetric members of the EH family of modes.

For weakly guiding modes, $\Delta = (n_1 - n_2)/n_1$ is small compared with unity. The characteristic equation for HE_{mp} modes is

$$k_{t1}a\frac{J_m(k_{t1}a)}{J_{m-1}(k_{t1}a)} = (jk_{t2}a)\frac{H_m^{(2)}(jk_{t2}a)}{H_{m-1}^{(2)}(jk_{t2}a)}, \tag{3.38}$$

where the subscript p refers to the pth root of the above equation. The characteristic equation for EH_{mp} modes is

$$k_{t1}a\frac{J_{m+2}(k_{t1}a)}{J_{m+1}(k_{t1}a)} = (jk_{t2}a)\frac{H_{m+2}^{(2)}(jk_{t2}a)}{H_{m+1}^{(2)}(jk_{t2}a)}. \tag{3.39}$$

Both the HE and EH modes exhibit nearly transverse field distribution. The longitudinal components have a phase shift of $\pi/2$ with respect to the transverse components; they remain small compared with the transverse field. The characteristic equation for HE_{mp} modes is the same as the characteristic equation for $\mathrm{EH}_{m-2,p}$ modes. Therefore, for weakly guiding fibers, any $\mathrm{HE}_{l+1,p}$ mode is degenerate with $\mathrm{EH}_{l-1,p}$ modes (i.e. they have the same propagation constants or effective index).

When we linearly combine the degenerate $\mathrm{HE}_{l+1,p}$ and $\mathrm{EH}_{l-1,p}$ modes, we obtain the linearly polarized LP_{lp} mode, which has the same effective index as the $\mathrm{HE}_{l+1,p}$ mode. The LP_{lp} mode has only E_x and H_y in the core and cladding; it is nearly uniformly polarized over the fiber cross-section. The LP_{01} mode is just the HE_{11} mode. Each LP mode occurs in four different versions: two orthogonal directions of polarization, each with $\cos l\theta$ and $\sin l\theta$ variations. Figure 3.9 shows the phase parameter B as a function of fiber parameter $V = ka\sqrt{n_1^2 - n_2^2}$ for low-order LP_{lp} modes. For a more detailed discussion of solutions of modes of step-index optical fibers, see ref. [7].

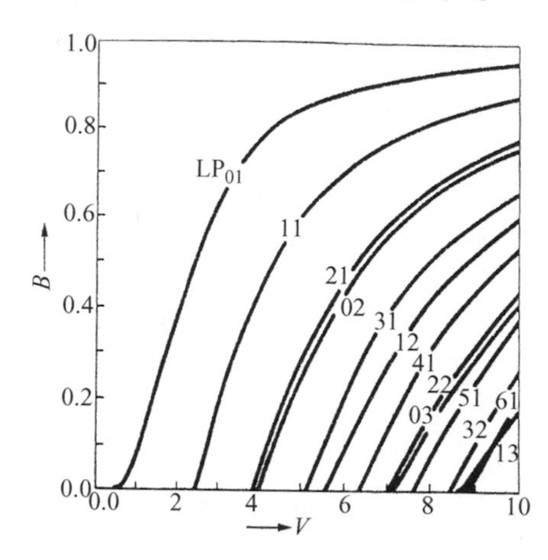

Figure 3.9. The phase parameter B of propagating modes in step index round fibers. The phase parameter B is related to the effective index n_{eff} of the propagating mode, $B = (n_{\mathrm{eff}}^2 - n_2^2)/(n_1^2 - n_2^2)$. The phase parameters B are shown as a function of the fiber material parameter V, $V = ka\sqrt{n_1^2 - n_2^2}$, for lower order LP_{1p} modes in weakly guiding fibers. This figure is taken from ref. [7] with copyright permission from Oxford University Press and the University of Minnesota Press.

3.6.3 Properties of optical fibers

Optical fibers are used primarily as transmission lines, often over many kilometers. There are very few devices made from fibers, the most prominent being fiber optical amplifiers and grating filters. Therefore, properties of modes in fibers that are most interesting for fiber communications are the number of propagating modes, the attenuation rate of the modes, the dispersion of the n_{eff}, the polarization of the excited mode and the change of state of polarization as the mode propagates. The low attenuation rates at 1.3 and 1.55 μm wavelengths dictate the operating wavelengths of optical fiber networks. In single-mode fibers, the indices and the core radius are controlled to cut off all the higher order modes. Thus, only the HE_{11} mode exists. However, the solution of n_{eff} from Eq. (3.38) clearly depends on λ. This is called the modal dispersion. On the other hand, the n_1 and n_2 also have slightly different values at different wavelengths. This is known as material dispersion. Dispersion causes the pulses of optical radiation to spread after propagating a long distance in the fiber, and it limits the data rate that can be transmitted through the fiber. In some fibers, material and mode dispersion cancel each other at specific wavelength such as 1.3 μm; they are called the zero-dispersion fibers. The effects of dispersion after a certain distance of propagation could also be canceled by propagating in another section of fiber with opposite dispersionî; this is called dispersion compensation. The polarization of the propagating mode is determined by the excitation source. However, the cylindrical fiber is degenerate in two orthogonal polarization directions. Any minute changes in uniformity caused by factors such as bending and stress cause the polarization of the radiation to rotate randomly in the fiber as it propagates. By means of intentional strain or ellipticity of the cross-section, polarization-maintaining fibers maintain the polarization of the radiation as it propagates.

3.6.4 Cladding modes

There are also cladding modes in optical fibers, corresponding to the continuous substrate and air modes in the planar and channel waveguides. They are excited whenever there are defects, bending of the fiber or dielectric discontinuity. Cladding modes are solutions of the boundary value equations when their effective indices are less than n_2. These modes do not decay exponentially away from the core. A typical single-mode fiber has a core about 10 μm in diameter, and the cladding has a diameter of the order of 100 μm. Thus there are many propagating cladding modes, with the effective indices very close to each other, resembling a continuous mode distribution. In the absence of the exponential decay factor, cladding modes have high attenuation. Their amplitude is very small at distances far from the discontinuity. Even for fibers with finite cladding thickness, cladding modes resemble closely the radiation modes in planar waveguides.

3.7 Excitation of guided wave modes

Usually, the guided wave mode (or modes) in an abruptly terminated planar waveguide, channel waveguide or weakly guiding optical fiber is excited at its end by illumination from a laser (or an abruptly terminated fiber). The second most common method of excitation is by phased matched interaction of the incident radiation with the guided wave, utilizing the evanescent tail of the mode in the lower index cladding, such as in a directional coupler for channel waveguides or a prism coupler for planar waveguides. We will discuss here the end excitation.

From the point of view of Gaussian optics, as discussed in Chapter 2, we could approximate the guided wave mode of a channel waveguide or optical fiber by a Gaussian beam. The radiation from the laser can also be approximated by another Gaussian beam. Efficient excitation of the guided wave mode is obtained when we use a lens to match the two Gaussian beams, as discussed in Section 2.4.7, provided that the direction of polarization is also matched.

In order to understand more thoroughly the excitation process, the modal expansion technique could be used to calculate the end excitation efficiency of a specific guided wave mode. Let us consider an example in which the laser is the source, and let the waveguide be excited by the laser. They are oriented along the z axis. Let the waveguide at $z > 0$ be abruptly terminated at $z = 0$. The laser radiation is incident on the waveguide from $z < 0$. At $z = 0$, the transverse electric field of the incident radiation coming from $z < 0$ may be expressed as

$$\underline{E_t}(x, y) = E_x(x, y)\underline{i_x} + E_y(x, y)\underline{i_y}, \tag{3.40}$$

where $\underline{i_x}$ and $\underline{i_y}$ are unit vectors in the x and y directions. For $z \leq 0$, $\underline{E_t}$ consists of just the incident laser mode and the reflected laser radiation. If we neglect the reflection, $\underline{E_t}$ is just the incident laser radiation. For $z \geq 0$, $\underline{E_t}$ consists of the guided wave modes and radiation (or cladding) modes of the waveguide. Both the x and y polarized electric fields must be continuous across the $z = 0$ plane.

For the channel waveguide at $z = 0$,

$$E_x = \sum_m A_m \psi_{x,m}(x, y) + \int_\beta b(\beta)\psi_x(\beta; x, y)\, d\beta \tag{3.41}$$

and

$$E_y = \sum_m C_m \psi_{y,m}(x, y) + \int_\beta d(\beta)\psi_y(\beta; x, y)\, d\beta. \tag{3.42}$$

Here, $\psi_{x,m}$ is the mth x polarized guided wave mode, $\psi_{y,m}$ is the mth y polarized guided wave mode, ψ_x is the x polarized radiation mode and ψ_y is the y polarized radiation mode. If the modes are orthogonal to each other (or non-overlapping) then we can multiply both sides of Eq. (3.41) by $\psi_{x,j}(x, y)$ and integrate with respect to x and y from $-\infty$ to $+\infty$. In that case, we obtain

$$|A_j|^2 = \frac{\left| \int_{-\infty}^{\infty} dx \int_{-\infty}^{\infty} dy \left[E_x \psi^*_{x,j}\right] \right|^2}{\left[\int_{-\infty}^{\infty} dx \int_{-\infty}^{\infty} dy |\psi_{x,j}|^2 \right]^2}. \tag{3.43}$$

The expression

$$\int_{-\infty}^{\infty} \int_{-\infty}^{\infty} E_x \psi^*_{x,j}\, dx\, dy$$

is called the overlap integral between the incident field and the jth order mode. The expression

$$|A_j|^2 \int_{-\infty}^{\infty} dx \int_{-\infty}^{\infty} dy |\psi_{x,j}|^2 \Big/ \int_{-\infty}^{\infty} dx \int_{-\infty}^{\infty} dy |E_t|^2$$

is the power efficiency for coupling the laser radiation into the x polarized jth guided wave mode. A similar expression is obtained for coupling into the y polarized guided wave mode, $|C_j|^2$.

When modes in both polarizations are excited, the total polarization of the radiation in waveguides or fibers will be position dependent because of the difference of the phase velocity of different modes. The coupling is sensitive with respect to geometrical, strain or bending perturbations. Clearly, radiation (or cladding) modes are also excited at $z = 0$. However, they radiate (or attenuate) away over a short distance. Therefore, most of the time, only the excitation of guided waves is of practical interest.

References

1 W. S. C. Chang, M. W. Muller and F. J. Rosenbaum, "Integrated Optics," in *Laser Applications*, vol. 2, ed. M. Ross, New York, Academic Press, 1974
2 P. M. Morse and H. Feshback, *Methods of Theoretical Physics*, Chapter 11, New York, McGraw-Hill, 1953
3 P. M. Morse and H. Feshback, *Methods of Theoretical Physics*, Section 7.2, New York, McGraw-Hill, 1953
4 C. Dragone, "Efficient N×N Star Coupler Using Fourier Optics," *Journal of Lightwave Technology*, **7**, 1989, 479
5 D. Marcuse, *Theory of Dielectric Optical Waveguides*, Chapter 1, New York, Academic Press, 1974
6 M. K. Smit and C. van Dam, "PHASAR-Based WDM-Devices, Principles, Design and Applications," *IEEE Journal of Selected Topics in Quantum Electronics*, **2**, 1996, 236
7 H.-G. Unger, *Planar Optical Waveguides and Fibers*, Chapter 5, Oxford, Oxford University Press, 1977

4

Guided wave interactions and photonic devices

In order to understand optical fiber communication components and systems, we need to know how laser radiation functions in photonic devices. The operation of many important photonic devices is based on the interactions of several guided waves. We have already discussed the electromagnetic analysis of the individual modes in planar and channel waveguides in Chapter 3. From that discussion, it is clear that solving Maxwell's equations simultaneously for several modes or waveguides is too difficult. There are only approximate and numerical solutions. In this chapter, we will first learn special electromagnetic techniques for analyzing the interactions of guided waves. Based on these techniques, practical devices such as the grating filter, the directional coupler, the acousto-optical deflector, the Mach–Zehnder modulator and the multimode interference coupler will be discussed. The analysis techniques are very similar to those techniques used in microwaves, except we do not have metallic boundaries in optical waveguides, only open dielectric structures.

The special mathematical techniques to be presented here include the perturbation method, the coupled mode analysis and the super-mode analysis (see also ref. [1]). In guided wave devices, the amplitude of radiation modes is usually negligible at any reasonable distance from the discontinuity. Thus, in these analyses, the radiation modes such as the substrate and air modes in waveguides and the cladding modes in fibers are neglected. They are important only when radiation loss must be accounted for in the vicinity of any dielectric discontinuity. The radiation modes are also important in special situations such as in a prism coupler, in which a radiation beam excites a guided wave over a long interaction distance, or vice versa [2].

There are three types of interactions which are the basis of the operation of most guided wave photonic devices. (1) The adiabatic transition of guided wave modes in waveguide or fiber structures from one cross-section to another cross-section as the modes propagate. An example of this type of interaction is the

Y-branch that splits one channel waveguide into two channel waveguides. The combination of two symmetrical Y-branches back to back with two phase shifting channel waveguides interconnecting them constitutes the well known Mach–Zehnder interferometer modulator and switch. (2) The phase matched interaction between two guided wave modes over a specific distance. An example of photonic devices based on this type of interaction is the directional coupler in channel waveguides or fibers. (3) Interaction of guided wave modes through periodic perturbation of the optical waveguide. An example of this is the grating filter in channel waveguides (or optical fibers) or the acousto-optical deflector (or scanner) in planar waveguides.

In the following sections, we will present first the perturbation analysis and derive the coupled mode equations. The perturbation analysis will also be used to find the super-modes of waveguide structures involving more than one waveguide. The discussion on the application of the analysis to photonic devices will be organized according to different types of interactions utilized to achieve device operations. (1) Guided wave interactions in the same single-mode waveguide. Grating filters and acousto-optic deflectors and analyzers are examples illustrating this type of interaction. (2) Guided wave interactions in parallel waveguides. A directional coupler is discussed to illustrate such interactions. Modal analysis involving super-modes will also be presented to analyze this type of interaction. (3) Guided wave interactions in waveguide structures that employ adiabatic transitions. Modal analysis using super-modes will be used mostly for analyzing such structures. The Mach–Zehnder interferometer (and modulator) is presented to illustrate this type of interaction. (4) Mode interference in multimode waveguides. An example is the multimode interference coupler.

4.1 Perturbation analysis

4.1.1 Fields and modes in a generalized waveguide

In any waveguide or fiber which has a transverse index variation independent of z (i.e. independent of the position along its longitudinal direction), the Maxwell equations for its electric and magnetic fields, $\underline{E}(x, y, z)$ and $\underline{H}(x, y, z)$, propagating along the z axis, can be explicitly expressed in terms of the longitudinal ($\underline{E_z}$, $\underline{H_z}$) and transverse ($\underline{E_t}$, $\underline{H_t}$) fields as follows. Let

$$\underline{E} = [E_x\underline{i_x} + E_y\underline{i_y}] + E_z\underline{i_z} = \underline{E_t} + E_z\underline{i_z} = \underline{E}(x, y)\,e^{-j\beta z}e^{j\omega t},$$

$$\underline{H} = [H_x\underline{i_x} + H_y\underline{i_y}] + H_z\underline{i_z} = \underline{H_t} + H_z\underline{i_z} = \underline{H}(x, y)\,e^{-j\beta z}e^{j\omega t},$$

$$\nabla = \left[\frac{\partial}{\partial x}\underline{i_x} + \frac{\partial}{\partial y}\underline{i_y}\right] + \frac{\partial}{\partial z}\underline{i_z} = \nabla_t + \frac{\partial}{\partial z}\underline{i_z},$$

then

$$\nabla_t \times \underline{E_t} = -j\omega\mu H_z \underline{i_z}, \tag{4.1a}$$

$$\nabla_t \times \underline{H_t} = j\omega\varepsilon(x, y) E_z \underline{i_z}, \tag{4.1b}$$

$$\nabla_t \times E_z \underline{i_z} - j\beta \underline{i_z} \times \underline{E_t} = -j\omega\mu \underline{H_t}, \tag{4.1c}$$

$$\nabla_t \times H_z \underline{i_z} - j\beta \underline{i_z} \times \underline{H_t} = j\omega\varepsilon(x, y) \underline{E_t}. \tag{4.1d}$$

Equations (4.1a) and (4.1b) imply that the transverse fields can be obtained directly from the longitudinal fields, or vice versa. One only needs to give either set of them to specify the field.

The nth guided wave mode, given by $\underline{e_n}$ and $\underline{h_n}$, is the nth discrete eigen value solution of $\underline{E}$ and $\underline{H}$ in the above vector wave equations plus the condition of the continuity of tangential electric and magnetic fields across all boundaries. In Chapter 3, we discussed two types of the solution of Maxwell's equations, the modes in the planar and channel waveguides and the modes in step-index round optical fibers. Outside the step-index round fibers and planar waveguides, there are no analytical solutions for the eigen value equation of the general waveguide. There are only approximate and numerical solutions. Nevertheless, in view of the properties of the modes discussed in Chapter 3, we expect the following properties of the $\underline{e_n}$ and $\underline{h_n}$ modes for any general waveguide with constant cross-section in z.

(1) The magnitude of the fields outside the higher index core or channel region decays exponentially away from the high-index region in lateral directions.
(2) The higher the order of the mode, the slower the exponential decay rate.
(3) The effective index $n_{\text{eff},n}$ ($n_{\text{eff},n} = \beta_n/k$) is less than the highest index in the core and larger than the index of the cladding or the substrate. n_{eff} is larger for a lower order mode.
(4) Most importantly, it can be shown from the theory of differential equations that the guided wave modes are orthogonal to each other and to the radiation substrate or cladding modes. Mathematically this is expressed for channel guided wave modes as

$$\iint_S (\underline{e_{t,m}} \times \underline{h^*_{t,n}}) \cdot \underline{i_z}\, ds = \int_{-\infty}^{\infty}\int_{-\infty}^{\infty} (\underline{e_{t,m}} \times \underline{h^*_{t,n}}) \cdot \underline{i_z}\, dx\, dy = 0 \quad \text{for } n \neq m, \tag{4.2}$$

where the surface integral is carried out over the entire transverse cross-section S that extends to $\pm\infty$. The guided wave modes and all the radiation modes constitute a complete set of modes so that any field can be represented as a superposition of the modes.

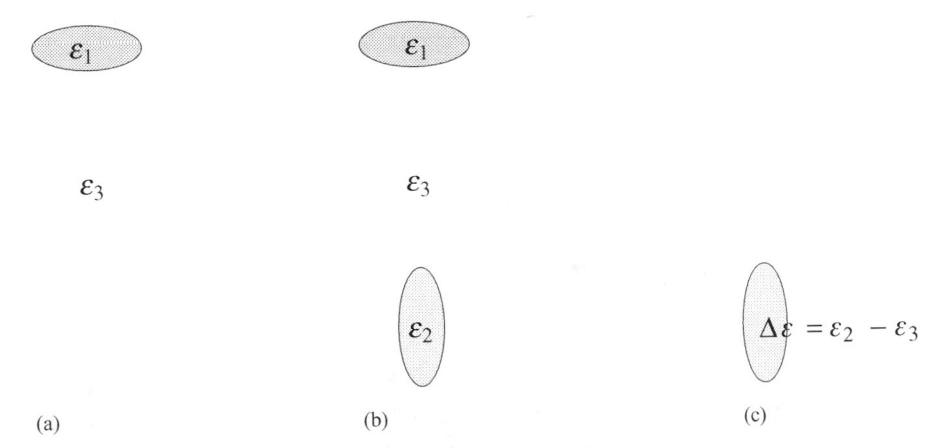

Figure 4.1. Index profile of a perturbation to a waveguide. (a) The permittivity variation, $\varepsilon(x, y)$ of the unperturbed waveguide. (b) The permittivity variation, $\varepsilon'(x, y)$ of the perturbed waveguide. (c) The permittivity perturbation, $\Delta\varepsilon$. The index profile of the original waveguide structure is shown in (a). An additional material with dielectric constant ε_2 is placed in the vicinity of the waveguide core as shown in (b). The net perturbation of the additional material to the original waveguide structure is shown as $\Delta\varepsilon$ in (c).

Moreover, the channel guided wave modes are normalized, i.e.

$$\frac{1}{2}\,\mathrm{Re}\left[\iint_S (\underline{e_{t,n}} \times \underline{h_{t,n}^*}) \cdot \underline{i_z}\, ds\right] = 1. \tag{4.3}$$

For planar guided wave modes, the modes are also orthogonal and normalized in the x variation, as shown in Eq. (3.7). However, the integration in the y coordinate is absent. The normalization means that the power carried by the mth normalized planar guided wave mode is one watt per unit distance (i.e. meter) in the y direction.

4.1.2 Perturbation analysis

Consider the two waveguide structures shown in Figs. 4.1(a) and (b). Let $\underline{E}$ and $\underline{H}$ be the guided wave solutions of Eqs. (4.1) for the waveguide structure with index profile $\varepsilon(x, y)$ shown in Fig. 4.1(a). Let $\underline{E'}$ and $\underline{H'}$ be the guided wave solutions of Eqs. (4.1) for the waveguide structure with $\varepsilon'(x, y)$ shown in Fig. 4.1(b). The two structures differ in the dielectric perturbation $\Delta\varepsilon$ shown by Fig. 4.1(c), where $\Delta\varepsilon(x, y) = \varepsilon'(x, y) - \varepsilon(x, y)$. Let us assume that $\underline{E}$ and $\underline{H}$ are already known. The guided wave modes of the structure shown in Fig. 4.1(b) are the perturbation of the guided wave modes of the structure in Fig. 4.1(a) due to the $\Delta\varepsilon$. The perturbation analysis allows us to calculate approximately the $\underline{E'}$ and $\underline{H'}$ from the $\underline{E}$ and $\underline{H}$, without solving Maxwell's equations. Perturbation analysis is applicable as long as $\Delta\varepsilon$ is either small or at a position reasonably far from the waveguide.

Mathematically, from vector calculus and Eqs. (4.1), we know that

$$\nabla \cdot [\underline{E}^* \times \underline{H}' + \underline{E}' \times \underline{H}^*] = -j\omega\, \Delta\varepsilon\, \underline{E}^* \cdot \underline{E}'.$$

Let us apply volume integration to both sides of this equation over a cylindrical volume V:

$$\iiint_V \nabla \cdot \lfloor \underline{E}^* \times \underline{H}' + \underline{E}' \times \underline{H}^* \rfloor dx\, dy\, dz = -j\omega \iiint_V \Delta\varepsilon \underline{E}' \cdot \underline{E}^* dx\, dy\, dz.$$

The cylinder has flat circular ends parallel to the xy plane, and has an infinitely large radius for the circular ends and a short length dz along the z axis. According to advanced calculus, the volume integration on the left hand side of this equation can be replaced by the surface integration of $\lfloor \underline{E}^* \times \underline{H}' + \underline{E}' \times \underline{H}^* \rfloor$ on the cylinder. The contribution of the surface integration over the cylindrical surface is zero because the guided wave fields $\underline{E}$ and $\underline{E}'$ have already decayed to zero at the surface. For a sufficiently small dz, $\underline{E}^* \cdot \underline{E}'$ is approximately a constant from z to $z + dz$. Therefore, we obtain

$$\iint_S \{[\underline{E}^* \times \underline{H}' + \underline{E}' \times \underline{H}^*]|_{z+dz} - [\underline{E}^* \times \underline{H}' + \underline{E}' \times \underline{H}^*]|_z\} \cdot \underline{i_z}\, ds$$

$$= -j\omega \left[\iint_S \Delta\varepsilon\, \underline{E}' \cdot \underline{E}^*\, ds \right] dz,$$

where S is the flat end surface of the cylinder oriented toward the $+z$ direction. In other words [1],

$$\iint_S \frac{\partial}{\partial z} [\underline{E_t^*} \times \underline{H_t'} + \underline{E_t'} \times \underline{H_t^*}] \cdot \underline{i_z}\, ds$$

$$= -j\omega \iint_S \Delta\varepsilon\,(x, y)\, \underline{E}' \cdot \underline{E}^*\, ds. \tag{4.4}$$

Mathematically, $\underline{E}'$ and $\underline{H}'$ can be represented by the superposition of any set of modes. They can be either the modes of the structure shown in Fig. 4.1(b) or the modes of the structure shown in Fig. 4.1(a). The two sets of the modes, ($\underline{E}$, $\underline{H}$) and ($\underline{E}'$, $\underline{H}'$), form a complete orthogonal set. From the perturbation analysis point of view, we are not interested in the exact modes of ($\underline{E}'$, $\underline{H}'$). We know they are close to the modes of ($\underline{E}$, $\underline{H}$). We only want to know how $\underline{E}'$ and $\underline{H}'$ are related to the $\Delta\varepsilon$ and the original ($\underline{E}$, $\underline{H}$).

In Eq. (4.4), let us express any $\underline{E}'$ and $\underline{H}'$ at any position z in terms of the modes of $(\underline{E}, \underline{H})$ as follows:

$$\underline{E}_{\mathrm{t}}'(x, y, z) = \sum_j a_j(z)\,\underline{e_{\mathrm{t},j}}(x, y)\,e^{-j\beta_j z}, \tag{4.5a}$$

$$\underline{H}_{\mathrm{t}}'(x, y, z) = \sum_j a_j(z)\,\underline{h_{\mathrm{t},j}}(x, y)\,e^{-j\beta_j z}. \tag{4.5b}$$

The radiation modes have been neglected in Eqs. (4.5). In general, the coefficients a_j will be different at different z. The variation of the a_j coefficient signifies that the $\underline{E}'$ and $\underline{H}'$ fields will vary as a function of z. Substituting Eqs. (4.5a) and (4.5b) into Eq. (4.4), letting $\underline{E}_{\mathrm{t}} = \underline{e_{\mathrm{t},n}}$ and $\underline{H}_{\mathrm{t}} = \underline{h_{\mathrm{t},n}}$, and utilizing the orthogonality and normalization relation in Eqs. (4.2) and (4.3), we obtain, for forward propagating waves:

$$\frac{da_n}{dz} = -j\sum_m a_m C_{m,n} e^{+j(\beta_n - \beta_m)z}, \tag{4.6}$$

$$C_{m,n} = \frac{\omega}{4}\iint_S \Delta\varepsilon\,(\underline{e_m}\cdot\underline{e_n^*})\,ds.$$

This is the basic result of the perturbation analysis [3]. It tells us how to find the a_j coefficients. Once we know the a_j coefficients, we know $\underline{E}'$ and $\underline{H}'$ from Eqs. (4.5a) and (4.5b). We will apply this result to different situations in the following sections.

4.1.3 Simple application of the perturbation analysis

In order to demonstrate the power of the results shown in Eq. (4.6), let us find the change in the propagation constant β_0 of a forward propagating guided wave mode caused by the addition of another dielectric material with index ε' in the vicinity of the original waveguide. Let the dielectric material be located at $\infty > x \geq L$ and $\infty > y > -\infty$. Let us apply this $\Delta\varepsilon$ to Eq. (4.6). If the original waveguide has only a single mode, $\underline{e_0}$, then we do not need to carry out the summation in Eq. (4.6). We obtain

$$\frac{da_0}{dz} = -ja_0\left[\frac{\omega}{4}\int_{-\infty}^{\infty}\int_{L}^{\infty}(\varepsilon' - \varepsilon_1)\underline{e_0}\cdot\underline{e_0^*}\,dx\,dy\right] = -j\Delta\beta\,a_0, \tag{4.7}$$

or

$$a_0 = Ae^{-j\Delta\beta z},$$

$$\Delta\beta = \frac{\omega}{4}(\varepsilon' - \varepsilon) \int_{-\infty}^{+\infty} \int_{L}^{+\infty} \underline{e_0} \cdot \underline{e_0^*} \, dx \, dy,$$

$$\underline{E_t'} = A\underline{e_0}(x, y) \, e^{j(\beta + \Delta\beta)z}.$$

Clearly the β_0 of the guided mode $\underline{e_0}$ is changed by the amount $\Delta\beta$. Notice that the perturbation analysis does not address how the field distribution of the original mode is affected by the perturbation. The perturbation analysis allows us to calculate $\Delta\beta$ without solving the differential equation.

4.2 Coupling of modes in the same waveguide, the grating filter and the acousto-optical deflector

Modes in different directions of propagation are independent solutions of the wave equations. For example, the independent modes can be the forward and backward propagating modes of the same order in a channel waveguide. They can also be the forward (or backward) propagating guided wave modes of different orders in the same channel waveguide. They can be planar guided wave modes in different directions of propagation in a planar waveguide. They can all be coupled by an appropriate $\Delta\varepsilon$ placed in the evanescent tail region. In the case of a prism coupler, there could even be the coupling of a guided wave mode to substrate, air or cladding modes [2]. Equations (4.4) and (4.6) are directly applicable in analyzing such interactions. However, the details will differ for different applications. We will discuss in this section the coupling of modes in different directions of propagation via two examples: (1) the grating filter in an optical waveguide or fiber and (2) the acousto-optical deflector or switch in a planar waveguide.

4.2.1 Grating filter in a single-mode waveguide

Grating filters are very important devices in wavelength division multiplexed (WDM) optical fiber communication networks. In such networks, signals are transmitted via optical carriers that have slightly different wavelengths. The purpose of a filter is to select a specific optical carrier (or a group of optical carriers within a specific band of wavelength) to direct it (or them) to a specific direction of propagation (e.g. reflection) [4].

A grating filter utilizes a perturbation of the channel waveguide by a periodic $\Delta\varepsilon$ to achieve the filtering function. In this example we will analyze a reflection filter.

The objectives of a grating filter are: (1) high and uniform reflection of incident waves in a single-mode waveguide within the selected wavelength band; (2) sharp reduction of reflection immediately outside the band; (3) high contrast ratio of the intensity of reflected optical carriers inside and outside the band.

Let us consider a grating layer which has a cosine variation of dielectric constant along the z direction, i.e. $\Delta\varepsilon$, and thickness d in the x direction and width w in the y direction. It is placed on top of a ridged channel waveguide of thickness t. An example of a ridged channel waveguide was shown in Fig. 3.6(b). Let us assume that the ridged waveguide has only a single mode.

Mathematically, let

$$\Delta\varepsilon = \Delta\varepsilon_0 \cos(Kz)\,\mathrm{rect}\left(\frac{2(x-H)}{d}\right)\mathrm{rect}\left(\frac{2y}{W}\right).$$

It has a periodicity $T = 2\pi/K$ in the z direction, a width W in the y direction, a thickness d in the x direction and a maximum change of dielectric constant $\Delta\varepsilon_0$. The $\Delta\varepsilon$ perturbation layer is centered at $x = H$, where $H \geq t + (d/2)$. It is a perturbation of the cladding refractive index n_3 of the channel waveguide. This mathematical expression is a simplified $\Delta\varepsilon$ of a practical grating that normally has a $\Delta\varepsilon$ described by a rectangular function of x and z.

Let the complex amplitude of the forward propagating guided wave mode be a_{f} and let the amplitude of the backward propagating mode at the same wavelength be a_{b}. Then the application of Eq. (4.4) to the field in the waveguide that has both the forward and the backward propagating mode yields

$$\underline{E_{\mathrm{t}}'}(x,y,z) = \left[a_{\mathrm{f}}(z)\,e^{-j\beta_0 z} + a_{\mathrm{b}}(z)\,e^{+j\beta_0 z}\right]\underline{e_{\mathrm{t.0}}}(x,y), \tag{4.8a}$$

$$\frac{da_{\mathrm{f}}}{dz} = -jC_{\mathrm{ff}}a_{\mathrm{f}} - jC_{\mathrm{bf}}a_{\mathrm{b}}e^{-j2\beta_0 z}, \tag{4.8b}$$

$$\frac{da_{\mathrm{b}}}{dz} = -jC_{\mathrm{bb}}a_{\mathrm{b}} - jC_{\mathrm{fb}}a_{\mathrm{f}}e^{j2\beta_0 z}, \tag{4.8c}$$

$$C_{\mathrm{ff}} = -C_{\mathrm{bb}} = -C_{\mathrm{fb}} = C_{\mathrm{bf}}$$

$$= \frac{\omega}{4}\left[\int_{H-\frac{d}{2}}^{H+\frac{d}{2}}\int_{-\frac{W}{2}}^{\frac{W}{2}} \Delta\varepsilon_0 |\underline{e_0}\cdot\underline{e_0^*}|\,dy\,dx\right]\left[\frac{1}{2}(e^{jKz}+e^{-jKz})\right], \tag{4.8d}$$

where there is a minus sign on C_{bb} and C_{fb} because, in the normalization of the modes shown in Eq. (4.3), the $\underline{i_z}$ is pointed toward the $+z$ direction. The $\underline{i_z}$ for the backward wave is pointing toward the $-z$ direction.

Clearly, a_{f} and a_{b} will only affect each other significantly along the z direction when the driving terms on the right hand side of Eqs. (4.8b) and (4.8c) have a slow z variation. Since the perturbation has a $\cos(Kz)$ variation, the maximum

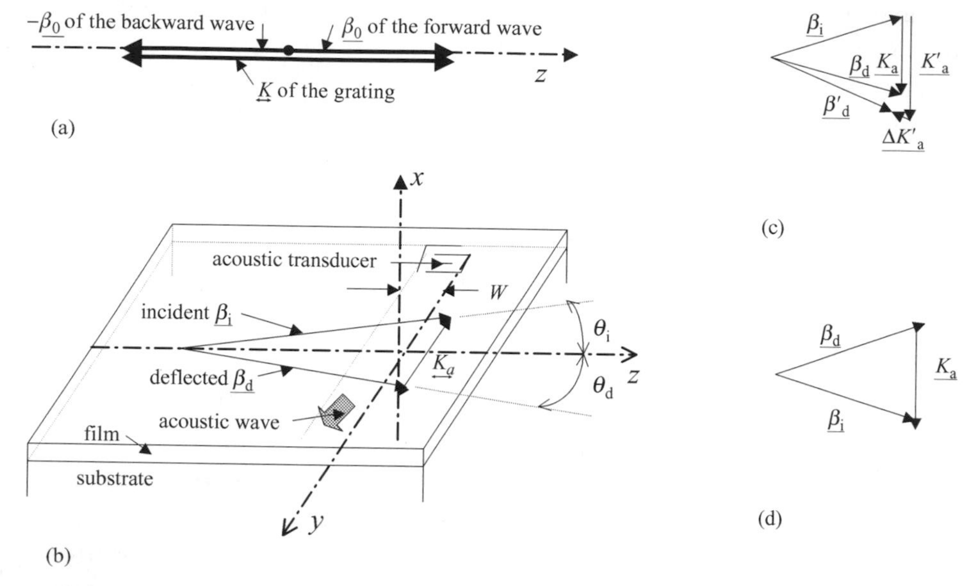

Figure 4.2. Phase matching of forward and backward waves and in acousto-optical deflection. (a) Matching of the propagation wave vectors of the forward and back ward waves by the K of the grating ($K = 2\pi/T$, T = periodicity) in the collinear z direction of a channel waveguide. (b) The matching of the propagation wave vectors, $\underline{\beta_i}$ and $\underline{\beta_d}$, of the incident and deflected planar optical guided waves in the θ_i and θ_d directions by the $\underline{K_a}$ of the grating, created by the surface acoustic wave. The width of the acoustic wave is W, which is also the interaction distance of the optical waves. The matching of β_i and β_d by $\underline{K_a}$ for (c) $\omega_d = \omega_i - \Omega$, and for (d) $\omega_d = \omega_i + \Omega$. When $\underline{K_a}$ is changed to $\underline{K_a'}$, a small mismatch is shown as $\Delta \underline{K_a'}$ in (c).

coupling between a_f and a_b will take place when $K = 2\beta_0$. This is known as the phase matching (or the Bragg) condition of the forward and backward propagating waves. When the Bragg condition is satisfied, the relationship among the β and the K is illustrated in Fig. 4.2(a), where the β_0 of the forward and backward propagating modes with $\exp(\pm j\beta_0 z)$ variations are represented by vectors with magnitude β_0 in the $\pm z$ directions. Since a cosine function is the sum of two exponential functions, $\underline{K}$ is represented as a bi-direction vector of magnitude K. If we designate λ_g as the free space wavelength at which the maximum coupling takes place, then the phase matching condition is satisfied when K is given by

$$K = \frac{4\pi\, n_{\text{eff}}}{\lambda_g}, \tag{4.9}$$

where n_{eff} is the effective index of the guided wave mode. When $K \approx 2\beta_0$, the terms involving C_{ff} and C_{bb} can be neglected in Eqs. (4.8b) and (4.8c).

For a reflection filter, we like to have large a_b when any carrier frequency (i.e. β) is within the desired wavelength band. Since β is inversely proportional to λ, Eq. (4.9) will not be satisfied simultaneously for all the carriers within the desired band. In order to analyze the grating properties as a function of wavelength for a given K, we need also to consider the solutions of Eqs. (4.8b) and (4.8c) under approximate phase matching conditions. Let

$$2\beta_0 - K = \delta_K. \tag{4.10}$$

Under this condition, we obtain from Eqs. (4.8b) and (4.8c)

$$\frac{da_f}{dz} = -j\frac{C_g}{2}a_b e^{j\delta_K z} \tag{4.11a}$$

and

$$\frac{da_b}{dz} = +j\frac{C_g}{2}a_f e^{-j\delta_K z}, \tag{4.11b}$$

where

$$C_g = \frac{\omega}{4}\int_{H-\frac{d}{2}}^{H+\frac{d}{2}}\int_{-\frac{W}{2}}^{\frac{W}{2}} \Delta\varepsilon_0 |\underline{e_0}|^2\, dy\, dx.$$

Equations (4.11) are known as the coupled mode equations between the forward and the backward propagating modes. We know the solutions for such differential equations are the familiar exponential functions, $e^{\gamma^+ z}$ and $e^{\gamma^- z}$. Specifically, the solutions of Eqs. (4.11) for the forward and backward propagating waves are

$$\left.\begin{aligned}
a_b(z) &= A_1 e^{\gamma^+ z} + A_2 e^{\gamma^- z},\\
a_f(z) &= -j\frac{2}{C_g}\left[A_1\gamma^+ e^{-\gamma^- z} + A_2\gamma^- e^{-\gamma^+ z}\right],\\
\gamma^+ &= -j\frac{\delta_K}{2} + Q,\\
\gamma^- &= -j\frac{\delta_K}{2} - Q,\\
Q &= \sqrt{\left(\frac{C_g}{2}\right)^2 - \left(\frac{\delta_K}{2}\right)^2}.
\end{aligned}\right\} \tag{4.12}$$

The A_1 and A_2 coefficients will be determined from boundary conditions at $z = 0$ and $z = L$.

For a grating that begins at $z = 0$ and terminates at $z = L$, a_b must be zero at $z = L$. Thus,

$$\left.\begin{aligned} A_2 &= -A_1 e^{2QL}, \\ a_b &= -A_1 2 e^{QL - j\left(\frac{\delta_K}{2} z\right)} \sinh[Q(L-z)], \\ a_f &= -jA_1 \frac{4}{C_g} e^{QL + j\left(\frac{\delta_K}{2} z\right)} \\ &\quad \times \left[j\frac{\delta_K}{2} \sinh(Q(L-z)) + Q \cosh(Q(L-z)) \right]. \end{aligned}\right\} \tag{4.13}$$

At $z = 0$, the ratio of the reflected power to the incident power is

$$\frac{|a_b(z=0)|^2}{|a_f(z=0)|^2} = \frac{\left(\frac{C_g}{2}\right)^2 \sinh^2 QL}{Q^2 \cosh^2 QL + (\delta_K/2)^2 \sinh^2 QL}. \tag{4.14}$$

At $z = L$, the ratio of the transmitted power in the forward propagating mode to the incident power of the forward mode at $z = 0$ is

$$\frac{|a_f(z=L)|^2}{|a_f(z=0)|^2} = \frac{Q^2}{Q^2 \cosh^2 QL + (\delta_K/2)^2 \sinh^2 QL}. \tag{4.15}$$

Since $|a_f(z=L)|^2 + |a_b(z=0)|^2 = |a_f(z=0)|^2$, the conservation of power of the incident, transmitted and reflected waves is obeyed. For a reflection filter, we want $|a_b(z=0)/a_f(z=0)|^2$ to be large within a desired band of wavelength and small outside this band.

Note that $|a_b(z=0)|$ is larger for larger L and for smaller δ_K/C_g. At $\lambda = \lambda_g$, δ_K is zero, and the grating reflection is a maximum. The maximum possible value of $|a_b(z=0)/a_f(z=0)|^2$ is unity. At $\delta_K = C_g$, there will not be any reflected wave. Let $\Delta\lambda_g$ be the wavelength deviation form λ_g such that, when $\lambda = \lambda_g \pm \Delta\lambda_g$, Q is zero. Then $2\Delta\lambda_g$ is the pass band of the filter, where

$$\Delta\lambda_g = \pm \frac{4\pi C_g n_{\text{eff}}}{K^2}. \tag{4.16}$$

In summary, one uses K to control the center wavelength λ_g at which the transmission of the forward propagating wave is blocked. One uses C_g to control the wavelength width $\Delta\lambda_g$ within which effective reflection occurs. The smaller the C_g, the narrower the range of the transmission wavelength. For a given transmission range, one uses L to control the magnitudes of the reflected and the transmitted waves. These are useful parameters for designing grating reflection filters [4].

The analysis presented here could be applied directly to analyze distributed Bragg reflectors (DBRs) and the distributed feedback (DFB) effect in semiconductor edge emitting lasers.

4.2.2 Acousto-optical deflector, frequency shifter, scanner and analyzer

An acousto-optical deflector (or scanner) is a device that deflects a planar guided wave mode in a planar waveguide into a different direction by a surface acoustic grating. The surface acoustic wave is generated from an electric signal applied to an acoustic transducer. After the acoustic wave passes the interaction region, it is absorbed. Thus there is no reflected acoustic wave. The strain from the acoustic wave creates a surface layer of traveling refractive index wave with periodic index variation, i.e. a surface layer of traveling grating. When the phase matching condition along both the lateral and the longitudinal direction is satisfied by the acoustic grating between the incident and the deflected planar guided wave, efficient diffraction occurs. Optical energy in the incident wave is transferred from the incident wave to the deflected wave, which has a slightly different direction of propagation from the incident wave [5].

Normally the acoustic transducer and the optical input coupler of the incident guided wave cannot be repositioned after the device has been made. Thus the directions of the incident optical guided wave and the acoustic wave are fixed. However, the periodicity of the grating is determined by the wavelength of the acoustic wave, which is determined by the frequency of the electrical signal applied to the transducer. The acoustic wavelength and frequency can be varied. For a given collimated incident guided wave, the direction of the deflected guided wave will vary according to the acoustic frequency.

Acousto-optic deflection has a number of applications. (1) When the acoustic frequency is scanned, the acousto-optical deflector is used as an optical scanner. (2) The optical frequency of the deflected beam is shifted from the frequency of the incident optical beam. Thus an acousto-optical deflector is used sometimes as a frequency shifter. (3) When the acoustic signal has a complex RF frequency spectrum, the optical energy deflected into various directions can be used to measure the power contained in various RF frequency components; such a device is known as an acousto-optical RF spectrum analyzer.

In the following, we will apply the coupled mode analysis to the acousto-optical deflector. (1) We will analyze the direction and the optical power of the deflected waves as the acoustic frequency and power are varied. The deflected beam will be shown to have an optical frequency shifted from the optical frequency of the incident beam by the acoustic frequency. (2) We will show that in order to obtain a high intensity of the deflected beam in an acousto-optical deflector, the product of the width of the acoustic beam and the coupling coefficient of the acoustic grating with the planar guided wave needs to have specific values. (3) We will show that, in acousto-optical spectrum analyzers, the intensity of a weakly deflected beam in a specific direction is proportional to the intensity of the acoustic wave at a specific RF frequency.

(1) Mathematical representation of acoustic grating and planar guided waves

Consider a single TE_0 mode planar waveguide. As we have discussed in Section 3.4, there could be planar guided modes propagating along different directions in the yz plane. The total guided wave is the summation of planar guided wave modes propagating in different directions. Let there be an acoustic surface wave propagating in the y direction in the planar waveguide, as illustrated in Fig. 4.2(b). The net effect of the acoustic wave is to create a periodic traveling wave of $\Delta\varepsilon$ in the y direction. Mathematically, a simplified acoustic $\Delta\varepsilon$ is described as

$$\Delta\varepsilon = \Delta\varepsilon_0 \cos(\underline{K_\mathrm{a}} \cdot \underline{\rho} - \Omega t)\, \mathrm{rect}\left(\frac{2x - t}{t}\right) \mathrm{rect}\left(\frac{2z - W}{W}\right), \tag{4.17}$$

where

$$\underline{K_\mathrm{a}} = K_\mathrm{a}\underline{i_y}, \quad \underline{\rho} = y\underline{i_y} + z\underline{i_z}.$$

$\underline{K_\mathrm{a}}$ is the vector representation of the propagation wave number for the acoustic wave; $2\pi/K_\mathrm{a}$ is the periodicity; Ω is the angular frequency of the acoustic wave; Ω/K_a is the acoustic velocity v. The rectangular functions, rect( ), designate an acoustic wave confined to the layer from $x = 0$ to $x = t$ and within a width W, from $z = 0$ to $z = W$.

Each planar TE_0 guided wave mode is designated by an angle θ_j which is the angle its direction of propagation makes with respect to the z axis. For small θ_j, the electric field of the TE_0 mode is still approximately polarized in the y direction. Therefore the total field of a summation of TE_0 modes can be expressed mathematically as

$$E'_y\underline{i_y} \approx \left[\sum_j a_j e^{-jn_0k_j\cos\theta_j z} e^{-jn_0k_j\sin\theta_j y}\right] E_{0,y}(x)\, e^{j\omega_j t}\underline{i_y}$$

$$\approx \left[\sum_j a_j e^{-j\underline{\beta_j}\cdot\underline{\rho}}\right] E_{0,y}(x)\, e^{j\omega_j t}\underline{i_y}, \tag{4.18}$$

where

$$\underline{\beta_j} = n_\mathrm{eff}k_j(\cos\theta_j\underline{i_z} + \sin\theta_j\underline{i_y}), \quad \underline{\rho} = z\underline{i_z} + y\underline{i_y}. \tag{4.19}$$

$E_{0,y}$ describes the x variation of the TE_0 mode. Note that, in anticipation of the traveling acoustic wave interaction which will couple incident and diffracted waves at slightly different frequencies, we have allowed the guided wave modes to be at slightly different frequencies. This point will be further clarified below.

(2) Acousto-optical interaction, the Bragg deflection and the frequency shift

Let us consider two specific planar TE_0 guided wave modes, propagating in the directions $+\theta$ (for $\underline{\beta_\mathrm{d}}$ of the deflected wave) and $-\theta$ (for $\underline{\beta_\mathrm{i}}$ of the incident wave) with respect to the z axis. The complex amplitudes for these modes are a_d and a_i. In this case, the acoustic $\Delta\varepsilon$ couples the incident wave, a_i, to the diffracted wave, a_d, as shown in Fig. 4.2(b). Equation (4.6), modified by the different frequency variations, is directly applicable to a_i and a_d. For the incident and the deflected modes, we obtain

$$\left.\begin{aligned}\frac{da_i}{dz} &= -ja_d C_a e^{+j(\underline{\beta_i}-\underline{\beta_d})\cdot\underline{\rho}} e^{j(\omega_d-\omega_i)t}\left[e^{j\underline{K_a}\cdot\underline{\rho}}e^{-j\Omega t} + e^{-j\underline{K_a}\cdot\underline{\rho}}e^{j\Omega t}\right],\\ \frac{da_d}{dz} &= -ja_i C_a e^{-j(\underline{\beta_i}-\underline{\beta_d})\cdot\underline{\rho}} e^{j(\omega_i-\omega_d)t}\left[e^{j\underline{K_a}\cdot\underline{\rho}}e^{-j\Omega t} + e^{-j\underline{K_a}\cdot\underline{\rho}}e^{j\Omega t}\right],\\ C_a &= \frac{\omega}{4}\Delta\varepsilon_0\int_0^t |\underline{e_0}\cdot\underline{e_0^*}|\,dx.\end{aligned}\right\}\tag{4.20}$$

Clearly, the phase matching condition for maximum interaction between a_i and a_d is

$$(\underline{\beta_i}-\underline{\beta_d})\cdot\underline{\rho} = \mp\underline{K_a}\cdot\underline{\rho} \quad \text{or} \quad \underline{\beta_d} = \underline{\beta_i}\pm\underline{K_a}. \tag{4.21}$$

This is known as the Bragg condition for acousto-optical deflection. In comparison with the grating filter, the phase matching condition expressed in Eq. (4.21) is a vector relation in the yz plane. The phase matching condition in the z direction is satisfied independently of the K_a value because of the balanced $+\theta$ and $-\theta$ orientations of the $\underline{\beta}$'s, and $|\underline{\beta_i}| = |\underline{\beta_d}| = n_{\text{eff}}k$. Here, in anticipation that $\omega_i \approx \omega_d$, we have taken the approximation $k_i = k_d = k$. Clearly, the magnitude of K_a determines the angular relationship between $\underline{\beta_i}$ and $\underline{\beta_d}$, i.e. the θ. In addition, according to Eqs. (4.20), the interaction is strong only when

$$\omega_d = \omega_i \mp \Omega. \tag{4.22}$$

Since Ω (in RF frequency) $\ll \omega_i$ and ω_d (in optical frequencies), $k_i = k_d = k$. The case using the upper signs in Eqs. (4.21) and (4.22) is illustrated in Fig. 4.2(c). The case using the lower signs is illustrated in Fig. 4.2(d). Note that the diffracted wave is at a slightly different optical frequency to the incident wave. This method is sometimes used to shift the optical frequency slightly from ω_i to ω_d.

(3) Deflection efficiency under the Bragg condition

Comparing Eqs. (4.20) with Eqs. (4.11), we notice the difference in the minus sign in the coupled mode equation. When the phase and frequency matching conditions are satisfied, the solution to Eqs. (4.20) is now a cos ($C_a z$) or sin ($C_a z$) variation. The exact form of the solution will again depend on the boundary conditions. Let a_i be the amplitude of the incident wave and let a_d be the amplitude of the diffracted wave. The interaction by the grating begins at $z = 0$ and ends at $z = W$. Thus, the boundary condition is "$a_i = A$ and $a_d = 0$ at $z = 0$." For this boundary condition and for the case shown in Fig. 4.2(c), the solution for the amplitude of the two planar guided waves is

$$a_i(z) = A\cos(C_a z),$$
$$a_d(z) = -jA\sin(C_a z). \tag{4.23}$$

The power diffraction efficiency, $|a_d(z=W)/a_i(z=0)|^2$, and the power transmission efficiency, $|a_i(z=W)/a_i(z=0)|^2$, are

$$\left|\frac{a_d(W)}{a_i(0)}\right|^2 = \sin^2(C_a W),$$
$$\left|\frac{a_i(W)}{a_i(0)}\right|^2 = \cos^2(C_a W), \tag{4.24}$$

and

$$\left|\frac{a_d(W)}{a_i(0)}\right|^2 + \left|\frac{a_i(W)}{a_i(0)}\right|^2 \equiv 1.$$

For applications such as the acousto-optical switch or optical frequency shifter, maximum diffraction efficiency is desired. In that case, we need $W = \pi/2C_a$. For devices which require only low efficiency acousto-optical diffraction, the fraction of the optical power diffracted into the new direction is linearly proportional to $\Delta\varepsilon_0^2$, which is often proportional to the acoustic power at the frequency Ω in the small signal approximation. Usually, Ω (in megahertz or gigahertz) $\ll \omega$, thus the small θ assumption used in Eq. (4.18) is justified.

(4) Deflections slightly off the Bragg angle – the optical scanner and the acousto-optical spectrum analyzer

In an acousto-optical scanner, the acoustic velocity remains the same under moderate variations of acoustic frequency. The propagation wave vector of the incident guided wave and the direction of the acoustic wave are also fixed. When the acoustic frequency shifts from Ω to Ω', K_a changes to K'_a. $\underline{\beta'_d}$ will now be oriented in a new direction to satisfy the phase matching condition in the y direction,

$$(\underline{\beta'_d} - \underline{\beta_i}) \cdot \underline{i_y} = K'_a,$$

or

$$\sin\theta'_d = -\sin|\theta_i| + \frac{K'_a}{n_{\text{eff}}k}. \tag{4.25}$$

The optical frequency of the diffracted wave will change to $\omega'_d = \omega_i - \Omega'$. β_i, K'_a and β'_d are illustrated in Fig. 4.2(c). The shift of θ'_d as a function of K'_a is the principal mechanism for controlling the direction of deflection in optical scanning. Waves with $\underline{\beta}$ that do not satisfy the phase matching condition in the y direction will have negligible amplitude. For incident planar guided wave modes that are reasonably wide in the y direction, the direction of $\underline{\beta_i}$ is well defined.

Let the electrical signal driving the acoustic transducer have different frequency components or sweep from one frequency to another. In that case, the phase matching condition, Eq. (4.21), is no longer satisfied exactly in the z direction for all frequencies. Specifically, for the case shown in Fig. 4.2(c),

$$\left.\begin{array}{c} \underline{\beta'_d} = \underline{\beta_i} + \underline{K'_a} + \Delta \underline{K'_a} \\ \text{or} \\ \Delta \underline{K'_a} = \Delta K'_a \underline{i_z} = n_{\text{eff}} k (\cos\theta'_d - \cos\theta_i) \underline{i_z}. \end{array}\right\} \quad (4.26)$$

When we include the $\Delta K'_a$, Eqs. (4.20) become

$$\left.\begin{aligned} \frac{da_i}{dz} &= -j a'_d C_a e^{j\Delta K'_a z}, \\ \frac{da'_d}{dz} &= -j a_i C_a e^{-j\Delta K'_a z}. \end{aligned}\right. \quad (4.27)$$

The solutions of Eqs. (4.27) for $|a'_d| = 0$ at $z = 0$ are

$$\left.\begin{aligned} a_i &= A e^{j\frac{\Delta K'_a}{2}z} \left[\cos\sqrt{C_a^2 + (\Delta K'_a/2)^2}\, z \right. \\ &\quad \left. - j \frac{(\Delta K'_a/2)}{\sqrt{C_a^2 + (\Delta K'_a/2)^2}} \sin\sqrt{C_a^2 + (\Delta K'_a/2)^2}\, z \right], \\ a'_d &= -\frac{-jC_a A}{\sqrt{C_a^2 + (\Delta K'_a/2)^2}} e^{-j\frac{\Delta K'_a}{2}z} \left[\sin\sqrt{C_a^2 + (\Delta K'_a/2)^2}\, z \right]. \end{aligned}\right\} \quad (4.28)$$

Note that, for small $[\sqrt{C_a^2 + (\Delta K'_a/2)^2}]W$, the intensity of the diffracted beam, $|a'_d|^2$, is independent of $\Delta K'_a$. For an optical scanner, this means that the intensity of the deflected beams will be uniform with respect to the deflection angle. In an acousto-optical spectral analyzer [6], when one measures the optical power deflected into different directions θ'_d, the detected optical power measures the RF power at the frequency Ω' applied to the transducer with efficiency independent of Ω'. This is an important feature for such applications.

When the acousto-optical deflector is used as a beam scanner or a spectral analyzer, there will be a guided wave lens to focus the deflected wave into a small spot at the focal plane of the lens. From our analysis of the generalized guided waves in planar waveguides, we know that the focused beam will have a finite spot size. Unless the size of the lens is smaller than the width of the guided wave, the spot width is determined from the width of the guided wave and the focal length of the lens. The total range of K_a that can be scanned and the size of the spot determine how many resolvable spots can be obtained in such a scanner or analyzer. The range of the Ω (or K_a) that can be scanned is determined by the acoustic property of the material and the acoustic transducer. The time response of such a scanner is determined by the transit time of the acoustic wave to travel from one edge of the guided wave beam to the other edge.

4.3 Propagation of modes in parallel waveguides – the coupled modes and the super-modes

The operation of a number of devices, including the directional coupler, is based on the mutual interactions of modes in two parallel waveguides via the evanescent field of the guided wave modes [7]. We can analyze such interactions using the coupled mode analysis of the modes of the individual waveguide as we have done in Section 4.2 [3, 8]. In addition, there is an alternative approach based on the modal analysis of the super-modes of the total two-waveguide structure. Both approaches will be discussed in the following.

For two infinitely long parallel waveguides with uniform distance of separation, the super-modes of the total structure can be found by perturbation analysis. When there are two coupled waveguides with a continuously varying distance of separation, we will approximate the original continuously varying structure by steps of local waveguides that have constant separation within each step. In other words, we will find the super-modes for each local section. Modal analyses of the super-modes can then be applied to the junctions between two adjacent steps. The device properties are determined from the cumulative effect of all the successive junctions.

4.3.1 Modes of two uncoupled parallel waveguides

Consider the two waveguides shown in Fig. 4.3(a). Let the distance of separation D between the two waveguides, A and B, be very large at first. In that case, the modes of A and B will not be affected by each other. The modes of the total structure, $\underline{e_{tn}}$ and $\underline{h_{tn}}$, are just the linear combination of the modes of individual waveguides, ($\underline{e_{An}}$, $\underline{h_{An}}$) and ($\underline{e_{Bn}}$, $\underline{h_{Bn}}$). The fields of the total structure can be expressed as the summation of all the modes of the waveguides A and B:

$$\underline{E} = \sum_n a_{An}\underline{e_{An}}e^{-j\beta_{An}z} + a_{Bn}\underline{e_{Bn}}e^{-j\beta_{Bn}z}, \tag{4.29a}$$

$$\underline{H} = \sum_n a_{An}\underline{h_{An}}e^{-j\beta_{An}z} + a_{Bn}\underline{h_{Bn}}e^{-j\beta_{Bn}z}. \tag{4.29b}$$

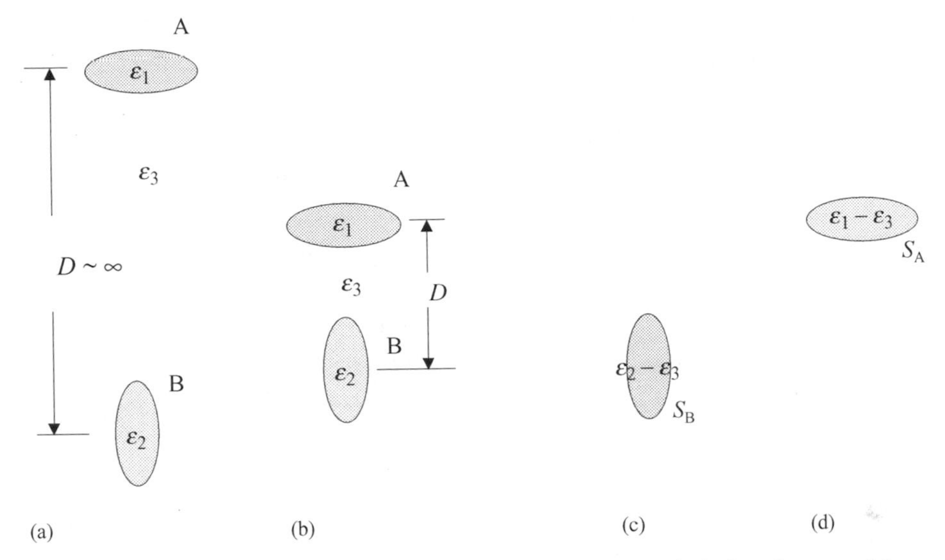

Figure 4.3. The perturbation of permittivity for the modes in isolated waveguides A and B. (a) The permittivity profile of two well separated waveguides, A and B, with core dielectric constants ε_1 and ε_2. (b) The permittivity profile of two neighboring waveguides, A and B, at separation D with core dielectric constants ε_1 and ε_2. (c) For A, the perturbation of ε_3 by ε_2 of waveguide B. (d) For B, the perturbation of ε_3 by ε_1 of waveguide A.

Here, the a coefficients are independent of z. Because of the evanescent decay of the fields, the overlap of the fields $(\underline{e_{An}}, \underline{h_{An}})$ with $(\underline{e_{Bn}}, \underline{h_{Bn}})$ is negligible, i.e.

$$\iint_S (\underline{e_{t,An}} \times \underline{h^*_{t,Bm}}) \cdot \underline{i_z}\, ds = 0. \tag{4.30}$$

In other words, the A and B modes can be considered as being orthogonal to each other.

4.3.2 Analysis of two coupled waveguides based on modes of individual waveguides

When the two waveguides are closer, but not very close, to each other, the perturbed fields, $\underline{E'}$ and $\underline{H'}$, can again be expressed as the summation of ($\underline{e_{An}}$ and $\underline{e_{Bn}}$) and ($\underline{h_{An}}$ and $\underline{h_{Bn}}$) as follows:

$$\left.\begin{aligned} \underline{E'} &= \sum_n a_{An}(z)\, \underline{e_{An}} e^{-j\beta_{An} z} + a_{Bn}(z)\, \underline{e_{Bn}} e^{-j\beta_{Bn} z}, \\ \underline{H'} &= \sum_n a_{An}(z)\, \underline{h_{An}} e^{-j\beta_{An} z} + a_{Bn}(z)\, \underline{h_{Bn}} e^{-j\beta_{Bn} z}, \end{aligned}\right\} \tag{4.31}$$

where the a coefficients are now functions of z. However, the effect of the perturbation created by the finite separation distance D will be different for A and B modes as shown below.

Consider now the two waveguides, A and B, separated by a finite distance D as shown in Fig. 4.3(b). For modes of waveguide A, the significant perturbation of the variation of the permittivity from the structure shown in Fig. 4.3(a) is the increase of permittivity from ε_3 to ε_2 at the position of waveguide B as shown in Fig. 4.3(c). For modes of waveguide B, the perturbation of the variation of the permittivity is shown in Fig. 4.3(d), which is the increase of permittivity from ε_3 to ε_1 at the position of waveguide A. Applying the result in Eq. (4.6) to this case, we obtain:

$$\begin{aligned}\frac{da_{\mathrm{A}n}}{dz} &= -j\left[C_{\mathrm{A}n,\mathrm{A}n}a_{\mathrm{A}n} + \sum_m C_{\mathrm{B}m,\mathrm{A}n}e^{j(\beta_{\mathrm{A}n}-\beta_{\mathrm{B}m})z}a_{\mathrm{B}m}\right],\\ \frac{da_{\mathrm{B}n}}{dz} &= -j\left[C_{\mathrm{B}n,\mathrm{B}n}a_{\mathrm{B}n} + \sum_m C_{\mathrm{A}m,\mathrm{B}n}e^{j(\beta_{\mathrm{B}n}-\beta_{\mathrm{A}m})z}a_{\mathrm{A}m}\right],\end{aligned} \tag{4.32}$$

where

$$C_{\mathrm{A}n,\mathrm{A}n} = \frac{\omega}{4}\iint_{S_\mathrm{B}} (\varepsilon_1-\varepsilon_3)[\underline{e_{\mathrm{A}n}}\cdot\underline{e^*_{\mathrm{A}n}}]\,ds,$$

$$C_{\mathrm{B}m,\mathrm{A}n} = \frac{\omega}{4}\iint_{S_\mathrm{B}} (\varepsilon_1-\varepsilon_3)[\underline{e_{\mathrm{B}m}}\cdot\underline{e^*_{\mathrm{A}n}}]\,ds,$$

$$C_{\mathrm{B}n,\mathrm{B}n} = \frac{\omega}{4}\iint_{S_\mathrm{A}} (\varepsilon_2-\varepsilon_3)[\underline{e_{\mathrm{B}n}}\cdot\underline{e^*_{\mathrm{B}n}}]\,ds,$$

$$C_{\mathrm{A}m,\mathrm{B}n} = \frac{\omega}{4}\iint_{S_\mathrm{A}} (\varepsilon_2-\varepsilon_3)[\underline{e_{\mathrm{A}m}}\cdot\underline{e^*_{\mathrm{B}n}}]\,ds.$$

Equations (4.32) denote the well-known coupled mode equation [3]. It is used extensively to analyze many waveguide devices. There are number of ways in which Eqs. (4.32) may be simplified. (1) Since there is evanescent decay of $\underline{e_{\mathrm{A}n}}$ before the field reaches S_B, $C_{\mathrm{A}n,\mathrm{A}n}$ is always much smaller than $C_{\mathrm{B}m,\mathrm{A}n}$. Similar comments can be made for $C_{\mathrm{B}n,\mathrm{B}n}$. Thus, $C_{\mathrm{A}n,\mathrm{A}n}$ and $C_{\mathrm{B}n,\mathrm{B}n}$ are often neglected in Eqs. (4.32) for reasonably large separation distance D, especially when the effect on $a_{\mathrm{A}n}$ and $a_{\mathrm{B}n}$ by the $C_{\mathrm{B}m,\mathrm{A}n}$ and $C_{\mathrm{A}m,\mathrm{B}n}$ is reasonably large. The example given in Section 4.1.3 illustrates the case when $C_{\mathrm{A}n,\mathrm{A}n}$ cannot be neglected. (2) When there is no $\underline{e_m}$ mode in the second waveguide, $C_{\mathrm{B}m,\mathrm{A}n}$ or $C_{\mathrm{A}m,\mathrm{B}n}$ will be zero. Either $C_{\mathrm{A}n,\mathrm{A}n}$ or $C_{\mathrm{B}n,\mathrm{B}n}$ is then used to calculate the slight change of the propagation wave number of the modes, as we have done in Section 4.1.3. (3) When there is more than one mode in waveguides A and B, there should also be more terms, such as $C_{\mathrm{A}n,\mathrm{A}j}$ and

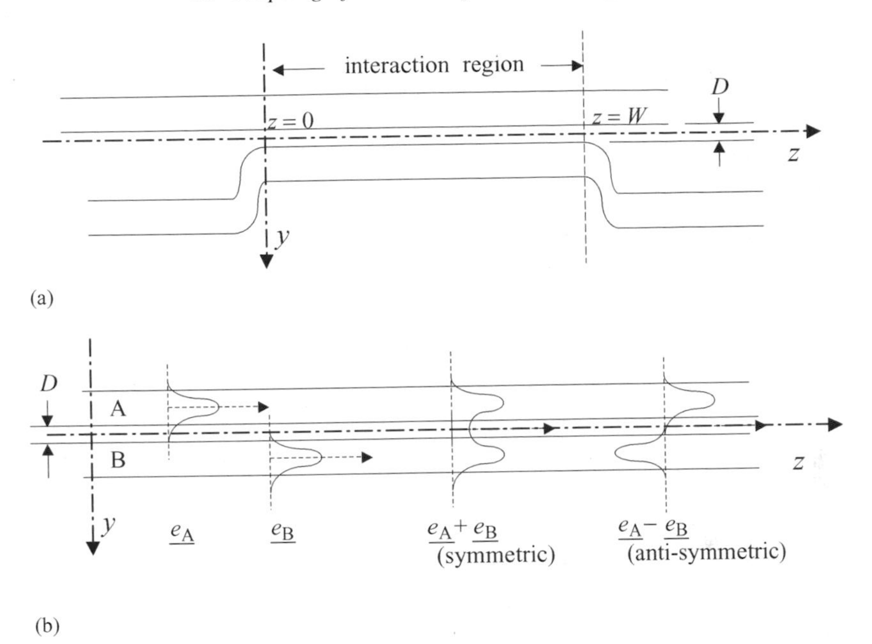

Figure 4.4. Top view of a directional coupler and the modes of two coupled identical waveguides. (a) Top view of two channel waveguides in a directional coupler. The interaction region is W. The separation distance of the two waveguides in the interaction region is D. (b) The field patterns of symmetric and anti-symmetric super-modes of the two coupled identical waveguides in the interaction region. $\underline{e_A}$ and $\underline{e_B}$ are field patterns of the modes of the isolated waveguides A and B.

$C_{\mathrm{B}n,\mathrm{B}j}$ in a more precise analysis. However, these C coefficients are even smaller than $C_{\mathrm{A}n,\mathrm{A}n}$ and $C_{\mathrm{B}n,\mathrm{B}n}$ because of the orthogonality of the unperturbed modes of the same waveguide. Therefore, those terms have not been included in Eqs. (4.32).

In the following Section 4.3.3, we will discuss an application of the coupled mode equations to a device called the directional coupler.

4.3.3 The directional coupler, viewed as coupled individual waveguide modes

A directional coupler has an interaction region that has two parallel identical channel waveguides (or fibers). A prescribed fraction of power in waveguide A is transferred into waveguide B within the interaction region and vice versa. The top view of a channel waveguide directional coupler is illustrated in Fig. 4.4(a). Within the interaction region, the waveguides are separated from each other by a distance D, which is usually of the order of or less than the evanescent decay length. The length of the interaction section is W. Outside the interaction region, the waveguides are well separated from each other without any further interaction [7]. Clearly, Eqs. (4.32) are directly applicable to the modes of the individual waveguides in the interaction region.

Let $\underline{e_A}$ and $\underline{e_B}$ be the modes of the two waveguides (or fibers) that are interacting with each other through their evanescent field in the interaction section. Let the two waveguides have cores with cross-sections S_A and S_B and dielectric constants ε_A and ε_B. The cores are surrounded by a cladding which has dielectric constant ε_3. Let the coupling region begin at $z = 0$ and end at $z = W$ as shown in Fig. 4.4(a). For mathematical convenience, the coupling is assumed to be uniform within this distance. Application of Eqs. (4.32) yields

$$\left.\begin{aligned}
\frac{da_A}{dz} &= -jC_{BA}e^{j\Delta\beta z}a_B(z),\\
\frac{da_B}{dz} &= -jC_{AB}e^{-j\Delta\beta z}a_A(z),\\
C_{AB} &= \frac{\omega}{4}\iint_{S_A}(\varepsilon_B - \varepsilon_3)[\underline{e_A}\cdot\underline{e_B^*}]\,ds,\\
C_{BA} &= \frac{\omega}{4}\iint_{S_B}(\varepsilon_A - \varepsilon_3)[\underline{e_B}\cdot\underline{e_A^*}]\,ds,\\
\Delta\beta &= \beta_A - \beta_B.
\end{aligned}\right\} \tag{4.33}$$

C_{AA} and C_{BB} have been neglected in anticipation of the large effects to be produced by C_{AB} and C_{BA} at small $\Delta\beta$. Solution of a_A and a_B will again depend on initial conditions. Let the initial conditions be $a_A = A$ and $a_B = 0$ at $z = 0$. Then, similarly to the solution for Eq. (4.27), we obtain

$$\left.\begin{aligned}
a_A &= Ae^{j\frac{\Delta\beta}{2}z}\left[\cos\left(\sqrt{C_{BA}C_{AB} + \left(\frac{\Delta\beta}{2}\right)^2}\,z\right)\right.\\
&\qquad\left. - j\frac{\left(\frac{\Delta\beta}{2}\right)}{\sqrt{C_{BA}C_{AB} + \left(\frac{\Delta\beta}{2}\right)^2}}\sin\left(\sqrt{C_{BA}C_{AB} + \left(\frac{\Delta\beta}{2}\right)^2}\,z\right)\right],\\
a_B &= \frac{-jC_{AB}A}{\sqrt{C_{BA}C_{AB} + \left(\frac{\Delta\beta}{2}\right)^2}}e^{-j\frac{\Delta\beta}{2}z}\sin\left(\sqrt{C_{BA}C_{AB} + \left(\frac{\Delta\beta}{2}\right)^2}\,z\right),
\end{aligned}\right\} \tag{4.34a}$$

for $0 \le z \le W$.

Similarly, if the boundary conditions are $a_B = B$ and $a_A = 0$ at $z = 0$, we obtain

$$\left.\begin{aligned} a_A &= \frac{-jC_{BA}B}{\sqrt{C_{BA}C_{AB} + \left(\frac{\Delta\beta}{2}\right)^2}} e^{+j\frac{\Delta\beta}{2}z} \sin\left(\sqrt{C_{BA}C_{AB} + \left(\frac{\Delta\beta}{2}\right)^2}\, z\right), \\ a_B &= Be^{-j\frac{\Delta\beta}{2}z}\left[\cos\left(\sqrt{C_{BA}C_{AB} + \left(\frac{\Delta\beta}{2}\right)^2}\, z\right)\right. \\ &\quad \left. + j\frac{\left(\frac{\Delta\beta}{2}\right)}{\sqrt{C_{BA}C_{AB} + \left(\frac{\Delta\beta}{2}\right)^2}} \sin\left(\sqrt{C_{BA}C_{AB} + \left(\frac{\Delta\beta}{2}\right)^2}\, z\right)\right], \end{aligned}\right\} \tag{4.34b}$$

for $0 \leq z \leq W$.

At $z = W$, the power transmitted from one waveguide to another and the power that remains in the original waveguide are calculated from a_B and a_A. Note that, unless $\Delta\beta = 0$, there cannot be full transfer of power from A to B. Substantial transfer of power from A to B (or vice versa) at $z = W$ can take place only when $\Delta\beta$ is small. $\beta_A = \beta_B$ is the phase matching condition for maximum transfer of power. As for all coupled mode interactions, the C coefficients, the W and the $\Delta\beta$ are used to control the net power transfer from A to B and from B to A. If W is too large, then a_A and a_B will exhibit oscillatory amplitudes as z progresses.

Conventionally, the directional coupler has two identical channel waveguides. In that case, $C_{BA} = C_{AB} = C$, and the ratio $|a_B|^2/|a_A|^2$ is the power distribution among the two waveguides. At $z = 0$, let there be an input power I_{in} in waveguide A and no input power in waveguide B. Then the output power I_{out} in waveguide B after an interaction distance W is given directly by Eqs. (4.34a):

$$\frac{I_{out}}{I_{in}} = \frac{1}{C^2 + \left(\frac{\Delta\beta}{2}\right)^2} \sin^2\left(\sqrt{C^2 + \left(\frac{\Delta\beta}{2}\right)^2}\, W\right). \tag{4.35}$$

If ε_A and ε_B are the dielectric constants of electro-optical materials, then β_A or β_B can be changed by the instantaneous electric field applied to the waveguide. A directional coupler modulator is a directional coupler with electro-optical control of $\Delta\beta$. Since it is the power transfer that will be affected by $\Delta\beta$, it is an intensity modulator. Furthermore, the power transfer is dependent on the interaction length W.

The discussion presented in this section is also the approach used commonly in the literature to discuss the directional coupler [3, 7, 8]. However, it is instructive to discuss the directional coupler in terms of the propagation of the super-modes in the total two-waveguide structure in Section 4.3.4. Such an approach has not been described in most optics books. It is very useful for understanding devices such as the Mach–Zehnder modulator.

4.3.4 Directional coupling, viewed as propagation of super-modes

The interaction region of a directional coupler could also be considered as a super-waveguide with a complicated cross-sectional variation of ε. There are super-modes for the total structure. However, we do not yet know what these super-modes are. Here, we will use Eqs. (4.32) to show that the modes of the total structure are just a symmetrical and anti-symmetrical combination of the modes of the uncoupled waveguides. Once we know the super-modes, the power transfer discussed in the previous section is just given by the superposition of super-modes as they propagate in the z direction. Let the two waveguides in Fig. 4.4(b) be identical. This is the classical example of a pair of coupled identical waveguides.

Mathematically, in terms of Eqs. (4.32), we have $\Delta\beta = 0$, $\varepsilon_{\mathrm{A}} = \varepsilon_{\mathrm{B}}$ and $C_{\mathrm{AB}} = C_{\mathrm{BA}} = C$. Let the boundary conditions at $z = 0$ be $a_{\mathrm{A}} = A$ and $a_{\mathrm{B}} = B$. Then, the solutions of Eqs. (4.32) are

$$\left.\begin{aligned} a_{\mathrm{A}}(z) &= \frac{1}{2}(A - B)e^{+jCz} + \frac{1}{2}(A + B)e^{-jCz}, \\ a_{\mathrm{B}}(z) &= \frac{1}{2}(B - A)e^{+jCz} + \frac{1}{2}(A + B)e^{-jCz}, \end{aligned}\right\} \tag{4.36}$$

where

$$C = \frac{\omega}{4} \iint_{S_{\mathrm{B}}} (\varepsilon_{\mathrm{A}} - \varepsilon_3)[\underline{e_{\mathrm{B}}} \cdot \underline{e_{\mathrm{A}}}]\, dS.$$

Substituting this result into Eq. (4.31), we obtain

$$\begin{aligned} \underline{E}' = {} & \frac{1}{\sqrt{2}}(A - B)\left[\frac{1}{\sqrt{2}}(\underline{e_{\mathrm{A}}} - \underline{e_{\mathrm{B}}})\right] e^{-j(\beta - C)z} \\ & + \frac{1}{\sqrt{2}}(A + B)\left[\frac{1}{\sqrt{2}}(\underline{e_{\mathrm{A}}} + \underline{e_{\mathrm{B}}})\right] e^{-j(\beta + C)z}. \end{aligned} \tag{4.37}$$

Therefore, the field is a superposition of two super-modes. The mode which consists of the symmetric combination, $\underline{e_{\mathrm{s}}} = (1/\sqrt{2})(\underline{e_{\mathrm{A}}} + \underline{e_{\mathrm{B}}})$, is a normalized symmetric eigen mode with $\beta_{\mathrm{s}} = \beta + C$. The mode which consists of the anti-symmetric combination, $\underline{e_{\mathrm{a}}} = (1/\sqrt{2})(\underline{e_{\mathrm{A}}} - \underline{e_{\mathrm{B}}})$, is an anti-symmetric eigen mode with $\beta_{\mathrm{a}} = \beta - C$. The symmetric mode $\underline{e_{\mathrm{S}}}$ is the lowest order mode of the entire structure with the highest effective index. The excitation of the super-modes depends on the initial condition. When $A = B$, only the symmetric mode is excited. When $A = -B$, only the anti-symmetric mode is excited. When $B = 0$ (or $A = 0$), the symmetric and the anti-symmetric modes are excited with equal amplitude, and this is the case we analyzed in the previous section for $B = 0$ at $z = 0$ and $\Delta\beta = 0$. Since the symmetric and the anti-symmetric modes do not have the same phase velocity, the relative phase between the two modes will oscillate as a function of the distance of propagation. Consequently, the intensity of the total field in waveguides A and B will be a function of z for $0 < z < W$. When $CW = \pi/2$, $a_A = 0$ at $z = W$. We would have transferred all the power from A at $z = 0$ to B at $z = W$. For $z > W$, the two waveguides are well separated from each other with $C = 0$. The symmetric and anti-symmetric modes have the same β as the modes of the individual waveguides. The power in waveguides A and B is independent of z. In summary, one can use either the summation of the symmetric and the anti-symmetric modes, $\underline{e_s}$ and $\underline{e_a}$, or the modes of the individual waveguides, $\underline{e_A}$ and $\underline{e_B}$, to represent the total field.

4.3.5 Super-modes of two coupled non-identical waveguides

We will follow an approach similar to the analysis of the symmetric and anti-symmetric super-modes to find the super-modes of two coupled non-identical waveguides. For initial amplitude either $A = 0$ or $B = 0$ at $z = 0$, the solution of the amplitudes of the modes of each individual waveguide, a_A and a_B, has already been given in Eqs. (4.34a) and (4.34b). When both A and B are non-zero at $z = 0$, we obtain, from Eqs. (4.31) and (4.32)

$$\begin{aligned}
\underline{E'} = &\left\{\left[\left(1 + \frac{\left(\frac{\Delta\beta}{2}\right)}{\sqrt{C_{BA}C_{AB} + \left(\frac{\Delta\beta}{2}\right)^2}}\right)\frac{A}{2}\underline{e_A} + \left(1 - \frac{\left(\frac{\Delta\beta}{2}\right)}{\sqrt{C_{BA}C_{AB} + \left(\frac{\Delta\beta}{2}\right)^2}}\right)\frac{B}{2}\underline{e_B}\right]\right. \\
&\left. + \left[\frac{C_{BA}\frac{B}{2}\underline{e_A} + C_{AB}\frac{A}{2}\underline{e_B}}{\sqrt{C_{BA}C_{AB} + \left(\frac{\Delta\beta}{2}\right)^2}}\right]\right\} \exp\left[-j\left(\frac{\beta_A + \beta_B}{2} + \sqrt{C_{BA}C_{AB} + \left(\frac{\Delta\beta}{2}\right)^2}\right)z\right] \\
&+ \left\{\left[\left(1 - \frac{\left(\frac{\Delta\beta}{2}\right)}{\sqrt{C_{BA}C_{AB} + \left(\frac{\Delta\beta}{2}\right)^2}}\right)\frac{A}{2}\underline{e_A} + \left(1 + \frac{\left(\frac{\Delta\beta}{2}\right)}{\sqrt{C_{BA}C_{AB} + \left(\frac{\Delta\beta}{2}\right)^2}}\right)\frac{B}{2}\underline{e_B}\right]\right. \\
&\left. - \left[\frac{C_{BA}\frac{B}{2}\underline{e_A} + C_{AB}\frac{A}{2}\underline{e_B}}{\sqrt{C_{BA}C_{AB} + \left(\frac{\Delta\beta}{2}\right)^2}}\right]\right\} \exp\left[-j\left(\frac{\beta_A + \beta_B}{2} - \sqrt{C_{BA}C_{AB} + \left(\frac{\Delta\beta}{2}\right)^2}\right)z\right].
\end{aligned} \tag{4.38}$$

It is clear that the two super-modes have propagation wave numbers

$$\left.\begin{aligned}\beta_1 &= \frac{\beta_A+\beta_B}{2} + \sqrt{C_{BA}C_{AB} + \left(\frac{\Delta\beta}{2}\right)^2},\\ \beta_2 &= \frac{\beta_A+\beta_B}{2} - \sqrt{C_{BA}C_{AB} + \left(\frac{\Delta\beta}{2}\right)^2}.\end{aligned}\right\} \tag{4.39}$$

When $\beta_A = \beta_B$, $\Delta\beta = 0$ and $C_{AB} = C_{BA}$, we again obtain the symmetric and the anti-symmetric modes. When $C_{AB} = C_{BA} = 0$, the super-modes become just the waveguide modes of the isolated waveguide. It is interesting to note that, in principle, the coupled waveguide should have orthogonal modes with different propagation wave numbers. For the case of coupled non-identical waveguides, the perturbation analysis gave us the field variation only in terms of $\underline{e_A}$ and $\underline{e_B}$. The relative magnitude of the mix of $\underline{e_A}$ and $\underline{e_B}$ that forms the super-modes will vary depending upon C_{BA}, C_{AB} and $\Delta\beta$, as well as the initial conditions A and B. The super-modes can be quite asymmetrical. When the mix of $\underline{e_A}$ and $\underline{e_B}$ and the coupling of two asymmetrical waveguides vary in the direction of propagation, such as in the case of crossing channel waveguides, the device will have very interesting power transfer characteristics [9].

4.4 Propagation of super-modes in adiabatic branching waveguides and the Mach–Zehnder interferometer

In this section, we will analyze components utilizing the super-modes and the modal analysis. A new concept, the adiabatic transition, will be introduced.

4.4.1 Adiabatic Y-branch transition

Consider the transition for a guided wave mode propagating from waveguide C into waveguide D, as shown in Fig. 4.5(a). Let waveguide C be a single-mode waveguide and let waveguide D be a multimode waveguide. As the waveguide cross-section expands, the second mode emerges at $z = z_1$ (i.e. there exists a second mode in the electromagnetic solution of an infinitely long waveguide that has the same transverse dielectric index variation as the cross-sectional index variation at $z = z_1$). The third mode emerges at $z = z_2$, etc. The transition section can be approximated by many steps of local waveguides that have constant cross-section within each step, as shown in Fig. 4.5(b). At each junction of two adjacent steps, modal analysis can be used to calculate the excitation of the modes in the next step by the modes in the previous step. For adiabatic transition in the forward direction, the steps are so small that only the lowest order mode is excited in the next section by the lowest order

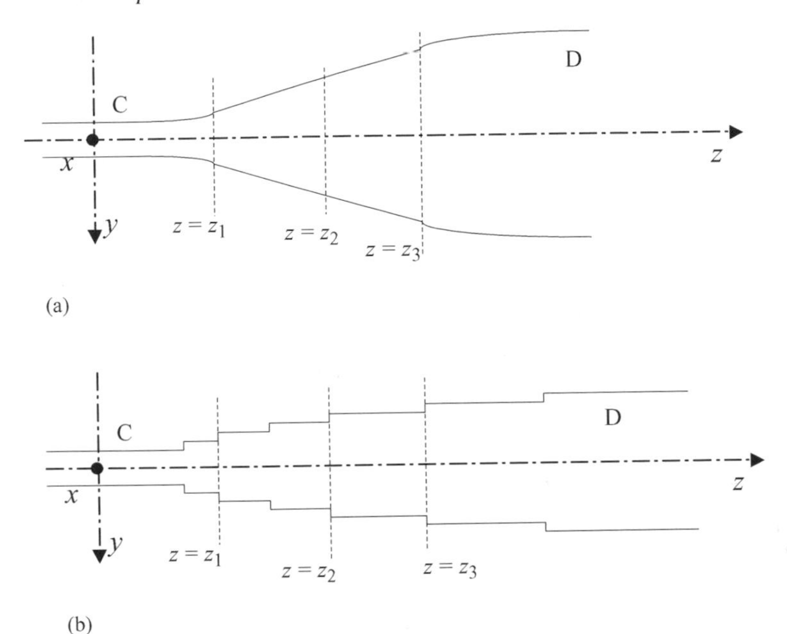

Figure 4.5. Top view of an adiabatic transition and its step approximation. (a) The transition from a single-mode channel waveguide to a multimode channel waveguide. (b) The step approximation of the transition. The local section of the waveguide within each step has a profile of dielectric constant independent of z. The second mode exists for $z > z_1$. The third mode exists for $z > z_2$. The fourth mode exists for $z > z_3$.

mode in the previous section. A negligible amount of power is coupled into higher order modes and radiation modes. Therefore, in a truly adiabatic transition only, the lowest order mode is excited in the multimode output waveguide by the lowest order mode in the input section, and there is no power loss. Conversion of power into higher order modes will occur when the tapering is not sufficiently adiabatic or when scattering occurs. The same conclusion can be drawn for propagation of the lowest order mode in the reverse direction, i.e. from D to C.

Let us now consider a reverse transition where the incident field has several modes. Whenever a higher order mode is excited at D, it will not be transmitted to C. The power in this higher order mode will be transferred into the radiation modes at the z position where this mode is cut off. Only the power in the lowest order mode will be transmitted from D to C.

4.4.2 Super-mode analysis of wave propagation in a symmetric Y-branch

A guided wave component used frequently in fiber and channel waveguide devices is a symmetric Y-branch that connects one single-mode channel waveguide to two single-mode channel waveguides. Its top view in the yz plane is illustrated in

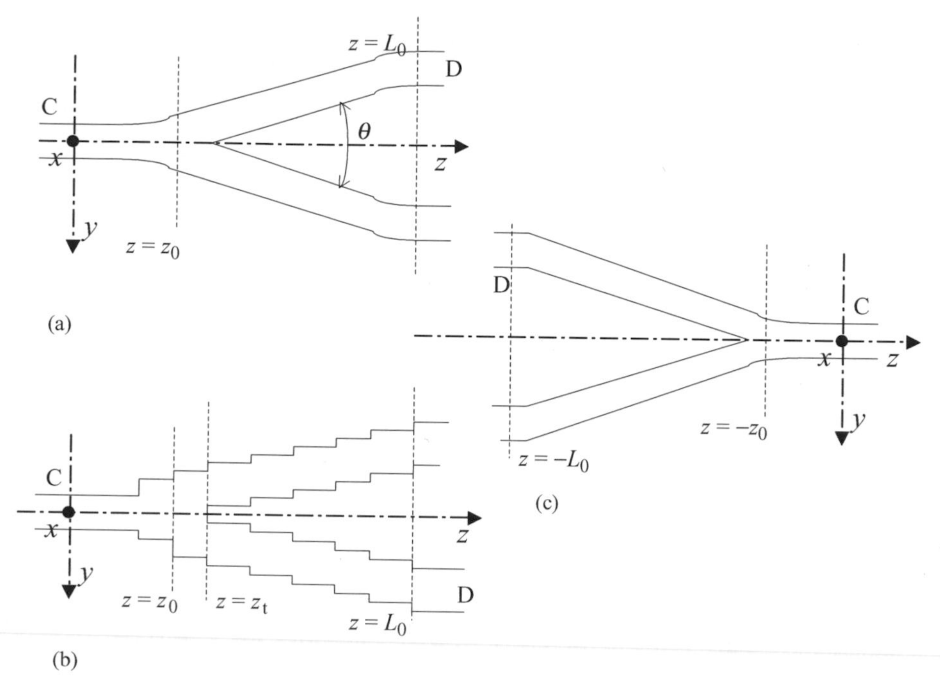

Figure 4.6. Top view of a symmetric Y-branch coupler. (a) Symmetric 3 dB coupler that splits equally the power in the input channel waveguide at C into two identical channel waveguides at *D*. For a symmetric Y-branch, the modes at D always have symmetrical amplitude and phase. (b) The step approximation of the Y-branch coupler. (c) Reverse symmetric coupler that combines the fields from two input waveguides into a single-mode output waveguide. Whether the power from the two input waveguides will be transmitted into the mode of the output waveguide or radiated into the substrate depends on the relative phase and amplitude of the optical fields at the input of the two channel waveguides in a reverse Y-branch.

Fig. 4.6(a). It is symmetric with respect to the xz plane in the y direction. The waveguides at $z > L_0$ have constant separation and cross-sectional profile in the y direction. The index profile in the x direction within the channel waveguide is uniform for the entire device. The objective of such a device is to split the power in the original waveguide at C equally into two waveguides at D, where they are well separated from each other. It is an adiabatic transition when the angle of the branching, θ, is sufficiently small that the scattering and conversion loss from $z = 0$ to $z = L_0$ can be neglected. Ideally, a symmetric Y-branch should function like a 3 dB coupler from the input to both outputs.

The Y-branch coupler can be analyzed as follows. In Fig. 4.6(a), the input waveguide has a single TE_0 mode at $z < 0$. The waveguide width in the y direction begins to broaden at $z > 0$. At $z > z_0$, the waveguide (or the split waveguides) has two modes. At $z \approx L_0$, each isolated waveguide has a single TE_0 mode, $\underline{e_A}$

and $\underline{e_B}$. Thus the two super-modes are the symmetric mode, $(1/\sqrt{2})(\underline{e_A} + \underline{e_B})$, and the anti-symmetric mode, $(1/\sqrt{2})(\underline{e_A} - \underline{e_B})$, discussed in Section 4.3.4. From the symmetry point of view, no anti-symmetric mode should be excited in an adiabatic transition. We expect the output mode to be a symmetric mode. In the absence of an anti-symmetric mode, an equal amount of optical power is carried into the two individual waveguides.

We can reach the same conclusion by examining the transition region in detail. For the transition region $0 < z < L_0$, the Y-branch has a step approximation as shown in Fig. 4.6(b). Within each step we will have local modes. Let z_t be the vertex of the waveguide split. For $z < z_t$, the local modes are the eigen solutions of Eq. (4.1) for the index profile of that step. For $z > z_t$, the local modes are the super-modes given in Eqs. (4.37). The local modes evolve from the input TE_0 mode at $z = 0$ to the super-modes at $z = L_0$. For a given z position z_0, the waveguide begins to have two modes for $z > z_0$. The first-order mode is a symmetric mode and the second-order mode is an anti-symmetric mode. At each step junction in Fig. 4.6(b), the excitation of the modes in the next step can be calculated by modal analysis, as discussed in Section 3.6. For adiabatic transitions, the mismatch between the fields of the incident and the transmitted symmetric mode is so small at each step that the transmission coefficient to the symmetric mode into the next step is unity, and there is negligible coupling to the radiation and higher order modes. The end result is that no anti-symmetric mode is excited over the entire Y-branch.

In the reverse situation shown in Fig. 4.6(c), when the incident field is the lowest order symmetric mode of the double waveguides, it is transmitted without loss to the output waveguide as the TE_0 mode. However, if the incident mode is an anti-symmetric mode, it will continue to propagate as the anti-symmetric mode from $z = -L_0$ to its cut-off point. Let the cut-off point be $z = -z_0$. At z just before $-z_0$, the anti-symmetric mode will be very close to cut-off, with a very long evanescent tail, and its n_{eff} is very close to the effective index of a cladding or substrate mode. As z approaches $-z_0$, the anti-symmetric mode begins to transfer its energy into the radiation mode in the cladding or the substrate. Because of the small overlap integral between the anti-symmetric and the TE_0 mode, the TE_0 mode will not be excited by the anti-symmetric mode. Similar comments can be made for any higher order mode excited at $z < -L_0$. It will be coupled to radiation modes at its cut-off point. In summary, only the power in the lowest order symmetric mode will be transferred to the TE_0 mode at the output.

4.4.3 Analysis of wave propagation in an asymmetric Y-branch

Note that the analysis based on symmetric and anti-symmetric modes applies only to symmetrical adiabatic Y-branches. When the branching angle is large in non-adiabatic transitions, mode conversion will occur at step junctions. When the

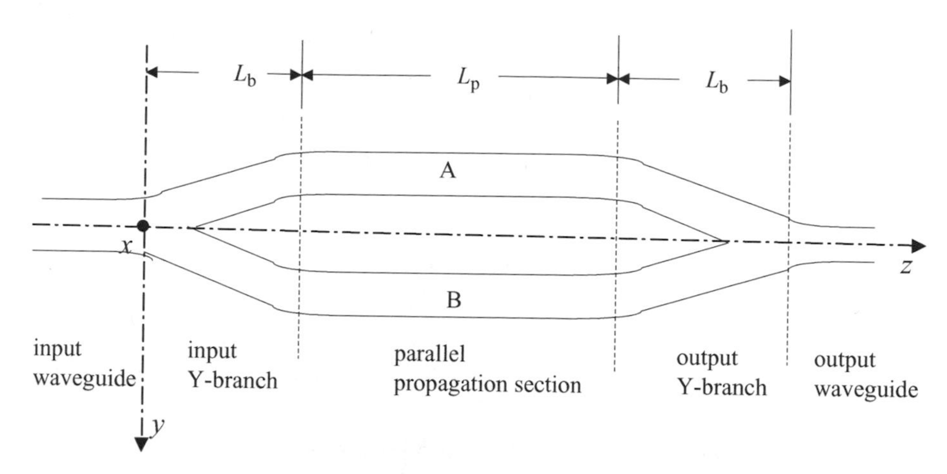

Figure 4.7. Top view of a channel waveguide Mach–Zehnder interferometer. Two waveguides, A and B, connect the input symmetrical Y-branch 3 dB coupler to the output reverse symmetrical Y-branch coupler. Waveguides A and B are well separated from and not coupled to each other. However, the index of waveguides A and B may be changed by electro-optic effects so that either the symmetric super-mode or the anti-symmetric super-mode, or a mixture of them, may appear at the input of the reverse symmetrical Y-branch coupler. Only the power in the symmetric mode will be transmitted to the symmetric single mode of the output waveguide.

branches are not symmetrical, the local super-modes could have very asymmetrical electromagnetic field profiles as we have discussed in Section 4.3. Conversion among super-modes might occur at each step junction. The output, i.e. the cumulative effect, will depend on initial conditions, the branching angle, the index profile and the asymmetry of the Y-branch. An asymmetrical Y-branch will behave sometimes as a power divider and sometimes as a mode splitter or converter. Numerical analysis based on modal analysis of the super-modes in the step approximation is required to find the answer. In general, a larger branching angle will lead to a power divider, while a small branching angle will lead to a mode splitter [10].

4.4.4 Mach–Zehnder interferometer

The Mach–Zehnder interferometer consists of two symmetric Y-branches back to back, connected by two parallel channel waveguides that are well separated from each other so that they are uncoupled; see Fig. 4.7. Similar devices can be made from optical fibers. The objective of the input Y-branch in the Mach–Zehnder interferometer is to excite equally the individual mode of the two waveguides immediately after the input Y-branch.

Let the input be a TE_0 mode with amplitude A at $z = 0$. At the exit of the input Y-branch at $z = L_b$, the amplitude of the symmetric mode is $A\exp(j\varphi)$; φ is the phase shift due to the propagation from $z = 0$ to $z = L_b$. In terms of the modes of the individual waveguides, the amplitudes are $(1/\sqrt{2})Ae^{j\phi}\underline{e_A}$ and $(1/\sqrt{2})Ae^{j\phi}\underline{e_B}$. When the two parallel waveguides are identical and have equal length L_p, the input to the output Y-branch at $z = L_b + L_p$ is $(1/\sqrt{2})Ae^{j\phi}e^{-j\beta_A L_p}(\underline{e_A} + \underline{e_B})$. Such a symmetric mode will yield an output $Ae^{j2\phi}e^{-j\beta_A L_p}$ at $z = 2L_b + L_p$.

When the two parallel waveguides in the propagation section have slightly different effective index or propagation wave number, β_A and β_B, the input to the output Y-branch is

$$\frac{1}{\sqrt{2}}Ae^{j\phi}e^{-j\beta_A L_p}\left(\underline{e_A} + \underline{e_B}e^{-j(\beta_B-\beta_A)L_p}\right). \tag{4.40}$$

In other words, there is a mixture of symmetric mode, $\underline{e_s}$, and anti-symmetric mode, $\underline{e_a}$, at $z = L_b + L_p$. When $\Delta\beta L_p = (\beta_B - \beta_A)L_p = \pm\pi$ or $(2n \pm 1)\pi$, where n is an integer, then the input to the output Y-branch is an anti-symmetric mode. In this case, the output TE_0 mode at $z = 2L_b + L_p$ will have zero amplitude. The power in the anti-symmetric mode was transferred into the radiation modes.

When the waveguides are made from electro-optical materials, the change in n_{eff} of each waveguide will be proportional to the electric field (i.e. the voltage applied to the electrode across the waveguide). When we calculate the power transmitted to the output based on the amplitude of the symmetric mode at $z = L_b + L_p$, we obtain

$$\frac{I_{out}}{I_{in}} = \frac{1}{2}[1 + \cos(\Delta\beta L_p)] = \frac{1}{2}\left[1 + \cos\left(\frac{\pi}{V_\pi}V\right)\right], \tag{4.41}$$

where V is the electrical voltage applied to the modulator that produces the $\Delta\beta$ and V_π is the voltage that will yield $\Delta\beta L_p = \pi$. Such a device is called a Mach–Zehnder modulator [11].

From the super-mode analysis point of view, a symmetric super-mode (and no anti-symmetric mode) is excited at the exit of the input Y-branch at $z = L_b$. When there is $\Delta\beta$ (i.e. $\beta_A - \beta_B$) created by the applied V, it becomes a mixture of symmetric and anti-symmetric modes as the optical wave propagates. After propagating a distance L_p in the parallel section, the mix of the anti-symmetric mode and symmetric mode at $z = L_p + L_b$ will depend on $\Delta\beta L_p$. For example, the mode at $z = L_p + L_b$ is an anti-symmetric mode when $\Delta\beta L_p = \pi$. Since the anti-symmetric mode will not be transmitted to the output waveguide, the power transmitted to the output waveguide is controlled by $\Delta\beta$.

The super-mode analysis is very important in order to understand the Mach–Zehnder modulator in depth. For example, when the attenuation of one of the waveguides is very large, e.g. waveguide B, then the input to the output Y-branch is

$$\frac{1}{\sqrt{2}} A e^{j\phi} e^{-jB_A L_p} \underline{e_A} = \frac{A}{2}(\underline{e_s} + \underline{e_a}) e^{j\phi} e^{-j\beta_A L_p}.$$

Since only $\underline{e_s}$ will be transmitted, the amplitude of the TE_0 mode at the output is $(A/2)e^{2j\phi}e^{-j\beta_A L_p}$. In other words, only one-quarter of the input power is transmitted and three-quarters of the input power is attenuated and radiated into the cladding or the substrate.

4.5 Propagation in multimode waveguides and multimode interference couplers

The interference of modes in a multimode waveguide has interesting and important applications. A multimode interference coupler consists of a section of a multimode channel waveguide, abruptly terminated at both ends. A number of access channel waveguides (usually single-mode) may be connected to it at the beginning and at the end. Such devices are generally referred to as $N \times M$ multimode interference (MMI) couplers, where N and M are the numbers of input and output waveguides, respectively [12].

Figure 4.8(a) illustrates a multimode interference coupler with two input and two output access waveguides. The multimode section is shown here as a step-index ridge waveguide with width W and length L. It is single-mode in the depth direction x and multimode ($n \geq 3$) in the lateral direction y. The objective of such a multimode coupler, similar to the star coupler, is to couple specific amounts of power from the input access waveguides into the output access waveguides. However, it is much more compact than the star coupler discussed in Section 3.4.3. Its operation is based on the interference of the propagating modes. We intend to show here that, based on the interference pattern of the modes excited by the input access waveguides, we could obtain specific distributions of the power in the output access waveguides at specific positions of z.

Let the multimode waveguide be a ridge waveguide, as shown in Fig. 3.6(b). For the planar waveguide mode (i.e. for very large W) in the core (i.e. in the ridge), it has just a single TE mode in the x direction with an effective index n_{e1}. The cladding region, outside the ridge, also has a planar waveguide mode with an effective index n_{e2}; $n_{e1} > n_{e2}$. Figure 4.8(b) illustrates the profile of the effective index of the planar TE_0 modes in the y direction. The channel guided wave modes in the core can be found by the effective index method discussed in Section 3.4 or by other numerical methods. Figure 4.8(c) illustrates the effective mode width, W_e, and the lateral field variation in the y direction for the first few modes.

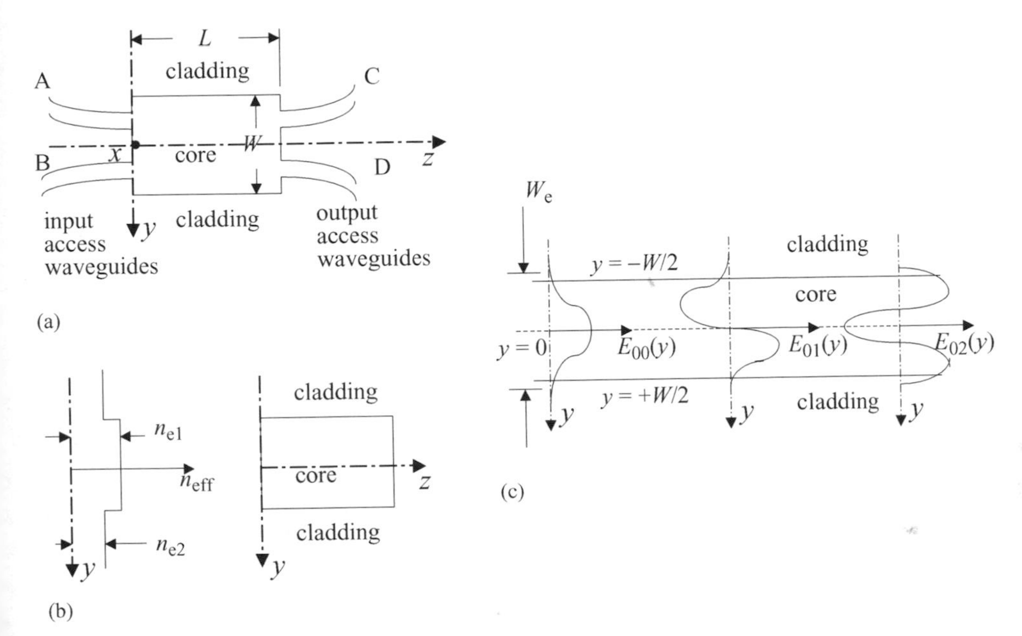

Figure 4.8. Multimode interference coupler. (a) Top view of a 2 × 2 multimode interference coupler. The multimode waveguide is of length L and width W. (b) Effective index profile of the multimode waveguide. (c) Field patterns, as a function of y, of the lowest order modes in the multimode waveguide.

Before we discuss the interference pattern of the modes, let us discuss the properties of the individual modes. For well guided modes, it has been shown in the literature [12] that the solution of the transcendental equation, Eq. (3.31), can be approximated by

$$\tan\left[(h_n/k)\frac{kW_e}{2}\right] \approx \infty.$$

Here, W_e is an effective width of the ridge, and $W_e > W$. W_e is usually taken to be the effective width of the lowest order mode $m = 0$ in the x direction and $n = 0$ in the y direction. In that case,

$$h_n = \frac{(n+1)\pi}{W_e}$$

and

$$\beta_{0n}^2 = n_{e1}^2 k^2 - h_n^2,$$
$$\beta_{0n} \approx n_{e1} k - \frac{(n+1)^2 \pi \lambda}{4 n_{e1} W_e^2}. \tag{4.42}$$

Equation (4.42) predicts that the propagation constants of the various lateral order modes will have a quadratic dependence on n. By defining L_π as the beat length (i.e. the propagation length in which the phase difference of two modes is π) between the $n = 0$ and $n = 1$ modes, we obtain

$$\left.\begin{aligned} L_\pi &= \frac{\pi}{\beta_{00} - \beta_{01}}, \\ \beta_{00} - \beta_{0n} &= \frac{n\,(n+2)\,\pi}{3L_\pi}. \end{aligned}\right\} \tag{4.43}$$

Let us now examine the total field of all the modes. As we have discussed in Section 3.5, the y variation of any input field at $z = 0$, $E_0(y, z = 0)$, can be expressed as a summation of the E_{0n} modes. Thus,

$$E_0(y, 0) = \sum_{n=0}^{n=N-1} C_n E_{0n}(y), \tag{4.44a}$$

$$E_0(y, z) = \sum_{n=0}^{n=N-1} \left\{ C_n E_{0n}(y)\, e^{\left[j \frac{n(n+2)\pi}{3L_\pi} z \right]} \right\} e^{-j\beta_{00} z}, \tag{4.44b}$$

$$E_{0n}(y) = A \sin(h_n y). \tag{4.44c}$$

Any input field at $z = 0$ will be repeated or mirrored at $z = L$, whenever

$$\exp\left[j \frac{n\,(n+2)\,\pi}{3L_\pi} L \right] = 1 \tag{4.45a}$$

or

$$\exp\left[j \frac{n\,(n+2)\,\pi}{3L_\pi} L \right] = (-1)^n . \tag{4.45b}$$

When the condition in Eq. (4.45a) is satisfied, the field at $z = L$ is a direct replica of the input field. When the condition in Eq. (4.45b) is satisfied, the even modes will have the same phase as the input, but the odd modes will have a negative phase, producing a mirror image of the input field. For the 2×2 coupler shown in Fig. 4.8(a), this means that power in input A will be transferred to output C when Eq. (4.45a) is satisfied. Power in input A will be transferred to output D when Eq. (4.45b) is satisfied.

More extensive use of the mode interference pattern can be obtained when we analyze it in detail in the following manner. Figure 4.8(c) shows that the y variation of the field of a well guided multimode channel waveguide mode resembles the lowest order sine terms of a Fourier series in y within the period $y = -W_e/2$ to $y = +W_e/2$. However, there are only a finite number of sine Fourier series terms in our modes. In order to recognize the more complex interference patterns, let us now extend the expression for the modes outside of the range $-W_e/2$ to $W_e/2$ in a periodic manner so that we can take advantage of our knowledge of Fourier series. Since these modes have a half-cycle sine variation within $-W_e/2 < y < W_e/2$, the

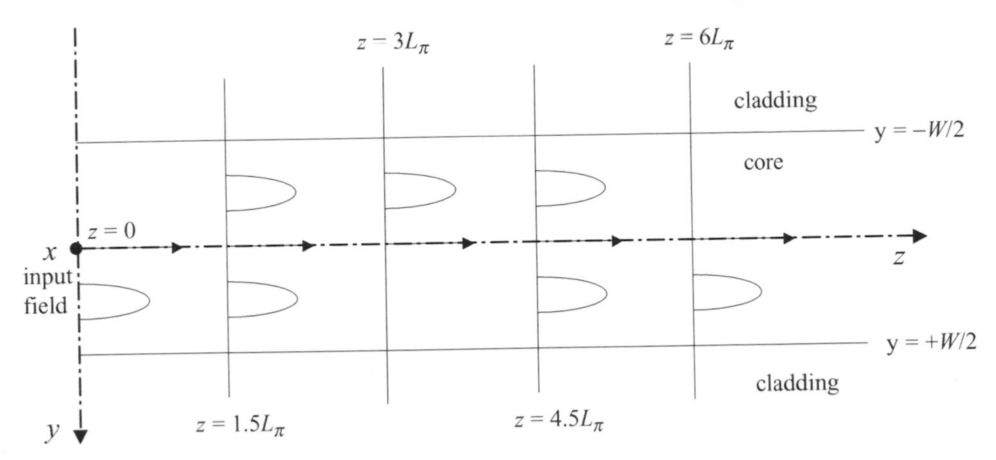

Figure 4.9. Images of the input field at various distances in a multimode interference coupler. The input field imposed at $z = 0$ can be decomposed into a field which is a summation of all the modes. The total field profile of the summation of these modes will yield a two-fold image of the input at $z = 1.5L_\pi$ and at $z = 4.5L_\pi$, a mirror single image at $z = 3L_\pi$ and a direct single image at $z = 6L_\pi$.

extended mode in $-3W_e/2 < y < -W_e/2$ and in $W_e/2 < y < 3W_e/2$ should be anti-symmetric with respect to the mode in $-W_e/2 < y < W_e/2$. Similar extensions can be made beyond $y > |3W_e/2|$. In other words, we will now consider the total extended field over all y coordinates, including the periodic extension of the fields outside the multimode waveguide region. The extended input field from all the input access waveguides (periodically repeated outside the region $y = -W_e/2$ to $W_e/2$) could then be expressed as a summation of these Fourier terms. Equations (4.44) show that at a distance L later the relative phase among the Fourier terms is changed. Different multifold images can be formed within the period, ranged from $-W_e/2$ to $W_e/2$, by manipulating these phase terms. As an example, let us consider $L = 3pL_\pi/2$, where p is an odd integer. Then,

$$E_0\left(y, \frac{3pL_\pi}{2}\right) = \sum_{n\ \text{even}} C_n E_{0n}(y) + \sum_{n\ \text{odd}} (-j)^p C_n E_{0n}(y)$$
$$= \frac{1 + (-j)^p}{2} E_0(y, 0) + \frac{1 - (-j)^p}{2} E_0(-y, 0). \quad (4.46)$$

The second line of Eq. (4.46) represents a pair of images of E_0 in quadrature and with amplitudes $1/\sqrt{2}$, at distances $z = 3L_\pi/2, 9L_\pi/2, \ldots$ The replicated, the mirrored and the double images of E_0 at various z distances are illustrated in Fig. 4.9. Clearly, we have a 3 dB power splitter from input B into output waveguides C and D at $z = 3L_\pi/2$ and $z = 9L_\pi/2$. We have transferred the power from B to C (called the cross-state) when $z = 3L_\pi$, and from B to D (called the through-state) when $z = 6L_\pi$. A 2×2 InGaAsP MMI cross-coupler has been made with

$W = 8$ μm and $L = 500$ μm, which gives an excess loss of 0.4 to 0.7 dB and an extinction ratio of 28 dB, and a 3 dB splitter with $L = 250$ μm, and imbalances between C and D well below 0.1 dB [13].

The actual design of an MMI coupler must take into account the number of input and output access waveguides, the number of modes in the multimode waveguide, the relative phase and amplitude of the incident modes in the input access waveguides and the position and width of access waveguides [13].

References

1 D. L. Lee, *Electromagnetic Principles of Integrated Optics*, Chapter 8, New York, John Wiley and Sons, 1986

2 R. Ulrich and P. K. Tien, "Theory of Prism-Film Coupler and Guide," *Journal of the Optical Society of America*, **60**, 1970, 1325

3 A. Yariv, "Coupled Mode Theory for Guided Wave Optics," *IEEE Journal of Quantum Electronics*, **QE-9**, 1973, 919

4 D. C. Flanders, H. Kogelnik, R. V. Schmidt and C. V. Shank "Grating Filters for Thin Film Optical Waveguides," *Applied Physics Letters*, **24**, 1974, 194

5 A. Korpel, "Acousto-optics, A Review of Fundamentals," *Proceedings of IEEE*, **69**, 1981, 48

6 D. L. Hecht, "Spectrum Analysis Using Acousto-optic Devices," *Proceedings of SPIE*, **90**, 1976, 148

7 S. Kurazono, K. Iwasaki and N. Kumagai, "New Optical Modulator Consisting of Coupled Optical Waveguides," *Electronic Communication Japan*, **55**, 1972, 103

8 H. Kogelnik and R. V. Schmidt, "Switched Directional Couplers with Alternating $\Delta\beta$," *IEEE Journal of Quantum Electronics*, **QE-12**, 1976, 396

9 G. B. Betts and W. S. C. Chang, "Crossing Channel Waveguide Electrooptic Modulators," *IEEE Journal of Quantum Electronics*, **QE-22**, 1986, 1027

10 W. K. Burns and A. F. Milton, "Mode Conversion in Planar Dielectric Separating Waveguides," *IEEE Journal of Quantum Electronics*, **QE-11**, 1975, 32

11 R. C. Alferness, "Waveguide Electrooptic Modulators," *IEEE Transactions Microwave Theory and Technique*, **MTT-30**, 1982, 1121

12 N. S. Kapany and J. J. Burke, *Optical Waveguides*, New York, Academic Press, 1972

13 L. B. Soldano and E. C. M. Pennings, "Optical Multi-mode Interference Devices based on Self-Imaging: Principles and Applications," *Journal of Lightwave Technology*, **13**, 1995, 615

5

Macroscopic properties of materials from stimulated emission and absorption

All optical oscillators and amplifiers can be analyzed as an electromagnetic structure, such as a cavity or a transmission line, which contains an amplifying medium. The operating characteristics of lasers are governed by both the electromagnetic properties of the structures and the properties of the amplifying medium.

Stimulated emission and absorption of radiation from energy states are the physical basis of amplification in all laser materials. In order to study lasers, it is necessary to understand the energy levels of the amplifying medium and the stimulated transitions involving them. Quantum mechanical analysis of the energy states and the stimulated emission and absorption is presented in most books on lasers. However, quantum mechanical analyses cover only the analysis of individual atoms. We need to relate the effects of the quantum mechanical interactions to the macroscopic properties of the materials so that properties of the lasers may be analyzed. Therefore, the macroscopic susceptibility of materials related to stimulated emission and absorption is the focus of discussion in this chapter.

It is not the purpose of this book to teach quantum mechanics. There are already many excellent books on this topic [1, 2, 3, 4]. The readers are assumed to have a fundamental knowledge of quantum mechanics. However, it is necessary first to review some of the major steps in order to understand precisely the notation used and the meaning of the results. This is presented in Section 5.1. The traditional semi-classical quantum mechanical analysis is also the easiest way to demonstrate stimulated emission and absorption. Therefore, we will present the traditional analysis of stimulated emission and absorption in Section 5.2. Sections 5.1 and 5.2 may be particularly helpful to engineering students who are not as familiar with quantum mechanics as physics students. For an in-depth discussion of laser characteristics, we need to analyze the collective effects of many atoms and the effect on the radiation field. In other words, we need to know the statistically averaged expectation value of the polarization of the atomic particles induced quantum mechanically by the electric field. Thus, in our quantum mechanical review in Section 5.2 we will stress the understanding of the expectation values of specific interactions.

In Section 5.3, we will introduce the quantum statistical concept of the density matrix so that we can analyze the macroscopic averaged susceptibility of all the atoms. When the quantum statistical analysis of macroscopic susceptibility is combined with the traditional analysis of stimulated transitions, we will then appreciate fully the significance of quantum mechanical interactions. Based on the macroscopic susceptibility, we will discuss in Section 5.4 the distinctive properties of the homogeneously and inhomogeneously broadened transitions that are very important in understanding the behavior of specific lasers. The non-linear macroscopic properties of lasers depend on the nature of the broadening. Saturation of the macroscopic susceptibility will be very important when we discuss properties of lasers in Chapter 6.

5.1 Brief review of basic quantum mechanics

In the following we will present (1) a brief review of elementary quantum mechanics using the differential operator representation that we are familiar with, (2) a discussion of the expectation value, (3) a brief summary of energy eigen value and energy states, and (4) a summary of matrix representation. See refs. [1], [2], [3], and [4] for more extensive reviews of quantum mechanics.

5.1.1 Brief summary of the elementary principles of quantum mechanics

In quantum mechanics, the state of a particle is described by its wave function ψ and a dynamic observable is represented by a Hermitian operator A. Some examples of dynamic observables in the form of Hermitian differential operators are as follows:

$$\begin{aligned} &\text{linear momentum:} && \underline{p} \to -j\hbar\nabla, \\ &\text{potential:} && V \to V, \\ &\text{total energy:} && H \to \frac{\hbar^2}{2m}\nabla^2 + V. \end{aligned}$$

The wave function ψ is a solution of the Schrödinger equation,

$$H\psi = j\hbar\frac{\partial\psi}{\partial t}. \tag{5.1}$$

H is the total Hamiltonian operator (i.e. the total energy operator) for the particle (or particles) and ψ is normalized, $\int\int\int \psi^*\psi \, dx \, dy \, dz = 1$. Since ψ is a function of x, y, z and t, the quantum mechanics presented in this manner is called the differential operator representation.

ψ has two meanings. (1) The probability density of finding this particle at $\underline{r}$ is $\psi(\underline{r}, t)\psi^*(\underline{r}, t)$. When ψ approaches zero at large $\underline{r}$, the implication is that the particle is confined to regions where $\psi\psi^*$ is large. (2) ψ can also be expanded as a summation of the eigen states of any dynamic observable. The square of the magnitude of the expansion coefficient has the meaning of the probability for the particle to be in that eigen state.

Let us present the above concept more formally. Let the dynamic observable A have a set of eigen values a_n and eigen functions v_n. In general, any ψ can be expanded in terms of the eigen functions $v_n(\underline{r})$ as follows:

$$\psi(\underline{r}, t) = \sum_n C_n(t) v_n(\underline{r}), \tag{5.2a}$$

where

$$A v_n = a_n v_n(\underline{r}); \tag{5.2b}$$

a_n is the nth eigen value of A and v_n is the eigen state of A associated with a_n.

The measurement of A on a particle which is in the state ψ will yield any one of the eigen values, a_n. The probability that the measured value is a specific a_n for a given ψ is $|C_n|^2$.

In laser analysis, the most frequently used set of eigen functions are the energy eigen states of the particles representing the optically active material. Stimulated transitions can take place between different pairs of energy states. In that case, A is the energy operator of the particle, a_n in Eqs. (5.2a) and (5.2b) is just the value of the nth energy level, and $|C_n|^2$ is the probability that the particle is in the nth energy state.

5.1.2 Expectation value

Repeated measurements of A on a particle in the state ψ yield its average value. The average value of any dynamic observable A, called the expectation value, is

$$\langle A \rangle = \int \psi^* A \psi \, dv, \tag{5.3}$$

e.g.

$$\langle p_x \rangle = -j\hbar \int \psi^* \frac{\partial \psi}{\partial t} \, dv.$$

We can differentiate $\langle A \rangle$ in Eq. (5.3) directly with respect to time. In that case, $\partial\psi/\partial t$ and $\partial\psi^*/\partial t$ are given by Eq. (5.1). Therefore, the equation of motion for $\langle A \rangle$ is

$$\frac{d\langle A \rangle}{dt} = \left\langle \frac{\partial A}{\partial t} \right\rangle + \frac{j}{\hbar} \langle [H, A] \rangle, \tag{5.4}$$

where $[H, A] = HA - AH$ is the commutator of H and A.

$\langle A \rangle$ is not a quantity commonly emphasized in many quantum mechanics books, which are concerned mostly with the behavior of individual atoms or particles. However, we are very interested in $\langle A \rangle$ in laser materials because we want to know the averaged properties over many atoms.

In terms of the set of eigen states v_n of A, $\langle A \rangle$ for a particle (or particles) in state ψ could also be expressed as

$$\langle A \rangle = \int \psi^* A \psi \, dv = \sum_n a_n |C_n|^2. \tag{5.5}$$

5.1.3 Summary of energy eigen values and energy states

Each dynamic observable has its own eigen functions and eigen values. For the total energy operator H,

$$H u_{E_j} = E_j \, u_{E_j}(\underline{r}). \tag{5.6}$$

Here E_j is the jth energy eigen value and u_{E_j} is the energy eigen function for the energy value E_j. All the u_E form an orthonormal complete set. If a particle is definitely in the state u_{E_j}, then

$$\psi = u_{E_j}(\underline{r}) e^{-j\frac{E}{\hbar}t} \tag{5.7a}$$

and

$$\langle H \rangle = \left\langle j\hbar \frac{\partial}{\partial t} \right\rangle = E_j. \tag{5.7b}$$

The best known energy operators discussed in quantum mechanics text books are the Hamiltonian operators for the harmonic oscillator and for the hydrogen atom. The eigen equations for these two cases are:

$$-\frac{\hbar^2}{2m} \frac{d^2}{dx^2} u_E + \frac{1}{2} k x^2 u_E = E u_E$$

and

$$-\frac{\hbar^2}{2m} \nabla^2 u_E + V(r) u_E = E u_E.$$

In terms of the energy eigen states, the expectation value of energy for any ψ (see Eqs. (5.3) and (5.5)) is the average energy of the particle. For stimulated emission or absorption, we are looking for any change of the expansion coefficients, $|C_n|^2$, which has the significance of the change of the probability of the particle in various energy states.

5.1.4 Summary of the matrix representation

The matrix representation of any operator A can be understood most easily by representing ψ in terms of different sets of eigen functions in the differential operator representation. Let

$$\psi = \sum_n C_n v_n.$$

Then ψ can also be represented as a column matrix on the basis of the set of v_n functions,

$$\begin{Vmatrix} C_1 \\ C_2 \\ C_3 \\ \vdots \\ C_i \\ \vdots \end{Vmatrix}.$$

ψ^* is a row matrix,

$$\begin{Vmatrix} C_1^* & C_2^* & C_3^* & \cdots & \cdots \end{Vmatrix}.$$

When A operates on ψ, it creates a new state $\phi = A\psi$. We can represent both the ψ and the ϕ on the same set of basis functions.

We could also express both ψ and ϕ in terms of a new basis set of eigen functions w_j. Then we obtain

$$\left.\begin{aligned} &\phi = \sum_i b_i w_i, \quad \psi = \sum_j C_j w_j, \quad \phi = A\psi, \\ &C_j = \int w_j^* \psi \, dv, \\ &b_i = \int w_i^* \phi \, dv = \int w_i^* A \left[\sum_j c_j w_j \right] dv = \sum_j A_{ij} C_j, \end{aligned}\right\} \tag{5.8}$$

with

$$A_{ij} = \int w_i^* A w_j \, dv.$$

Or, on the basis of w_j,

$$\begin{Vmatrix} b_1 \\ b_2 \\ b_3 \\ \vdots \end{Vmatrix} = \begin{Vmatrix} A_{11} & A_{12} & A_{13} & \cdots \\ A_{21} & A_{22} & A_{23} & \cdots \\ A_{31} & A_{32} & \cdots & \\ \vdots & & & \end{Vmatrix} \begin{Vmatrix} C_1 \\ C_2 \\ C_3 \\ \vdots \end{Vmatrix}, \tag{5.9}$$

where $\|A\|$ is a Hermitian matrix. $\|A\|$ will appear to be different when it is represented on the basis of a different set of eigen functions. If $\|A\|$ is expressed on the basis of v_n, which is the set of eigen functions of A, then

$$Av_n = a_n v_n, \tag{5.10a}$$

$$\int v_n^* A v_j \, dv = a_j \delta_{ij}, \tag{5.10b}$$

$$\|A\| = \begin{Vmatrix} A_{11} & 0 & 0 & \cdots \\ 0 & A_{22} & 0 & \\ 0 & 0 & A_{33} & \\ \vdots & & & \ddots \end{Vmatrix}. \tag{5.10c}$$

$\|A\|$ is now a diagonal matrix.

There are four reasons for learning the matrix representation. (1) It is used in many references. (2) There is a matrix representation to all the quantum mechanical equations. However, the basis function does not need to be expressed as a function of x, y and z. For example, the basis functions for the spin angular momentum are not functions of x, y and z. It is necessary to use matrix representation to analyze interactions involving spin. (3) The equation of motion of matrix elements would allow us to find solutions of matrix elements without knowing explicitly the basis functions. (4) When we can find the matrix of any dynamic observable A in diagonal form, we know its eigen values, and the basis functions for the diagonal representation are the eigen states. This is a powerful practical tool used to calculate energy levels. For those whose interests are only to read the references, the brief summary presented in the preceding paragraphs is sufficient. More discussions on matrix representation are presented in the following paragraphs to enable us to review matrix diagonalization and the equation of motion.

We will first discuss how $\|A\|$ transforms from one basis set of eigen functions to another basis set of basis functions. More specifically, we note that any basis function of the first set can, itself, be represented as a summation of the basis functions in the second set, and vice versa.

Let

$$v_j = \sum_n S_{jn}^* w_n;$$

then

$$S_{jn} = \int w_n v_j^* \, dv, \quad w_n = \sum_j S_{jn} v_j.$$

In terms of v_j,

$$A_{ij} = \int v_i^* A v_j \, dv = \int \left(\sum_m S_{im} w_m^* \right) A \sum_n S_{jn}^* w_n \, dv$$
$$= \sum_m \sum_n S_{im} \left[\int w_m^* A w_n \, dv \right] S_{jn}^*. \qquad (5.11)$$

Or,

$$\|A(\text{on the } v_n \text{ basis})\| = \|S\| \, \|A(\text{on the } w_n \text{ basis})\| \, \|\widetilde{S}\|.$$

$\|\widetilde{S}\|$ is the complex conjugate transposed matrix of $\|S\|$. $\|S\|$ is a unitary matrix; $\|\widetilde{S}\| = \|S^{-1}\|$. There are well known mathematical techniques, called matrix diagonalization, that can be utilized to solve for $\|S\|$ that will yield a diagonal $\|A\|$. In other words, when we find the $\|S\|$ which will transform $\|A\|$ into the diagonal form, we will have found the eigen values of the operator A. This is an important practical technique which allows us to find the energy levels of important materials. In order for the A matrix to be diagonalized, A must be a Hermitian operator.

Notice that the matrix representing any dynamic observable can be written without explicitly expressing the basis functions used to represent the dynamic observable as mathematical functions. What this means is that if we can write a $\|B\|$ on any basis, we can mathematically solve for $\|S\|$ that will make it diagonal, without ever knowing how to write the original basis set explicitly. For this reason, in the literature, it is common to express the matrix element B_{ij} as $\langle i|B|j\rangle$ without explicitly writing the integral expression as shown in Eq. (5.8)

If we take the time derivative of the matrix element B_{ij} of a dynamic observable B as we defined in Eq. (5.8), we obtain the Heisenberg equation of motion,

$$\frac{d}{dt}\langle i|B|j\rangle = \langle i| \left(\frac{\partial B}{\partial t} \right) |j\rangle + \frac{j}{\hbar}\langle i|[H, B]|j\rangle. \qquad (5.12)$$

This is an important equation. It will allow us to calculate the time variation of many important interaction Hamiltonians. In matrix representation, we will regard Eq. (5.12) as the equation of motion of B_{ij} without deriving it from Eq. (5.8) and without requiring it to have basis functions which are eigen functions of differential operators.

5.2 Time dependent perturbation analysis of ψ and the induced transition probability

The time dependence of ψ can be expressed in terms of the time dependence of $|C_n|^2$. Let ψ be expressed in terms of the energy states of a particle in a material. Let us consider the transitions between energy states of the particles present in an optically active laser material. The probability of emission of a quantity of energy implies a decrease in the probability of finding such particles in a higher energy state and an increase in the probability of finding them in a lower energy state. Conversely, when the probability of finding the particle in a higher energy state is increased while the probability of being in a lower energy state is reduced, this implies the probability of absorbing a quantity of energy. If the change in ψ is produced in response to the presence of electromagnetic radiation, it is a stimulated transition. More extensive reviews of time dependent solutions of the Schrödinger equation are presented in refs. [1] to [4].

In the following, we will follow a traditional presentation commonly used in many other books on lasers. We will first describe the mathematical formalism of the time dependent analysis of the interaction with electromagnetic radiation, followed by a discussion of the various approximations of the interaction Hamiltonian. The electric dipole approximation is the first-order approximation, followed by magnetic dipole and electric quadrupole interactions as second-order approximations. When the electromagnetic radiation has harmonic time variations (i.e. $\cos \omega t$), we will show that the traditional transition probability per unit time will be large only under the special circumstance where the energy difference between the two energy levels is equal to $h\nu$ of the radiation; ν is the electromagnetic frequency, $\omega = 2\pi\nu$. The transition probability will also be proportional to the radiation intensity.

The semi-classical analysis is the simplest way to describe the induced transitions and the emission and absorption processes. It explains clearly that the stimulated transition can yield emission and absorption. The analysis presented in this section is the same as that presented in many books on lasers. However, those books do not tell us how the electromagnetic behavior of the laser is determined by the stimulated transition. We need to know the macroscopic material properties produced by the stimulated transition. The macroscopic susceptibility approach will be presented in Section 5.3.

5.2.1 Time dependent perturbation formulation

The method that we will use to calculate ψ as a function of time is called the time dependent solution of the Schrödinger equation. Following most quantum mechanical textbooks, we will analyze ψ in terms of a superposition of the energy states of an atomic particle in the absence of radiation. The interaction of the radiation with the particle is expressed by an interaction Hamiltonian in which the electromagnetic radiation is expressed as a classical field. Since the radiation field is not quantized, this is called a semi-classical analysis. The interaction Hamiltonian is treated as a perturbation to the Hamiltonian of the particle in the absence of radiation.

Let H_0 be the total energy Hamiltonian of the particle in a material without any electromagnetic radiation; we will not solve the energy eigen value problem, but rather we will assume that its energy eigen values and energy states have already been found such that

$$H_0 \, u_n = E_n \, u_n. \tag{5.13}$$

The total Hamiltonian H for the Schrödinger equation consists of two parts, H_0 and H', where H' is the interaction Hamiltonian of the particle with the radiation field:

$$H = H_0 + H'. \tag{5.14}$$

In order to solve Eqs. (5.1), we shall assume

$$\psi = \sum a_n \, (t) \, u_n e^{-jE_n t/\hbar}.$$

If H' is zero, then the a_n are just constants. Substituting this form of ψ into Eq. (5.1), we obtain

$$\sum_n u_n \left[a_n \left(-\frac{jE_n}{\hbar} \right) e^{-jE_n t/\hbar} + \frac{da_n}{dt} e^{-jE_n t/\hbar} \right]$$
$$= -\frac{j}{\hbar} \sum_n a_n (H_0 + H') u_n e^{-jE_n t/\hbar}.$$

Multiplying by u_k^* and integrating both sides, we obtain, for each k,

$$\frac{da_k}{dt} = -\frac{j}{\hbar} \sum_n a_n H'_{kn} e^{j\omega_{kn} t}, \tag{5.15}$$

with

$$H'_{kn} = \int u_k^* H' u_n \, dv = \langle E_k | \, H' \, | E_m \rangle \, ,$$
$$\omega_{kn} = \frac{E_k - E_n}{\hbar}.$$

Up to this point the analysis is exact. However, we have not simplified very much the task of solving the Schrödinger equation. In order to find the a_k coefficients, we need to solve the infinite set of equations given in Eq. (5.15).

Simplification can be achieved by considering that the effect of H' operating on any wave function is small, i.e. H'_{kn} is small. We recognize that the change of a_n

due to a small H' will also be small. Moreover, there will be a first-order change of a_n, a second-order change of a_n, etc. We could evaluate Eq. (5.15) easily if we could calculate just the first-order perturbation effect of H' from the zeroth-order known solution and the second-order perturbation effect from the first- and zeroth-order solutions. In order to do this, we need to group quantities of equal orders of magnitude together and to require that they equal each other in the same order. To identify terms which belong to the same order of magnitude, we introduce a parameter λ into H and a_n as follows:

$$H = H_0 + \lambda H'$$

and

$$a_n = a_n^{(0)} + \lambda a_n^{(1)} + \lambda^2 a_n^{(2)} + \lambda^3 a_n^{(3)} + \cdots$$

When we substitute the above expressions into Eq. (5.15), we obtain

$$\frac{da_k^{(0)}}{dt} + \lambda \frac{da_k^{(1)}}{dt} + \lambda^2 \frac{da_k^{(2)}}{dt} + \cdots = -\frac{j}{\hbar} \sum_n \left(a_n^{(0)} + \lambda a_n^{(1)} + \lambda^2 a_n^{(2)} + \cdots\right) \lambda H'_{kn} e^{j\omega_{kn} t}.$$

A first-order solution is produced by requiring those terms with λ power to equal each other, the second-order effect is produced by terms with λ^2. Collecting terms with equal power of λ to equal to each other, we have

$$\left.\begin{aligned} \frac{da_k^{(0)}}{dt} &= 0, \\ \frac{da_k^{(1)}}{dt} &= -\frac{j}{\hbar} \sum_n a_n^{(0)} H'_{kn} e^{j\omega_{kn} t}, \\ \frac{da_k^{(2)}}{dt} &= -\frac{j}{\hbar} \sum_n a_n^{(1)} H'_{kn} e^{j\omega_{kn} t}, \ \ldots \end{aligned}\right\} \tag{5.16}$$

When all the $a_k^{(1)}$ satisfy Eqs. (5.16), the Schrödinger equation for the H given in Eq. (5.14) is satisfied for all values of λ. The significance of Eqs. (5.16) is that we can obtain solutions of the next-order terms based upon the information we already have. The solutions for all $a_k^{(0)}$ are clearly constants. Thus, the $a_k^{(0)}$ are just the initial values (i.e. the value of a_k before the electromagnetic radiation is applied.).

Let the particle be initially in the state u_m, which has energy E_m, then $a_m^{(0)} = 1$ and all other $a_n^{(0)} = 0$ for $n \neq m$. We obtain

$$\frac{da_k^{(1)}}{dt} = -\frac{j}{\hbar} H'_{km} e^{j\omega_{km} t}. \tag{5.17}$$

The purpose of the above derivation is now clear. Equation (5.17) allows us to calculate, in a very simple manner, the change in $|a_k^{(1)}|^2$ with respect to time. We need to know its magnitude in order to tell whether the particle is losing or gaining energy. The key quantity is the matrix element, $|H'_{km}|^2$.

5.2.2 Electric and magnetic dipole and electric quadrupole approximations

In most books on quantum mechanics, when the electromagnetic radiation is considered as a classical quantity given by the vector potential $\underline{A}$, the Hamiltonian for the interaction of the particle with the electromagnetic field is given by ref. [5] (see also Appendix 5 of ref. [1]):

$$H' = \frac{je\hbar}{m} \underline{A}(\underline{r}, t) \cdot \nabla, \tag{5.18}$$

where m is the mass of the particle. $\underline{A}$ is, in general, a function of the position of the particle, $\underline{r}$. Since r is small, we can write, in descending order of magnitude,

$$A(\underline{r}, t) = \underline{A}|_{r=0} + \underline{r} \cdot (\nabla \underline{A})|_{r=0} + \cdots. \tag{5.19}$$

Using the first term of the series, we obtain [5]

$$H' = -e\underline{E}(t) \cdot \underline{r}, \tag{5.20}$$

where $\underline{E}$ is the electric field. This H' has the form of the energy of an electric dipole. Thus the matrix element H'_{km} using Eq. (5.20) is called the electric dipole matrix element. It can be shown that the second-order term in Eq. (5.19) gives the magnetic dipole and electric quadrupole matrix elements. They are normally much smaller than the electric dipole matrix element. They are important only when $|H'_{km}|$ in the electric dipole approximation is zero. In that case, we say the electric dipole transition is forbidden. Transitions in which only the electric quadrupole or magnetic dipole terms are non-vanishing are frequently observed experimentally in spectroscopy. However, they are not important in lasers because they correspond to weak transitions. No laser has been known to oscillate in magnetic dipole or electric quadrupole transitions.

5.2.3 Perturbation analysis for an electromagnetic field with harmonic time variation

$\underline{E}$ typically has a harmonic time variation. If we further assume that $\underline{E}$ is polarized in the y direction, then we obtain

$$H'(\underline{r}, t) = H'(\underline{r})e^{-j\omega t} + H'(\underline{r})^* e^{j\omega t}, \tag{5.21a}$$

$$H'(\underline{r}) = -\frac{1}{2} e E_y y. \tag{5.21b}$$

Note that we do not use just the $-eE_y y \exp(-j\omega t)$ term for the time variation because H' must be a Hermitian operator, $H'_{kn} = H'^*_{nk}$.

As a special case, consider that E is turned on at $t = 0$ and $E = 0$ for $t < 0$. Then,

$$a_k^{(1)}(t) = \int_0^t \left(-\frac{j}{\hbar}\right) H'_{km}(t') e^{j\omega_{km}t'} dt'$$
$$= \hbar^{-1}\left[H'_{km}\frac{e^{j(\omega_{km}-\omega)t}-1}{\omega_{km}-\omega} + H'^*_{mk}\frac{e^{j(\omega_{km}+\omega)t}-1}{\omega_{km}+\omega}\right]. \tag{5.22}$$

The probability for the particle to be in the state u_k is $|a_k^{(1)}|^2$. It is likely to be very small unless $|\omega_{km}| \approx \omega$. When $|\omega_{km}| \approx \omega$, Eq. (5.22) can be written approximately as

$$|a_k^{(1)}|^2 = \frac{4|H'_{km}|^2}{\hbar^2}\left\{\frac{\sin^2\left[\frac{1}{2}(\omega_{km}-\omega)t\right]}{(\omega_{km}-\omega)^2} + \frac{\sin^2\left[\frac{1}{2}(\omega_{km}+\omega)t\right]}{(\omega_{km}+\omega)^2}\right\},$$

where the cross product terms have been neglected. The first term is large only when $E_k > E_m$ and $E_k - E_m \approx \hbar\omega$, while the second term is large only when $E_k < E_m$ and $E_m - E_k \approx \hbar\omega$. Thus, the time harmonic perturbation of E can cause both an upward transition, $E_k > E_m$, and a downward transition, $E_k < E_m$. It is a resonant transition because the energy difference $|E_k - E_m|$ must equal approximately $\hbar\omega$, the photon energy of the radiation field. Furthermore, at sufficiently large t, an applicable approximation is

$$\frac{\sin^2\left[\frac{1}{2}(\alpha)t\right]}{\alpha^2} \approx \frac{\pi t}{2}\delta(\alpha)$$

for $\alpha \approx 0$. Therefore,

$$|a_k^{(1)}|^2 = \frac{2\pi}{\hbar^2}|H'_{km}|^2\left[\delta(\omega_{km}-\omega) + \delta(\omega_{km}+\omega)\right]t$$
$$= \frac{2\pi}{\hbar}|H'_{km}|^2\left[\delta(E_k - E_m - \hbar\omega) + \delta(E_m - E_k - \hbar\omega)\right]t. \tag{5.23}$$

The significance of the result given in Eq. (5.23) is that (1) the change of ψ in the first-order approximation, i.e. $|a_k^{(1)}|$, is proportional to the intensity of the radiation field, (2) the magnitude of $|a_k^{(1)}|$ is small unless the matrix element for the electric dipole transition is large, and (3) the energy difference, $|E_k - E_m|$, is equal approximately to $h\nu$ of the radiation. For this reason, such transitions, i.e. the change of $|a_k^{(1)}|^2$ with respect to time, are called induced transitions. Since $h\nu$ is commonly recognized as the energy of the photon of the radiation, and since the total energy of both the particle and the radiation must be conserved, we attribute the absorption process to the particle gaining energy equal to $E_k - E_m$, while the radiation loses its energy equal to $h\nu$. Conversely, in the emission process, the radiation gains a photon while the particle loses its energy equal to $E_m - E_k$. Actually, such a relationship cannot be proven in the semi-classical derivation of $|a_k^{(1)}|$. The concept of the photon comes from the quantized analysis of the radiation field. Such an energy balance emerges naturally from the analysis of induced and spontaneous transitions involving the quantized radiation field.

5.2.4 Induced transition probability between two energy eigen states

There are two practical situations in which Eq. (5.23) will be applied frequently.

(1) Let there be a group of states near E_k, and let the number of states be g_k and the normalized density of states per unit of ω_{km} be $\rho(\omega_{km})$. Then

$$\begin{aligned} |a_k^{(1)}|^2 &= \frac{2\pi t}{\hbar^2} \int_{-\infty}^{+\infty} |H'_{km}|^2 [\delta(\omega_{km} - \omega) + \delta(\omega_{km} + \omega)] g_k \rho(\omega_{km})\, d\omega_{km} \\ &\approx g_k \frac{2\pi |H'_{km}|^2}{\hbar^2} \rho(\omega)\, t, \end{aligned} \tag{5.24}$$

where

$$\int \rho(\omega_{km})\, d\omega_{km} = 1.$$

(2) The location of the energy level E_m is uncertain, and it can be described only by the probability function $f(E_m)\, dE_m$ of being found between E_m and $E_m + dE_m$. Therefore we multiply $|a_k^{(1)}|^2$ in Eq. (5.23) by $f(E_m)\, dE_m$. After integration, we obtain

$$|a_k^{(1)}|^2 = \frac{2\pi |H'_{km}|^2}{\hbar} f(h\nu)\, t,$$

where

$$\int_{-\infty}^{+\infty} f(h\nu)\, d(h\nu) = 1; \tag{5.25}$$

$f(h\nu)$ is a probability distribution function centered about the energy $h\nu_0$.

We can also replace $f(E_m)$ by a corresponding line shape function $g(\nu)$ in the frequency domain,

$$g(\nu) = 2\pi\hbar\, f(h\nu),$$

$$\int_{-\infty}^{\infty} g(\nu)\, d\nu_k = 1,$$

$$|a_k^{(1)}|^2 = \frac{1}{\hbar^2} |H'_{km}|^2 g(\nu) t. \tag{5.26}$$

A typical normalized line shape function called the Lorentz line shape function is

$$g(\nu) = \frac{(\Delta\nu/2\pi)}{(\nu - \nu_0)^2 + (\Delta\nu/2)^2}, \tag{5.27}$$

where ν_0 is the center frequency of the transition. At $\nu = \nu_0$, $g(\nu_0) = 2/\pi\Delta\nu$.

The transition probability per unit time, W_{mk}, is traditionally defined as

$$W_{mk} = \frac{d}{dt}|a_k^{(1)}|^2. \tag{5.28}$$

Therefore, the transition probability W_{km} is independent of time for either case (1) or case (2). W_{km} is zero when $E = 0$. It is dependent on the matrix element of H'_{km} between states u_k and u_m. This is known as the *golden rule* of induced transitions. The W_{km} for an upward transition is the same as for a downward transition. If we accept the concept that when downward transition takes place the particle emits a photon, and a photon is absorbed for the upward transition, then W is the induced emission and the absorption transition probability per unit time for the particle. When H'_{km} is zero for the electric dipole approximation, we say that the electric dipole transition between states u_m and u_k is forbidden.

Two conditions were used in obtaining the results expressed in Eqs. (5.24)–(5.26): (1) $2\pi/t$ is small compared with the width of $\rho(\omega_{km})$ or $1/t \ll \Delta\nu$, so that Eq. (5.24) and Eq. (5.26) are valid; (2) $|a_k^{(1)}| \ll 1$ at $\omega = \omega_{km}$ in Eq. (5.22), so that the perturbation procedure is applicable. Putting the two conditions together, we obtain

$$\frac{|H'_{km}|}{\hbar} \ll \frac{1}{t} \ll \Delta\nu. \tag{5.29}$$

Note again that if the original energy state u_m has $E_m > E_k$, then we have an induced emission, whereas for $E_m < E_k$ we have an induced absorption.

5.3 Macroscopic susceptibilty and the density matrix

Macroscopically, what we can measure is the susceptibility created by the quantum mechanical interactions of many particles in response to an applied electromagnetic field. In this section we will first temporarily set aside all the discussions presented in Section 5.2. Instead, we will show how to calculate this susceptibility, and then we reconcile it with the results obtained in Section 5.2 at the end of this section.

In an isotropic and homogeneous medium, the product of the electric susceptibility and the vacuum electric permittivity represents the proportionality constant between the electric field and the electric polarization. The polarization is equal to the averaged number of particles per unit volume, N, times the statistically averaged electric dipole moment per particle. Therefore our method for calculating the susceptibility consists in calculating the statistically averaged electric dipole moment (in response to the applied electric field) for a particle in a state ψ.

5.3.1 Polarization and the density matrix

We have described previously the expectation value of a dynamic observable such as the dipole polarization of a single particle. According to Eq. (5.3), if

$$\psi = \sum_n C_n(t)\, u_n(\underline{r}),$$

then

$$\langle A \rangle = \sum_{m,n} C_m^* A_{mn} C_n.$$

If we statistically average the $\langle A \rangle$ over different particles, i.e. $\overline{\langle A \rangle}$, we obtain

$$\overline{\langle A \rangle} = \sum_{m,n} \overline{C_m^* C_n} A_{mn} = \sum_{m,n} \rho_{nm} A_{mn} = \sum_n (\|\rho\|\,\|A\|)_{nn} = \mathrm{tr}(\|\rho\|\,\|A\|), \tag{5.30a}$$

where

$$\|\rho\| = \text{density matrix}, \qquad \rho_{nm} = \overline{C_m^* C_n}. \tag{5.30b}$$

In our case, we are interested in the statistically averaged dipole moment μ of the particles induced by the electric field, i.e. $\overline{\langle A \rangle} = \overline{\langle \mu \rangle}$. We obtain the ρ_{nm} by solving the time variation of ρ_{nm} as follows:

$$\frac{\partial \rho_{nm}}{\partial t} = \overline{\left(\frac{\partial C_m^*}{\partial t}\right) C_n} + \overline{C_m^* \frac{\partial C_n}{\partial t}}. \tag{5.31}$$

Since ψ satisfies Eq. (5.1), we can obtain the time variation of $\|\rho\|$ from the time variation of C_n as follows:

$$j\hbar \sum_n \frac{\partial C_n}{\partial t} u_n(\underline{r}) = \sum_n C_n(t) H u_n(\underline{r}).$$

Multiplying by u_m^* and using the orthonormal properties of u_n, we obtain

$$\frac{\partial C_m}{\partial t} = \frac{1}{j\hbar} \sum_n C_n H_{mn}.$$

Therefore,

$$\frac{\partial \rho_{nm}}{\partial t} = \frac{j}{\hbar} \left[\sum_k [\rho_{nk} H_{km} - H_{nk} \rho_{km}] \right]$$

or

$$\frac{\partial}{\partial t}\|\rho\| = \frac{j}{\hbar}[\|\rho\|\,\|H\| - \|H\|\,\|\rho\|]\,. \tag{5.32}$$

Equation (5.32) will be used to solve for $\|\rho\|$. The H, including both the Hamiltonian of the particle and its interaction with the radiation field, was already given in Eq. (5.14). More extensive discussions on the density matrix are given in refs. [6] and [7].

Following the simplification used in Eqs. (5.21b), we consider the electric field E and the dipole to be polarized in the y direction without any loss in generality. Then $\mu = ey$. For the sake of simplicity, we will also only carry out the analysis for the dipole, i.e. $\overline{\langle\mu\rangle}$, of a particle which has only two energy levels, E_2 and E_1, where $E_2 > E_1$. Moreover, we note that because ey is an odd function while $u_k^* u_k$ is always an even function, we obtain the following properties for the matrix element of ey:

(1)
$$\mu_{11} = \mu_{22} = 0. \tag{5.33}$$

(2) Because of the Hermitian property of μ, $\mu_{12} = \mu_{21} = \mu$. When there are N particles per unit volume, we obtain

$$H'_{21} = -E\,(t)\,\mu_{21} = -E\,(t)\,\mu, \tag{5.34a}$$

$$\overline{\langle\mu\rangle} = \mu\,(\rho_{12} + \rho_{21})\,, \tag{5.34b}$$

$$\underline{P} = \varepsilon_0 \chi \underline{E} = N\overline{\langle\mu\rangle}, \tag{5.34c}$$

where E is the electric field in the y direction.

5.3.2 Equation of motion of the density matrix elements

From Eq. (5.32), the equation for ρ_{21} is

$$\begin{aligned}
\frac{d\rho_{21}}{dt} &= -\frac{j}{\hbar}\left\{\sum_k (H_0 + H')_{2k}\rho_{k1} - \sum_k \rho_{2k}(H_0 + H')_{k1}\right\} \\
&= -\frac{j}{\hbar}[H'_{21}\rho_{11} + E_2\rho_{21} - E_1\rho_{21} - \rho_{22}H'_{21}] \\
&= -\frac{j}{\hbar}[-E(t)\mu_{21}\,(\rho_{11} - \rho_{22}) + (E_2 - E_1)\,\rho_{21}]\,.
\end{aligned} \tag{5.35}$$

Similarly,

$$\frac{d\rho_{22}}{dt} = -j\frac{1}{\hbar}[\rho_{21}H'_{12} - H'_{21}\rho_{12}] = -j\frac{E}{\hbar}\mu(\rho_{21} - \rho_{21}^*)$$

and

$$\frac{d}{dt}(\rho_{11} - \rho_{22}) = 2j\frac{\mu}{\hbar}E(t)(\rho_{21} - \rho_{21}^*). \tag{5.36}$$

From Eqs. (5.35) and (5.36), we could obtain a solution for $\|\rho\|$ as a function of an applied electric field E. However, two modifications are needed before we can use the solution.

The first is that Eq. (5.35) is mathematically very similar to the equations of resonant circuits. ρ_{21} has a forced solution and a natural solution. Two observations can be made about the solutions of such an equation. (a) ρ_{21} is small unless $E(t)$ has a harmonic time variation $\exp(j\omega t)$ with $\omega \approx \omega_{21}$. The forced solution always has the time variation of the radiation field which is $\exp(-j\omega t)$. (b) At $E = 0$, the natural solution has the form of a constant times $\exp(-j\omega_{21}t)$, where $\omega_{21} = (E_2 - E_1)/\hbar$. If E is turned off, then ρ_{21} will continue to have the $\exp(j\omega_{21}t)$ variation forever. This result is incorrect because we expect ρ_{21} eventually to decay to zero at thermal equilibrium. This result is obtained because Eq. (5.35) represents ρ without including any interaction of the particles with the surroundings. In reality, particles interact with their neighbors. The neighboring particles can have a mutual exchange of their energy state without changing their total energy, i.e. the individual C_k coefficients are exchanged so that ρ decays eventually to zero without affecting the total energy of the particles. Through such exchange interactions, the ψ will return eventually to a random distribution in the absence of the radiation field, i.e. $\overline{C_m^* C_n} = 0$, for $m \neq n$. For this reason, a relaxation term is added to Eq. (5.35) to obtain

$$\frac{d\rho_{21}}{dt} = -j\omega_{21}\rho_{21} + j\frac{\mu}{\hbar}(\rho_{11} - \rho_{22})\,E(t) - \frac{\rho_{21}}{T_2}. \tag{5.37}$$

Here, T_2 is called the transverse relaxation time, and $1/T_2$ represents the rate at which the $\overline{C_m^* C_n}$ $(m \neq n)$ is relaxed to zero from exchange interactions among neighbors. T_2 is the relaxation time of the off-diagonal elements of the density matrix. It does not affect the total energy of the particles.

The second modification is that Eq. (5.36) predicts that ρ_{11} and ρ_{22} are constants after E becomes zero. Here we have neglected that the particles will also continue to exchange energy with their surroundings through mechanisms such as thermal excitation. Eventually, ρ_{11} and ρ_{22} should return to their thermal equilibrium distribution. Therefore, another decay constant is added to Eq. (5.36):

$$\frac{d}{dt}(\rho_{11} - \rho_{22}) = \frac{2j\mu E(t)}{\hbar}(\rho_{21} - \rho_{21}^*) - \frac{(\rho_{11} - \rho_{22}) - (\rho_{11} - \rho_{22})_0}{\tau}. \tag{5.38}$$

Here, τ is called the longitudinal relaxation time; $1/\tau$ is the rate at which the total energy of the particles returns to thermal equilibrium; $(\rho_{11} - \rho_{22})_0$ is the thermal equilibrium value of $(\rho_{11} - \rho_{22})$; and τ is the relaxation time of the diagonal elements of the density matrix.

5.3.3 Solutions for the density matrix elements

We will now solve for the matrix parameters ρ_{21} and $\rho_{11} - \rho_{22}$ according to Eqs. (5.37) and (5.38). Let

$$E(t) = E_0 \cos \omega t = \frac{E_0}{2}(e^{j\omega t} + e^{-j\omega t}) \tag{5.39a}$$

and

$$\rho_{21}(t) = \sigma_{21}(t)\, e^{-j\omega t} . \tag{5.39b}$$

Then Eqs. (5.37) and (5.38) become

$$\frac{d\sigma_{21}}{dt} = j(\omega - \omega_{21})\sigma_{21} + \frac{j\mu E_0}{2\hbar}(\rho_{11} - \rho_{22}) - \frac{\sigma_{21}}{T_2}$$

and

$$\frac{d}{dt}(\rho_{11} - \rho_{22}) = \frac{j\mu E_0}{\hbar}(\sigma_{21} - \sigma_{21}^*) - \frac{(\rho_{11} \quad \rho_{22}) - (\rho_{11} - \rho_{22}^*)}{\tau} .$$

At the steady state (i.e. when d/dt in the above equations is zero), then ignoring terms that do not have the $\exp(-j\omega t)$ variation, we obtain

$$\begin{aligned} \mathrm{Re}(\sigma_{21}) &= \frac{(\omega_{21} - \omega)\Omega T_2^2(\rho_{11} - \rho_{22})}{1 + (\omega - \omega_{21})^2 T_2^2} \\ &= \frac{(\omega_{21} - \omega) T^2 \Omega(\rho_{11} - \rho_{22})_0}{1 + (\omega - \omega_{21})^2 T_2^2 + 4\Omega^2 T_2 \tau} , \end{aligned} \tag{5.40a}$$

$$\begin{aligned} \mathrm{Im}(\sigma_{21}) &= \frac{\Omega T_2(\rho_{11} - \rho_{22})}{1 + (\omega - \omega_{21})^2 T_2^2} \\ &= \frac{\Omega T_2\,(\rho_{11} - \rho_{22})_0}{1 + (\omega - \omega_{21})^2\, T_2^2 + 4\Omega^2 T_2 \tau} , \end{aligned} \tag{5.40b}$$

$$(\rho_{11} - \rho_{22}) = (\rho_{11} - \rho_{22})_0 \frac{1 + (\omega - \omega_{21})^2\, T_2^2}{1 + (\omega - \omega_{21})^2\, T_2^2 + 4\Omega^2 T_2 \tau} , \tag{5.40c}$$

where $\Omega = \mu\, E_0/2\hbar$ and $\sigma_{21} = \sigma_{12}^*$.

For the special case of a material with only two energy levels, $\rho_{11} > \rho_{22}$ in thermal equilibrium with $E_1 < E_2$, there is only net stimulated absorption. For optically active laser materials, there are other energy levels and pumping methods that are available to create a $(\rho_{11} - \rho_{22})_0 > 0$. Then we may have net stimulated emission. These methods will be discussed in Chapter 6.

5.3.4 Susceptibility

Results obtained from Eqs. (5.34) and (5.40) allow the calculation of the polarization P as follows. Let

$$E(t) = \mathrm{Re}[E_0 e^{j\omega t}] \quad \text{and} \quad \chi = \chi' - j\chi''.$$

Then

$$\begin{aligned} P(t) &= \mathrm{Re}[\varepsilon_0 \chi\, E_0 e^{j\omega t}] = E_0 \varepsilon_0 (\chi' \cos\omega t + \chi'' \sin\omega t) \\ &= N\mu(\rho_{12} + \rho_{21}) = N\mu[\sigma_{21}^* e^{j\omega t} + \sigma_{21} e^{-j\omega t}] \\ &= 2N\mu\, \mathrm{Re}(\sigma_{21}) \cos\omega t + 2N\mu\, \mathrm{Im}(\sigma_{21}) \sin\omega t. \end{aligned}$$

Hence,

$$\begin{aligned} \chi'' &= \Delta N_0 \frac{\mu^2 T_2}{\varepsilon_0 \hbar} \frac{1}{1 + 4\Omega^2 T_2 \tau + (\omega - \omega_{21})^2 T_2^2} \\ &= \Delta N(\nu) \frac{\mu^2 T_2}{\varepsilon_0 \hbar} \frac{1}{1 + (\omega - \omega_{21})^2 T_2^2} = \Delta N(\nu) \frac{\mu^2}{2\varepsilon_0 \hbar} g(\nu) \end{aligned} \tag{5.41a}$$

and

$$\begin{aligned} \chi' &= \Delta N_0 \frac{\mu^2 T_2^2}{\varepsilon_0 \hbar} \frac{(\omega_{21} - \omega) T_2}{1 + 4\Omega^2 T_2 \tau + (\omega - \omega_{21})^2 T_2^2} \\ &= \Delta N(\nu) \frac{\mu^2 T_2^2}{\varepsilon_0 \hbar} \frac{(\omega_{21} - \omega) T_2}{1 + (\omega - \omega_{21})^2 T_2^2} \\ &= \Delta N(\nu) \frac{\mu^2}{2\varepsilon_0 \hbar} \left(\frac{1}{\pi\, \Delta\nu} \right) (\omega_{21} - \omega)\, g(\nu), \end{aligned} \tag{5.41b}$$

where $\Delta N = N(\rho_{11} - \rho_{22})$, $\Delta N_0 = N(\rho_{11} - \rho_{22})_0$,

$$g(\nu) = \frac{(\Delta\nu/2\pi)}{(\nu - \nu_{21})^2 + (\Delta\nu/2)^2},$$

and $2\pi\nu = \omega$, where $\Delta\nu$ = the line width at half width half maximum $= 1/(\pi\, T_2)$, and

$$\frac{\chi'}{\chi''} = \frac{2}{\Delta\nu}(\nu_{21} - \nu) = 2\pi\, T_2 (\nu_{21} - \nu). \tag{5.41c}$$

Here, $g(\nu)$ is the Lorentz line shape function given in Eq. (5.27). Figure 5.1 illustrates the χ' and χ'' as a function of the radiation frequency ν. Note in particular the resonance effect at $\nu \approx \nu_{21}$ and the change of sign of χ' as ν is scanned past ν_{21}. See also refs. [1] and [8] for discussions of the susceptibility due to induced transitions.

Independently of this calculation and according to the general theory of solid state physics (see Appendix 1 of ref. [1] for a general derivation), all χ'' and χ' are

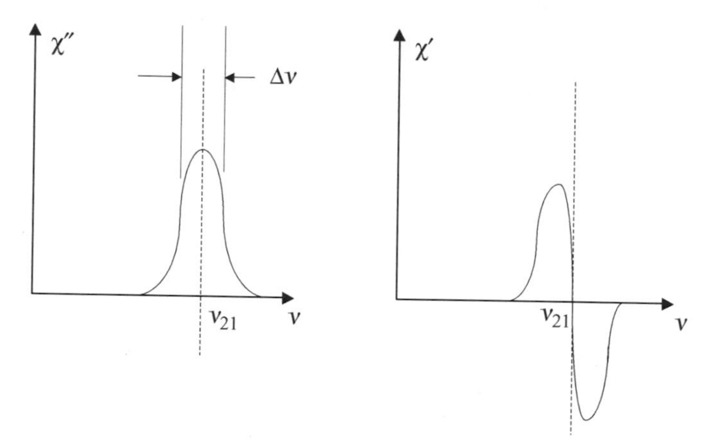

Figure 5.1. Unsaturated χ'' and χ' for a transition between two energy states at E_1 and E_2. The χ'' has a peak at ν_{21}, while the χ' has a zero at ν_{21}. The line width of the transition is $\Delta\nu$. χ'' and χ' are related by the Kramers–Kronig relations in Eqs. (5.42).

always related by the Kramers–Kronig relations as follows:

$$\chi' = \frac{1}{\pi} PV \int_{-\infty}^{\infty} \frac{\chi''(\omega')}{\omega' - \omega} d\omega'$$
$$\chi'' = -\frac{1}{\pi} PV \int_{-\infty}^{\infty} \frac{\chi'(\omega')}{\omega' - \omega} d\omega'. \tag{5.42}$$

where PV stands for the Cauchy principal value of the integral that follows. Clearly, the χ' and χ'' given in Eqs. (5.41c) satisfy Eqs. (5.42). The practical significance of the Kramers–Kronig relations is that, whereas χ'' can often be measured experimentally, χ' can be calculated from the experimentally measured χ''.

5.3.5 Significance of the susceptibility

The impact of χ' and χ'' on electromagnetic waves propagating in such a material can be most clearly illustrated via an example. For a plane wave in a medium with susceptibility χ we have

$$E(z,t) = \mathrm{Re}\left[E e^{j(\omega t - k'z)}\right],$$
$$k' = \omega\sqrt{\mu_0 \varepsilon'},$$
$$\varepsilon' = \varepsilon\,(\text{material}) + \varepsilon_0 \chi\,(\text{induced transition}),$$
$$\varepsilon = n^2 \varepsilon_0,$$
$$k' \approx \omega\sqrt{\mu_0 \varepsilon}\left[1 + \frac{\varepsilon_0}{2\varepsilon}(\chi' - j\chi'')\right] \approx k_0\left(1 + \frac{\chi'}{2n^2}\right) - jk_0\left(\frac{\chi''}{2n^2}\right);$$

therefore

$$E(z,t) = \text{Re}\left[E e^{j\omega t} e^{-jk_0\left(1+\frac{\chi'}{2n^2}\right)z} e^{\left(\frac{\gamma z}{2}\right)} \right], \tag{5.43}$$

where

$$\gamma = -\frac{k_0 \chi''}{n^2},$$

$$I(z) = \frac{\frac{1}{2}}{\sqrt{\mu_0/\varepsilon}} |E|^2 = I_0 e^{\gamma z},$$

$$\frac{dI}{dz} = \gamma I.$$

Therefore a negative χ'' will signify a growing plane wave, and a positive χ'' will signify a decaying plane wave. On the other hand, χ' affects the phase velocity, or the effective index, of the plane wave propagating in this medium. In other words, the amplification properties of the amplitude of electromagnetic modes in lasers will be given by χ'', while the phase of the modes will be affected by χ'. Since all laser oscillators require a cavity resonance, the phase velocity affects the resonance frequency of the laser.

5.3.6 Comparison of the analysis of χ with the quantum mechanical analysis of induced transitions

The quantum mechanical analysis in Section 5.2 showed that, for a particle initially in the state with energy E_m, the probability of finding the particle in the state with energy E_k (initially empty) is proportional to time due to the presence of the radiation. The rate of the induced transition, i.e. W_{mk}, is large only when the frequency of the radiation field ν is approximately $|E_k - E_m|$. W_{mk} is proportional both to the intensity of the radiation and, in the-first-order approximation, to the magnitude squared of the electric dipole matrix element of the two energy states U_k and U_m. The W_{mk} is the same no matter whether E_k is higher or lower than E_m. However, we cannot analyze the effect of the atomic particle on the radiation field by semi-classical quantum mechanical analysis. We can only make the conjecture that, based upon the principle of energy balance, a photon of the radiation energy is absorbed from the radiation field for $E_k > E_m$, and a photon of energy is emitted to the radiation field for $E_k < E_m$. On the other hand, the analysis of the susceptibilty of the material presented in this section did not tell us anything about what happened to the atomic particles. It described only the χ experienced by the radiation field. There are amplification (or attenuation) and change of phase velocity for an electromagnetic wave propagating through such media when $|E_k - E_m| \approx h\nu$. No comments can be made about the emission or absorption of a photon from atomic particles. Only when we combine the two analyses can we say that the atomic particles have a change in probability of being in the state with energy E_k while the radiation field experiences amplification (or attenuation) and a change of effective index.

A comparison of these two results is highly instructive. (1) The average energy per unit volume per unit time absorbed (or emitted) by atomic particles calculated according to the quantum mechanical results of transition probabilities. (2) The power per unit volume lost (or gained) by the radiation field due to the attenuation (or amplification) caused by χ''. For a propagating plane electromagnetic wave with a harmonic time variation, its time averaged power is related to its zero-to-peak electric field E_y by

$$I = \frac{1}{\sqrt{\mu_0/\varepsilon}} \frac{|E_y|^2}{2}.$$

According to Eq. (5.28), for a two energy level system, the net energy per unit volume per unit time gained by an atomic particle is

$$\begin{aligned}\Delta N W_{21} \hbar\omega_{21} &= \Delta N\, \hbar\omega \frac{1}{4\hbar^2} E_y^2 \mu^2 g(\nu) \\ &= \Delta N \sqrt{\frac{\mu_0}{\varepsilon}} \frac{\mu^2 \omega}{2\hbar} I g(\nu).\end{aligned}$$

According to Eq. (5.43), the power per unit volume lost (or gained) by the radiation field is

$$\begin{aligned}\gamma I &= -\frac{k_0}{n^2} \Delta N \frac{\mu^2}{2\varepsilon_0 \hbar} g(\nu) I \\ &= -\Delta N \sqrt{\frac{\mu_0}{\varepsilon}} \frac{\mu^2 \omega}{2\hbar} I g(\nu).\end{aligned}$$

Clearly the power absorbed (or emitted) by atomic particles is equal to the power lost (or gained) by the radiation field.

5.4 Homogeneously and inhomogeneously broadened transitions

In the discussion about susceptibility, we have shown in Eqs. (5.41) that χ'' will have a resonant response at $\nu \approx \nu_{21}$. The unsaturated (i.e. $4\Omega^2 T_2 \tau \ll 1$) full-width-half-maximum linewidth for this resonance is

$$\Delta\nu = 1/\pi\, T_2.$$

The stronger the interaction with neighboring atoms, the shorter the T_2 and the wider the linewidth. However, we have assumed that T_2 is the same for all particles. Thus the transition has a homogeneously broadened line. The result is derived for a single pair of energy levels, E_1 and E_2. This is equivalent to the case of induced transition probability described in Eq. (5.25) where there is uncertainty of E_m caused by the finite lifetime of the particles in that state. However, there are also situations in which different atomic particles see a slightly different surrounding environment. For example, the thermal energy of the particles may be different, or different particles might see varying quantities of crystalline electric field. We will also discuss in Chapter 7 that, in semiconductors, different electrons have different states. The total effective χ is then the summation of the individual components χ. In this case, the transition has an inhomogeneously broadened line shape. The case described by Eq. (5.24) could be considered as the quantum mechanical analog of an inhomogeneously broadened line where g_k and H'_{km} can be considered independent of ω_{km}. Transitions with inhomogeneously broadened lines typically do not have the Lorentz line shape shown by $g(\nu)$ in Eqs. (5.41). For an inhomogeneously broadened line, the use of a Lorentz line shape is a gross simplification.

Equations (5.41a) and (5.41b) also showed that the magnitude of Ω will affect χ' and χ''. Ω is proportional to the magnitude of the electric field E. This is called the saturation effect. The effect of saturation will be different for homogeneously and for inhomogeneously broadened lines. Therefore, in the following we will discuss separately the inhomogeneously and the homogeneously broadened lines. In particular, we have seen in Section 2.2 that the resonance frequencies of the longitudinal modes of the laser cavity are separated by $c/2D$. For sufficiently long D, the resonance frequency separation is smaller than the line width of the transition. There may be competition among adjacent modes caused by saturation. Therefore, we need to discuss how the saturation caused by a strong radiation at ν will affect the χ seen by another radiation at ν'. Homogeneous and inhomogeneous broadenings are also discussed in ref. [1].

5.4.1 Homogeneously broadened lines and their saturation

Consider first the homogeneously broadened line that is described by Eqs. (5.41a) and (5.41b). The susceptibility of a homogeneously broadened transition is dependent on the magnitude of $|E|^2$ or I. There are three cases to be considered.

(1) When $4\Omega^2 T_2 \tau$ is much smaller than unity,

$$\left.\begin{aligned}\chi'' &= \Delta N_0 \frac{\mu^2}{2\varepsilon_0 \hbar} g(\nu), \\ \chi' &= \Delta N_0 \frac{\mu^2}{2\varepsilon_0 \hbar} \frac{2(\nu_{21} - \nu)}{\Delta\nu} g(\nu),\end{aligned}\right\} \tag{5.44}$$

with

$$g(\nu) = \frac{(\Delta\nu/2\pi)}{(\nu - \nu_{21})^2 + (\Delta\nu/2)^2} \quad \text{and} \quad \Delta\nu = 1/\pi\, T_2.$$

In this case the intensity of the radiation has no effect on χ. We expect that this is what will be observed experimentally in a weak radiation field.

(2) In this case, $4\Omega^2 T_2 \tau$ is comparable to or larger than unity. According to Eqs. (5.41), if we write χ in the form of a normalized line shape function, we have

$$\left.\begin{aligned} \chi'' &= \Delta N_0 \frac{\mu^2}{2\varepsilon_0\hbar} \frac{1}{\sqrt{1+4\Omega^2 T_2\tau}} \frac{\left(\dfrac{\Delta\nu_s}{2\pi}\right)}{(\nu-\nu_{21})^2 + \left(\dfrac{\Delta\nu_s}{2}\right)^2}, \\ \chi' &= \Delta N_0 \frac{\mu^2}{2\varepsilon_0\hbar} \frac{2\pi\, T_2}{\sqrt{1+4\Omega^2 T_2\tau}} \frac{(\nu_{21}-\nu)\left(\dfrac{\Delta\nu_s}{2}\right)}{(\nu-\nu_{21})^2 + \left(\dfrac{\Delta\nu_s}{2}\right)^2}, \end{aligned}\right\} \tag{5.45}$$

with

$$\Delta\nu_s^2 = \frac{1}{\pi^2 T_2^2} + \frac{4\Omega^2\tau}{\pi^2 T_2} \quad \text{or} \quad \Delta\nu_s = \Delta\nu\sqrt{1+4\Omega^2 T_2\tau}. \tag{5.46a}$$

Thus the first effect of large E^2 is an increase of the linewidth to $\Delta\nu_s$. The second effect is a reduction of ΔN_0 to ΔN in Eqs. (5.41). According to Eq. (5.40c),

$$\begin{aligned} \Delta N &= \Delta N_0 \frac{1}{1 + \dfrac{4\Omega^2 T_2\tau}{1 + 4(\nu-\nu_{21})^2 (1/\Delta\nu^2)}} \\ &= \Delta N_0 \frac{1}{1 + \dfrac{I(\nu)}{I_s(\nu)}}. \end{aligned} \tag{5.46b}$$

Therefore, we can write

$$\gamma = \frac{\gamma_0(\nu)}{1 + \dfrac{I(\nu)}{I_s(\nu)}}. \tag{5.47}$$

Here, I_s is given by

$$I_s = \frac{1 + 4(\nu-\nu_{21})^2 \dfrac{1}{\Delta\nu^2}}{\dfrac{2\tau\mu^2}{\pi\,\Delta\nu\,\hbar^2}\sqrt{\dfrac{\mu_0}{\varepsilon}}}. \tag{5.48}$$

Equation (5.47) is a description of how the unsaturated $\gamma_0(\nu)$ will be changed by saturation. It involves a change in both magnitude and frequency variation. In the extreme, Ω is so large that γ and ΔN are reduced to zero. Equations (5.45) show that for the same $|E|^2$, the saturation effect is largest when $\nu = \nu_{21}$. The smaller the $\Delta\nu$, the sharper the optical frequency (or wavelength) dependence near resonance.

(3) Let a homogeneously broadened line be irradiated by a strong radiation at ν. If we now probe the particles with a second, weak radiation at ν', then the radiation at ν' will have a χ'' which will be given by

$$\chi'' = \frac{\mu^2}{2\varepsilon_0\hbar}\Delta N_0 \frac{1}{1 + \dfrac{4\Omega^2 T_2\tau}{1 + 4\pi^2 T_2^2(\nu - \nu_{21})^2}} \frac{(\Delta\nu/2\pi)}{(\nu' - \nu_{21})^2 + (\Delta\nu/2)^2} \tag{5.49a}$$

and

$$\chi' = \chi''\left[\frac{2}{\Delta\nu}(\nu_{21} - \nu')\right]. \tag{5.49b}$$

Note that, according to the last factor of Eq. (5.49a), the χ'' seen by the radiation at ν' has the unsaturated linewidth $\Delta\nu$. The saturation effect of the intense radiation at ν is only to reduce ΔN, not to broaden the line shape for the radiation at ν'. Figure 5.2(a) illustrates the difference among the χ'' as a function of ν for an unsaturated transition and the saturated χ'' for a radiation field at its own frequency ν. Figure 5.2(b) illustrates the cases of the χ'' for a weak radiation field at ν', where the saturation is caused by either a strong radiation at ν_1 or a strong radiation at ν_2, where the strong radiation has identical intensity. Since $|\nu_1 - \nu_{21}| < |\nu_{21} - \nu_2|$, the saturation effect of the strong radiation at ν_1 is larger.

5.4.2 Inhomogeneously broadened lines and their saturation

A very different saturation effect is created in inhomogeneously broadened transitions. Typically, particles in a solid medium see a slightly different crystalline field. The particles in a gaseous medium may have different thermal velocities. In this case, different particles may have slightly different ν_{21}. The total χ'' at ν is the sum of the contributions of all the component χ'' from different sets of particles with different center atomic transition frequency ν_ξ. Let the normalized distribution of the component transitions with center frequency ν_ξ be $p(\nu_\xi)$. Similarly to the case of homogeneous broadening, we will consider three situations: (1) no saturation, (2) saturation effect at the frequency of the strong radiation ν, and (3) the effect on χ at ν' while the saturation is created by a strong radiation at ν.

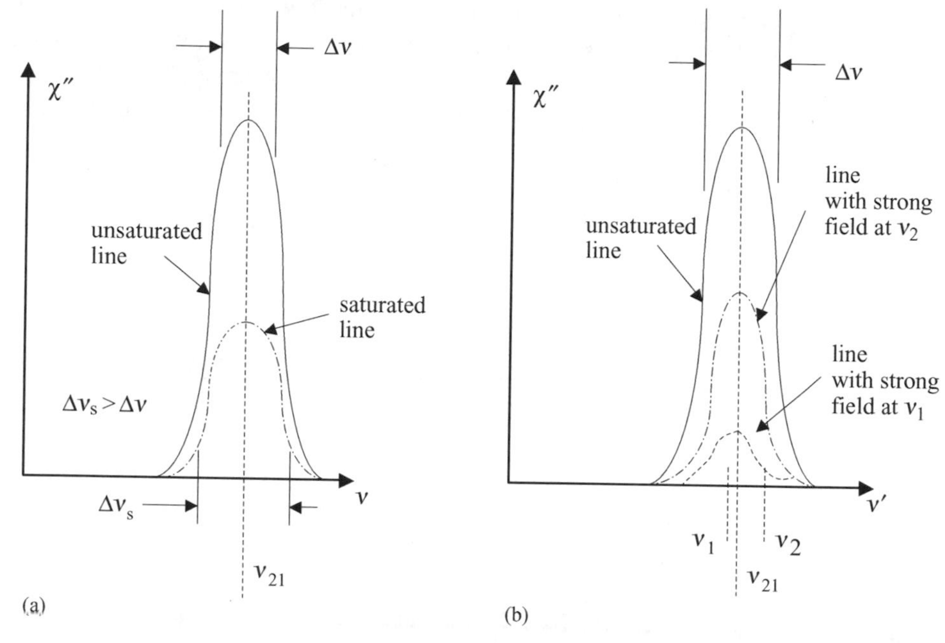

Figure 5.2. Saturation effect in a homogeneous broadened line. (a) The saturation of χ'' of a strong radiation as a function of its ν. For a strong radiation, the saturation effect of χ'' is exhibited as a reduction of the peak value and a broadening of the transition with a larger line width $\Delta\nu_s$. (b) The saturation of χ'' of a weak radiation as a function of its ν' when a strong radiation is at ν_1 or at ν_2. For a strong radiation at ν_2, the saturation effect of the χ'' of a weak radiation is exhibited as a reduction of the peak value with no change in its line width $\Delta\nu$. For a strong radiation with the same intensity at ν_1, where ν_1 is closer to the center of the transition ν_{21} than ν_2, the reduction of χ'' is even larger without any change of $\Delta\nu$.

When the radiation is weak so that the saturation effect can be neglected, we obtain for each component of χ centered about ν_ξ,

$$\chi''_\xi(\nu, \nu_\xi) = \frac{\Delta N_0 \mu^2}{2\varepsilon_0 \hbar} \frac{\dfrac{(\Delta\nu)}{2\pi}}{\left(\dfrac{\Delta\nu}{2}\right)^2 + (\nu - \nu_\xi)^2}.$$

Thus, the total χ' and χ'' are

$$\begin{aligned} \chi'' &= \frac{\Delta N_0 \mu^2 (\Delta\nu)}{4\pi\varepsilon_0\hbar} \int_{-\infty}^{\infty} \frac{p(\nu_\xi)}{\left(\dfrac{\Delta\nu}{2}\right)^2 + (\nu - \nu_\xi)^2} d\nu_\xi, \\ \chi' &= \frac{\Delta N_0 \mu^2}{2\pi\varepsilon_0\hbar} \int_{-\infty}^{+\infty} \frac{p(\nu_\xi)(\nu_\xi - \nu)}{\left(\dfrac{\Delta\nu}{2}\right)^2 + (\nu - \nu_\xi)^2} d\nu_\xi, \end{aligned} \tag{5.50}$$

where

$$\int_{-\infty}^{\infty} p(\nu_\xi)\, d\nu_\xi = 1.$$

For slowly varying $p(\nu_\xi)$ and small $\Delta\nu$, $1/[(\Delta\nu/2)^2 + (\nu - \nu_\xi)^2]$ in Eqs. (5.50) can be approximated as $(2\pi/\Delta\nu)\,\delta(\nu - \nu_\xi)$. Thus,

$$\chi'' = \frac{\Delta N_0 \mu^2}{2\,\varepsilon_0\,\hbar}\, p(\nu). \tag{5.51}$$

No simple answer exists for χ', since $p(\nu_\xi)$ can have various distribution functions and χ' might have a very complex line shape. However, the total effective χ' and χ'' are still related by the Kramers–Kronig relationship given in Eqs. (5.42).

Now let us consider the saturation effect. Let there be a strong radiation at ν. Then, the component of χ'' which has the center frequency ν_ξ will be saturated like any homogeneously broadened line, as given in Eqs. (5.41) and (5.45),

$$\chi''_\xi(\nu, \nu_\xi) = \frac{\Delta N_0 \mu^2}{2\varepsilon_0 \hbar} \frac{(\Delta\nu/2\pi)}{(\Delta\nu_s/2)^2 + (\nu - \nu_\xi)^2}.$$

The component χ''_ξ is broadened by the strong radiation like any homogeneously broadened line with $\Delta\nu_s > \Delta\nu$. Thus the total χ is

$$\left.\begin{aligned}
\chi'' &= \frac{\Delta N_0 \mu^2 (\Delta\nu)}{4\pi\,\varepsilon_0 \hbar} \int_{-\infty}^{\infty} \frac{p(\nu_\xi)}{\left(\frac{\Delta\nu_s}{2}\right)^2 + (\nu - \nu_\xi)^2}\, d\nu_\xi, \\
\chi' &= \frac{\Delta N_0 \mu^2}{2\pi\,\varepsilon_0 \hbar} \int_{-\infty}^{+\infty} \frac{p(\nu_\xi)(\nu_\xi - \nu)}{\left(\frac{\Delta\nu_s}{2}\right)^2 + (\nu - \nu_\xi)^2}\, d\nu_\xi,
\end{aligned}\right\} \tag{5.52}$$

where

$$\int_{-\infty}^{\infty} p(\nu_\xi)\, d\nu_\xi = 1.$$

Comparing the χ'' results given in Eqs. (5.50) and (5.52), we see that the saturated transition has the same functional relation in the denominator of the integral as the unsaturated transition except for a larger $\Delta\nu_s$. For reasonably small $\Delta\nu_s$, $1/[(\Delta\nu_s/2)^2 + (\nu - \nu_\xi)] \approx (2\pi/\Delta\nu_s)\delta(\nu - \nu_\xi)$. Thus, for a slowly varying $p(\nu_\xi)$, the p function can be considered to be approximately a constant, and therefore can be taken outside the integral. In that case, we obtain for χ'' at frequency ν:

$$\chi''(\nu) = \frac{\Delta N_0 \mu^2 (\Delta\nu)\, p(\nu)}{4\pi\varepsilon_0 \hbar} \int_{-\infty}^{\infty} \frac{d\nu_\xi}{\left(\frac{\Delta\nu_s}{2}\right)^2 + (\nu - \nu_\xi)^2}$$

$$= \frac{\Delta N_0 \mu^2 p(\nu)}{2\varepsilon_0 \hbar} \frac{1}{\sqrt{1 + \frac{I}{I'_s}}}, \qquad (5.53)$$

$$\frac{I}{I'_s} = 4\Omega^2 T_2 \tau \quad \text{or} \quad I'_s = \frac{\hbar^2}{2T_2\, \tau \mu^2} \sqrt{\frac{\varepsilon}{\mu_0}}.$$

Note that, unlike the homogeneously broadened line, the line shape $p(\nu)$ is not affected by the saturation in the inhomogeneously broadened line.

For the third case, let an inhomogeneously broadened line be saturated by an intense radiation at ν. From Eqs. (5.40) and (5.41), we conclude that there is a reduction of ΔN at ν_ξ caused by this radiation,

$$\frac{\Delta N(\nu, \nu_\xi)}{\Delta N_0} = \frac{(\Delta\nu/2)^2 + (\nu - \nu)^2}{(\Delta\nu_s/2)^2 + (\nu - \nu_\xi)^2}.$$

For the second weak radiation at ν', the contribution to χ'' by the $\chi''(\nu_\xi)$ component is

$$\chi''_\xi(\nu') = \frac{\mu^2}{2\varepsilon_0 \hbar} \Delta N(\nu, \nu_\xi) \frac{(\Delta\nu/2\pi)}{(\nu' - \nu_\xi)^2 + (\Delta\nu/2)^2}.$$

Therefore, the total χ'' is

$$\chi''(\nu, \nu') = \frac{\Delta N_0 \mu^2}{2\varepsilon_0 \hbar} \int_{-\infty}^{\infty} p'(\nu_\xi) \frac{(\Delta\nu/2\pi)}{(\Delta\nu/2)^2 + (\nu' - \nu_\xi)^2}\, d\nu_\xi, \qquad (5.54)$$

where

$$p'(\nu_\xi) = p(\nu_\xi) \frac{(\Delta\nu/2)^2 + (\nu - \nu_\xi)^2}{(\Delta\nu_s/2)^2 + (\nu - \nu_\xi)^2}.$$

If we assume once more that $p'(\nu_\xi)$ is a slowly varying function compared with $1/[(\Delta\nu/2)^2 + (\nu' - \nu_\xi)^2]$, and as long as $|\nu - \nu'| > \Delta\nu/2$, then $p'(\nu_\xi)$ can be assumed to have a constant value, $p'(\nu')$, and p' can be taken out of the integral. In that case,

$$\chi''(\nu, \nu') = \frac{\Delta N_0 \mu^2}{2\varepsilon_0 \hbar} p'(\nu'). \qquad (5.55)$$

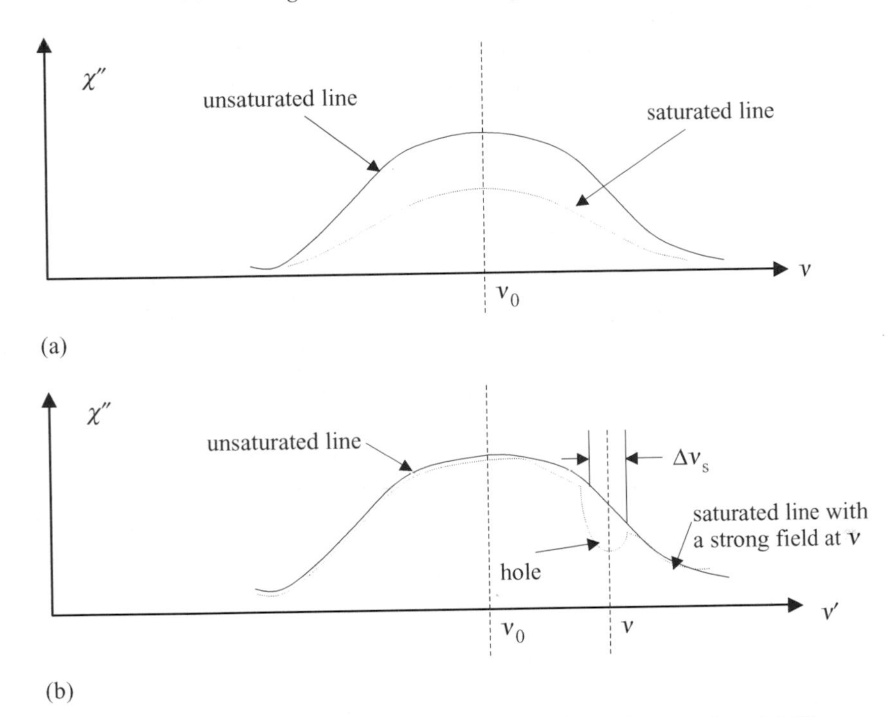

Figure 5.3. Saturation effect in an inhomogeneous broadened line. (a) The saturation of χ'' of a strong radiation as a function of its ν. ν_0 is the center frequency of the transition. For a small $\Delta\nu_s$ of the saturation of component transitions and a total transition that has a slowly varying unsaturated line shape, the magnitude of χ'' is reduced, but the line shape is the same as that of the unsaturated transition. (b) The saturation of χ'' of a weak radiation as a function of its ν' when a strong radiation is at ν. The saturation effect of χ'' is small when ν' is far away, $|\nu - \nu'| > \Delta\nu_s/2$. The saturation effect is large when ν' is close to ν. This uneven saturation effect is called the "hole burning" of the inhomogeneously broadened line.

Note how a homogeneously broadened line and an inhomogeneously broadened line saturate differently. (1) For a strong radiation interacting with a homogeneously broadened line, its γ is reduced from its unsaturated values according to Eq. (5.47), where I_s is dependent on frequency. Notice the change in the line shape, i.e. the broadening of the transition, as the saturation takes place. For a strong radiation interacting with an inhomogeneously broadened line, there is no change in the line shape prescribed by $p(\nu)$. Its γ is reduced by Eq. (5.53), which has a square root dependence on I. The I_s' is independent of frequency. (2) For a weak radiation at ν', the strong radiation at ν will reduce the value of γ of the homogeneously broadened line, but the line width for the radiation at ν' is the unsaturated $\Delta\nu$. For a weak radiation at ν' with a strong radiation at ν in an inhomogeneous broadened line, the reduction of γ (i.e. χ'') is not uniform, it is large only for those ν' close to ν. We know that the reduction is the largest at $\nu = \nu'$. At this frequency, the weak radiation sees the same χ'' as the strong radiation shown in Eq. (5.53). As ν moves away from ν', the reduction of γ from its unsaturated value is smaller. At $|\nu - \nu'| > \Delta\nu_s/2$, the $p(\nu')$ is hardly affected by the strong radiation. This is known as hole burning. We are unable to obtain an analytical expression for the line shape of the "hole" at $\nu \approx \nu'$ and within $\Delta\nu/2$ from ν. We know that only when $|\nu-\nu'|$ is smaller than $\Delta\nu_s$ will the saturation be significant. For this reason, $\Delta\nu_s$ is used as a measure of the hole width. Figure 5.3 illustrates the χ'' for three situations of the inhomogeneous broadened line: (1) the unsaturated line, (2) the saturated line for a radiation at its own frequency ν and (3) the response to a weak radiation at ν' with saturation caused by strong radiation at ν.

References

1 A. Yariv, *Quantum Electronics*, New York, John Wiley and Sons, 1989
2 R. T. Hecht, *Quantum Mechanics*, New York, Springer-Verlag, 2000
3 L. I. Schiff, *Quantum Mechanics*, New York, McGraw-Hill, 1968
4 W. Greiner, *Quantum Mechanics, An Introduction*, New York, Springer-Verlag, 1989
5 W. S. C. Chang, *Quantum Electronics*, Sections 5.4 and 5.5, Reading, MA, Addison-Wesley Publishing Co., 1969
6 L. E. Reichl, *A Modern Course in Statistical Physics*, Chapter 7, Sections E and F, Austin, University of Texas Press, 1980
7 R. K. Pathria, *Statistical Mechanics*, Chapter 5, New York, Pergamon Press, 1972
8 S. L. Chuang, *Physics of Optoelectronic Devices*, Appendix 1, New York, John Wiley and Sons, 1995

6

Solid state and gas laser amplifier and oscillator

6.1 Rate equation and population inversion

In a material which has only two energy levels, χ'' is always positive because $\rho_{11} > \rho_{22}$ for $E_1 < E_2$ at thermal equilibrium. Prior to the invention of lasers, there was no known method to achieve $\rho_{22} > \rho_{11}$. However, we now know that a negative χ'' in Eq. (5.41a) (i.e. $\rho_{22} > \rho_{11}$) can be achieved by pumping processes that are available in materials that have multiple energy levels, as described in the following.

Let there be many energy levels in the material under consideration, as shown in Fig. 6.1. Let there be a mechanism in which the populations at E_1 and E_2, i.e. $N_1 = N\rho_{11}$ and $N_2 = N\rho_{22}$, are increased by pumping from the ground state at pump rates R_1 and R_2. In solid state lasers, the pumping action may be provided by an intense optical radiation causing stimulated transition between the ground state and other higher energy states, where the particles in the higher energy states relax preferentially into the E_2 state. In gas lasers, the molecules in the ground state may be excited into higher energy states within a plasma discharge; particles in those higher energy states then relax preferentially to the E_2 state. Alternatively, collisions with particles of other gases may be utilized to increase the number of particles in the E_2 state. Various schemes to pump different lasers are reviewed in ref. [1]. In order to obtain amplification, it is necessary to have $R_2 \gg R_1$. In general, N_2 and N_1 can be calculated by the rate equation,

$$\begin{aligned} \frac{dN_2}{dt} &= R_2 - \frac{N_2}{t_2} - \left(N_2 - \frac{g_2}{g_1}N_1\right) g_1 W_i(\nu), \\ \frac{dN_1}{dt} &= R_1 - \frac{N_1}{t_1} + \frac{N_2}{t_{21}} + \left(N_2 - \frac{g_2}{g_1}N_1\right) g_2 W_i(\nu), \end{aligned} \tag{6.1}$$

where t_2 is the lifetime of particles in the upper level E_2 and t_1 is the lifetime of the particles in the lower level E_1; $1/t_{21}$ is the rate at which particles in the upper level E_2 make a transition to the lower level E_1; the total numbers of states at E_2 and E_1 are g_2 and g_1, called the degeneracies of E_2 and E_1, respectively; W_i is the

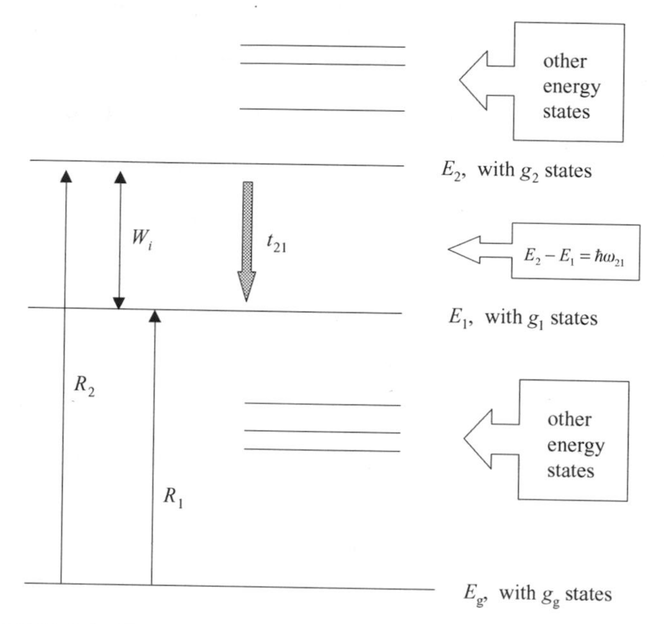

Figure 6.1. Pumping process in a material with many energy levels. This is a typical pumping scheme for a four-level laser. The four levels consist of the pump energy levels, the upper laser level E_2, the lower laser level E_1 and the ground level E_g. The rate at which the population of E_2 is increased is R_2. R_2 includes processes in which population from the ground state E_g is pumped to other energy states above E_2, called the pump energy levels, and then the population in those states is relaxed from pump energy levels to E_2. R_1 is the rate at which the population of E_1 is increased. Populations of the energy levels below E_1 usually have no effect on pumping.

transition probability, shown in Eq. (5.28) for $g_1 = g_2 = 1$; $1/t_2$ consists of $1/t_{21}$ and transition rates to other energy levels; $1/t_{21}$ consists of the transition rate $1/t_{\text{spont}}$ due to spontaneous radiation and the transition rate $(1/t)_{\text{nonrad}}$ due to non-radiative mechanisms. In other words,

$$\frac{1}{t_2} = \frac{1}{t_{21}} + \text{transition rates to other levels},$$
$$\frac{1}{t_{21}} = \frac{1}{t_{\text{spont}}} + \left(\frac{1}{t_{21}}\right)_{\text{nonrad}}.$$

At the steady state (i.e. $d/dt = 0$),

$$\Delta N_{\text{a}} = N_2 - \frac{g_2}{g_1} N_1 = \frac{R_2 t_2 - (R_1 + \delta\, R_2)\, t_1 \dfrac{g_2}{g_1}}{1 + \left[t_2 + (1-\delta)\, t_1 \dfrac{g_2}{g_1}\right] g_1 W_i},$$

$$\delta = \frac{t_2}{t_{21}} < 1.$$

When we compare the ΔN given in Eqs. (5.41) with the ΔN_a given in the above equation, we notice the difference in the negative sign and in g_2/g_1. We have only considered a single energy state in each energy level for the χ given in Eqs. (5.41). In addition, even for the case of $g_1 = g_2 = 1$, $\Delta N_a = -\Delta N$, because $\rho_{11} > \rho_{22}$ in ΔN while $\rho_{11} < \rho_{22}$ in ΔN_a. In other words, ΔN is used for the expression of χ under absorption, while ΔN_a is used for χ under amplification. The subscript "a" stands for amplification. In the absence of a strong radiation field at $\omega \sim \omega_{21}$, i.e. without saturation,

$$\Delta N_a = \frac{\Delta N_{0a}}{1 + \phi\, t_{21} g_1 W_i}, \tag{6.2}$$

where

$$\Delta N_{a0} = \left(N_2 - \frac{g_2}{g_1} N_1\right)_{a0} = R_2 t_2 - (R_1 + \delta\, R_2) t_1 \frac{g_2}{g_1},$$

$$\phi = \delta \left[1 + (1-\delta) \frac{t_1 g_2}{t_2 g_1}\right].$$

In the simple case of $R_1 = 0$ and $t_2 = t_{21}$,

$$\Delta N_{a0} = R_2 \left(t_2 - t_1 \frac{g_2}{g_1}\right).$$

The zero subscript designates the unsaturated value. For amplification of waves propagating through this medium, we need $\Delta N_{a0} > 0$. Whenever ΔN_{a0} is larger than zero, the population distribution, i.e. $N_2 - (g_2/g_1)N_1$, is said to be inverted. Rate equations are discussed in more detail in Chapter 7 of ref. [1]. References [1] and [2] are comprehensive general references on lasers.

6.2 Threshold condition for laser oscillation

Let us now consider a TEM_{00} mode propagating in a laser cavity. When we combine the results obtained in Section 2.2 and Eq. (5.43), we obtain the total phase shift and attenuation (or amplification) for making a round trip around the cavity:

$$e^{-j\theta} = e^{-2j[k'D - j\frac{\alpha}{2}D - \tan^{-1}(z_2/z_0) + \tan^{-1}(z_1/z_0)]} r_1 r_2 e^{-j(\theta_{m1}+\theta_{m2})}, \tag{6.3}$$

where D is the length of the cavity; k' is given in Section 5.3.5; $\alpha/2$ is the equivalent distributed amplitude decay rate per unit distance of propagation caused by both the diffraction loss per single pass and any other propagation loss mechanism such as scattering; r_1 and r_2 are the amplitude reflectivities of the two mirrors, and θ_{m1} and θ_{m2} are the reflection phase shifts at the two mirrors. However, for an amplifying medium, we now have a positive γ in Eq. (5.43), and we have

$$\gamma = -\frac{k_0 \chi_a''}{n^2},$$

and for homogeneous broadened lines,

$$\chi_a'' = -\Delta N_a(\nu) \frac{\mu^2}{2\varepsilon_0 \hbar} g(\nu),$$

$$\Delta N_a = N_2 - \frac{g_2}{g_1} N_1.$$

In order to oscillate, there are two threshold conditions: (1) the round trip phase shift must be integer multiples of 2π; (2) the total amplitude gain must be unity, or larger. In other words,

$$e^{-j\theta} = e^{-j2q\pi}, \tag{6.4a}$$

which means, for the real part of $e^{-j\theta}$, at the resonance frequency of the mode,

$$e^{(\gamma_t - \alpha)D} r_1 r_2 = 1 \tag{6.4b}$$

or

$$\gamma_t = \alpha - \frac{1}{D} \ln r_1 r_2, \tag{6.4c}$$

$$\gamma_t = \frac{k_0 \mu^2}{2n^2 \varepsilon_0 \hbar} (\Delta N_{a0})_t \, g(\nu), \tag{6.4d}$$

for homogenous broadened lines. γ_t is the value of γ when pumping has just produced enough population inversion, i.e. $(\Delta N_{a0})_t$, to satisfy the condition of oscillation. Clearly, for an inhomogeneous broadened line, $g(\nu)$ will be replaced by $p(\nu)$ from Eq. (5.51). Please also note that, before oscillation occurs, there is no saturation effect, i.e. $W_i \approx 0$ and $\Delta N_{a0} \approx \Delta N_a$.

Resonance occurs when the phase θ is $2q\pi$. This means that, at the frequency ν of the laser oscillation in the TEM_{00q} mode, we have

$$\frac{2\pi \nu n D}{c}\left[1 + \frac{\chi'}{2n^2}\right] - \tan^{-1}\frac{z_2}{z_0} + \tan^{-1}\frac{z_1}{z_0} + \frac{\theta_{m1} + \theta_{m2}}{2} = q\pi. \tag{6.5}$$

The equation for the resonance frequency of the TEM_{lmq} mode is similar to Eq. (6.5). The two arc tangent terms in that case are multiplied by $(l + m + 1)$.

When we calculated the resonance frequency of the cavity mode in Chapter 2, we did not include χ'. Let us call that the "cold" cavity resonance frequency ν_{00q} for the TEM_{00q} mode. For example, for $\chi' = 0$, the difference of the resonance frequencies of adjacent longitudinal modes (i.e. modes with $\Delta q = 1$) of the same lmth transverse order is the same. When we include χ', the ν of the laser oscillator will be close to, but not equal to, ν_{00q}. This shift in ν is called frequency pulling of the mode. The pulling on the resonance frequency can be expressed simply as follows:

$$\nu_{00q} = \frac{qc}{2nD} + \frac{c}{2\pi nD}\left(\tan^{-1}\frac{z_2}{z_0} - \tan^{-1}\frac{z_1}{z_0} - \frac{\theta_{m1}+\theta_{m2}}{2}\right),$$

$$\nu\left[1 + \frac{\chi_a'}{2n^2}\right] = \nu_{00q}.$$

From Eqs. (5.41) and (5.43), we have, for a homogeneous broadened line,

$$\frac{\chi_a'}{2n^2} = -\frac{(\nu_{21}-\nu)}{n^2\Delta\nu}\frac{n^2\gamma}{k_0}.$$

Therefore, at $\gamma = \gamma_t$ and assuming $\gamma_t(\nu) = \gamma_t(\nu_{00q})$,

$$\nu = \nu_{00q}\left[1 - \frac{\nu_{21}-\nu}{\Delta\nu}\frac{\gamma_t(\nu)}{k_0}\right]^{-1} = \nu_{00q} + (\nu_{21}-\nu)\frac{c\gamma_t(\nu_{00q})}{2\pi n\Delta\nu}$$

$$= \nu_{00q} - (\nu_{21}-\nu)\frac{1}{\Delta\nu}\frac{c\left[\alpha - \frac{1}{D}\ln(r_1 r_2)\right]}{2\pi n}. \tag{6.6}$$

It can be shown that the full width of the cold cavity optical resonance without quantum mechanical interaction is

$$\Delta\nu_{1/2} = \frac{\nu_q}{Q} = \frac{c\left[\alpha - \frac{1}{D}\ln(r_1 r_2)\right]}{2\pi n}. \tag{6.7}$$

Thus, the oscillation frequency of the laser is

$$\nu = \nu_{00q} - (\nu - \nu_{21})\frac{\Delta\nu_{1/2}}{\Delta\nu}. \tag{6.8}$$

The oscillation frequency of the TEM_{lmq} mode is also given by Eq. (6.8) when ν_{00q} is replaced by ν_{lmq} of the cold cavity. $\Delta\nu$ has been given in Eqs. (5.41).

6.3 Power and optimum coupling for CW laser oscillators with homogeneous broadened lines

For any optical resonant mode at frequency ν, if its γ is larger than γ_t, the amplitude of that mode will grow as it propagates within the cavity. Usually, the growth of the optical mode is initiated by noise (i.e. spontaneous emission). The magnitude of initial unsaturated γ depends on the intensity of the pump. The energy of all other modes which do not satisfy the condition $\gamma > \gamma_t$ will remain at the noise level. As the amplitude (E) of the optical wave in that mode grows, W_i will also increase. Equation (6.2) shows that ΔN (i.e. γ) will begin to saturate as W_i increases.

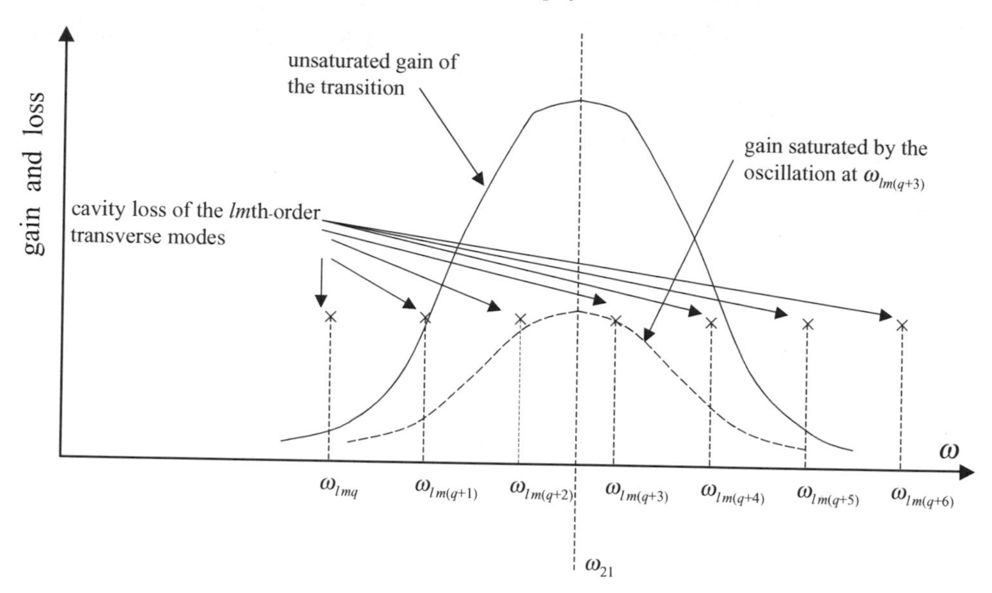

Figure 6.2. Saturated and unsaturated gain profile of a homogeneous transition and the losses of the *lm*th-order transverse modes. The unsaturated gain of the transition is shown together with the losses of the TEM_{lmq} modes that have different longitudinal orders. The $TEM_{lm(q+3)}$ mode is the mode to oscillate first because the unsaturated gain is the largest at its resonance frequency. Its oscillation creates a saturated steady state gain curve as illustrated. With the saturation, the losses of the other modes are now larger than the saturated gain, and they will not oscillate.

However, even with the saturation, as long as the γ is still larger than γ_t, the E of that mode and W_i will continue to increase. The saturation of γ will increase until γ is reduced to γ_t. At any time, if γ becomes less than γ_t, the saturation effect will reduce, and γ will increase again. Eventually, $\gamma(\text{saturated}) \equiv \gamma_t$, and equilibrium is reached. In other words, at the steady state, γ is pinned at γ_t. It follows that the mode that has the lowest α and the resonance frequency ν closest to the center of transition frequency will oscillate first.

The unsaturated gain profile of a homogeneously broadened line centered at $\omega_{21} = 2\pi\nu_{21}$ is depicted in Fig. 6.2. Let us assume that the TEM_{lm} modes have the lowest γ_t. All longitudinal modes of the same transverse order will have the same γ_t. The gains required for the oscillation of TEM_{lmq} at various longitudinal orders are marked as crosses at their resonance frequencies in Fig. 6.2. The oscillation will begin first in the mode with resonance frequency $\omega_{lm(q+3)}$, which is closest to the peak transition frequency ω_{21}, i.e. the mode with the largest unsaturated gain. Once this mode is in oscillation, saturation occurs. The saturation reduces the gain available to all other modes at resonance frequencies further away from ω_{21}. The saturated gain when the mode at $\omega_{lm(q+3)}$, is oscillating is also shown in Fig. 6.2. With saturation, the oscillation requirement $\gamma > \gamma_t$ can no longer be met for all other modes. In short, for a homogeneous broadened line, the mode with the resonance frequency closest to ω_{21} will have the largest unsaturated γ. The mode with the smallest γ_t and the largest unsaturated γ will oscillate first. The saturation effect created by the oscillation of this mode will be such that no other longitudinal or transverse mode will meet the condition (saturated γ) $> \gamma_t$. In other words, there is only one steady state mode that will oscillate in a strictly homogeneously broadened line.

We can analyze the internal field and the power output of laser oscillation in a homogeneously broadened line as follows. For the oscillating mode, where $\gamma = \gamma_t$ is required, we obtain, from Eq. (5.47),

$$\gamma D = \frac{\gamma_0 D}{1 + \frac{I}{I_s}} = \alpha D - \ln(r_1 r_2) = L_i + T,$$

where

$$\begin{aligned} \gamma_0 D = g_0 &= \text{unsaturated gain per pass}, \\ L_i &= \text{internal loss factor} = \alpha D, \\ T &= \text{mirror transmission} = -\ln(r_1 r_2), \end{aligned}$$

and I and I_s are explained in Eqs. (5.43), (5.47) and (5.48). Thus, the intensity I of the optical wave traveling inside the cavity is locked in by the saturation effect with

$$I = \left(\frac{g_0}{L_i + T} - 1 \right) I_s.$$

This is an important point. It means that, before laser oscillation can be achieved, we are concerned with obtaining $g_0/D > \gamma_t$. Once the oscillation has occurred, the I/I_s is determined by g_0, L and T. The net power per unit volume, P_e, emitted by stimulated emission, is γI, and the useful total output power, P_o is

$$\begin{aligned} P_o &= P_e \cdot \text{volume} \cdot \frac{T}{L_i + T} \\ &= \text{volume} \cdot \frac{(L_i + T)}{D} \left(\frac{g_0}{L_i + T} - 1 \right) I_s \cdot \frac{T}{(L_i + T)} \\ &= \frac{\text{volume} \cdot I_s}{D} \left(\frac{g_0}{L_i + T} - 1 \right) T. \end{aligned} \tag{6.9a}$$

Maximizing P_o with respect to T (i.e. setting $dP_o/dT = 0$) yields

$$T_{op} = -L_i + \sqrt{g_0 L_i}.$$

This result is, naturally, applicable only for homogeneously broadened lines because of the saturation relationship we have used here.

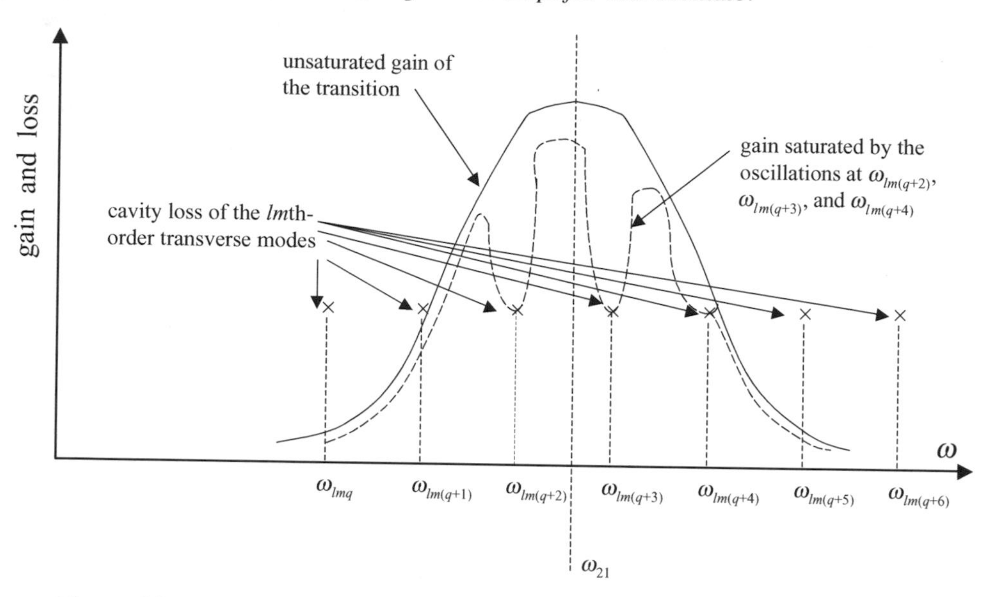

Figure 6.3. Saturated and unsaturated gain profiles of an inhomogeneous broadened transition and losses of the lmth-order transverse modes. There are three oscillating modes at $\omega_{lm(q+2)}$, $\omega_{lm(q+3)}$ and $\omega_{lm(q+4)}$. There are also three holes in the saturated gain curve. The gain of the saturated gain curve is pinned to the value of the loss of the cavity resonance mode at each hole.

6.4 Steady state oscillation in inhomogeneously broadened lines

The situation is very different in inhomogeneously broadened lines. Note the hole burning effect of a strong radiation at ν in an inhomogeneously broadened line depicted in Fig. 5.3(b). This implies that the saturation due to the oscillation of the mode at ν will not affect the gain of another mode at ν' with resonant frequency sufficiently separated from the oscillating mode. Therefore, for inhomogeneously broadened lines, there is likely to be a number of oscillating modes.

An example of the saturated profile of γ as a function of ω for an inhomogeneously broadened line is shown by the dashed curve in Fig. 6.3. We have again assumed that there is only one order of transverse mode, TEM_{lm}, which has a sufficiently low γ_t to be considered for oscillation. The γ_t of various longitudinal modes at $\omega_{lm(q+j)}$ are shown as crosses in Fig. 6.3. The resonance frequencies of the longitudinal modes are separated sufficiently far apart compared with $\Delta\nu_s$ that the saturation of one longitudinal mode does not reduce γ sufficiently to prevent the oscillation of the adjacent longitudinal mode. Consequently, there are three oscillating modes within the laser transition, and there are three holes in the profile of saturated γ. The γ curve is pinned to γ_t at each hole. However, those longitudinal modes at resonance frequencies smaller than $\omega_{lm(q+2)}$ and larger than $\omega_{lm(q+4)}$ will not oscillate because $\gamma < \gamma_t$ at those frequencies. If there are other modes that also have low γ_t and have resonance frequencies sufficiently close to these oscillating modes within $\Delta\nu_s$, there will then be competition among modes.

The optimum coupling of the cavity for maximizing its power output will also be different than the T_{op} given in Section 6.3 because the saturation of the inhomogeneously broadened line is different from that of the homogeneously broadened line. Consider the simple case where there is only one cavity resonance within the laser transition. Similarly to the example in Section 6.3, let

$$\gamma_0 D = g_0 = \text{unsaturated gain per pass},$$
$$L_i = \text{internal loss factor} = \alpha D,$$
$$T = \text{mirror transmission} = -\ln(r_1 r_2).$$

Then the saturation effect of γ is

$$\gamma D = \frac{\gamma_0 D}{\sqrt{1 + \frac{I}{I_s'}}} = L_i + T$$

or

$$\Delta N = \frac{\Delta N_0}{\sqrt{1 + \frac{I}{I_s'}}},$$

and hence

$$\frac{I}{I_s'} = \frac{\gamma_0^2 D^2 - (L_i + T)^2}{(L_i + T)^2},$$

where I_s' was explained in Eq. (5.53) and I was given in Eq. (5.43). The emitted power per unit volume, P_e, is γI. Therefore, the total output power from the laser cavity is

$$P_o = \text{volume} \cdot \gamma I \cdot \frac{T}{L_i + T} = \left(\frac{\text{volume} \cdot I_s'}{D}\right)\left[\left(\frac{\gamma_0 D}{L_i + T}\right)^2 - 1\right] T. \quad (6.9b)$$

In order to maximize P_o, we let $\partial P_o / \partial T = 0$. We obtain the equation

$$(\gamma_0 D)^2 (L_i - T_{op}) = (L_i + T_{op})^2.$$

The solution for T_{op} in the above equation will maximize P_o.

6.5 *Q*-switched lasers

The requirement that the steady state gain of the laser medium is locked to the threshold value limits the laser output. If the quality factor Q of laser resonant modes can be held first to a low value, then their oscillation threshold (i.e. the saturation of γ) will not be reached despite the very large γ which can be obtained by strong pumping. When this low Q factor is switched suddenly to the normal value at $t = 0$, then the initial γ greatly exceeds the γ_t of the cavity momentarily at that $t = 0_+$. Within a short period of time at $t > 0$, the amplitudes of a number of resonant modes, i.e. the total stored energy in the laser cavity, will build up quickly to a very large value. As the stored energy is building up to such a large amplitude, the initial population inversion is exhausted by the stimulated emission, i.e. the γ is reduced, and eventually γ drops below the threshold. Therefore, after the initial buildup, the stored energy of the cavity will begin to decay. During this transient period, the peak stored energy in the cavity will temporarily reach a very high value, and part of this stored energy is transmitted as the output. This is known as Q switching. Various practical methods of and variations from Q switching are discussed in Chapter 8 of ref. [1], and they are summarized in Section 20.1 of ref. [3]. It is a technique used to obtain a pulse of high laser power. There usually are many oscillating modes, because many modes have similar losses, and many modes have resonance frequencies well within the line width of the transition. It is impossible to discriminate one mode from another. Only modes with very large losses and modes with resonance frequencies far away from the center frequency of the transition will not be induced to oscillate.

From our discussion on the Q of the cavity, we know that

$$\frac{dE_{\text{cavity}}}{dt} = -\frac{1}{t_c} E_{\text{cavity}},$$

$$Q = -\frac{\omega E_{\text{cavity}}}{dE_{\text{cavity}}/dt} = \omega t_c = \frac{\omega n D}{c[\alpha D - \ln(r_1 r_2)]}.$$

Let the total number of photons in the oscillating modes be $\phi(t)$, the average mode volume be V, the total inverted number of particles be $n_p(t)$,

$$n_p(t) \equiv \left[N_2 - \frac{g_2}{g_1} N_1 \right] V,$$

and the average mode decay constant be t_c. At $t > 0$, the intensity of the contrapropagating optical waves will grow with distance, $dI/dz = \gamma I$. An observer traveling with the waves will see the growth of I in time as

$$\frac{dI}{dt} = \frac{dI}{dz}\frac{dz}{dt} = \gamma \frac{c}{n} I.$$

If the length of the laser amplifying medium is L ($<D$), then only a fraction of the photons is undergoing amplification, and the average growth rate is $(L/D)(\gamma c/n)$. Balancing the decay rate of the photons with the amplification rate, we obtain the following [4]:

$$\frac{d\phi}{dt} = \left[\frac{\gamma c L}{n D} - \frac{1}{t_c} \right] \phi.$$

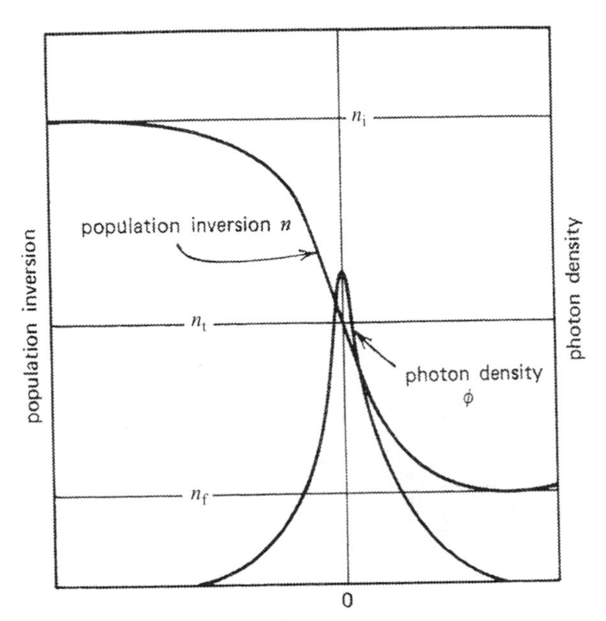

Figure 6.4. Population inversion and photon density during a giant pulse. The total inverted population n_p drops from the initial value n_i before the pulse to a final value n_f after the pulse. The total number of photons in oscillating modes ϕ rises from zero to its peak when n_p is at the threshold value n_t; then ϕ decays back to zero. Taken from ref. [4] with permission from the American Institute of Physics.

Note that, for $d\phi/dt = 0$, the threshold $\gamma_t = (nD/cLt_c)$ is just what balances the cavity decay with the gain. Using a normalized time, $\tau = t/t_c$, we can rewrite the above equation in a normalized form:

$$\frac{d\phi}{d\tau} = \left(\frac{\gamma}{\gamma_t} - 1\right)\phi.$$

Since γ is proportional to the population inversion,

$$\frac{d\phi}{d\tau} = \left(\frac{n_p}{n_t} - 1\right)\phi, \tag{6.10}$$

where $n_t = (\Delta N_a)_t V$ is the total inversion required at threshold and n_p is the total inverted number of atoms in the cavity at any instant of time t. The first term in Eq. (6.10) describes the rate of increase of the total number of photons in the cavity. Since each generated photon results from a decrease of total population inversion of $\Delta n_p = 2$ in a single transition, we obtain:

$$\frac{dn_p}{d\tau} = -2\phi\frac{n_p}{n_t}. \tag{6.11}$$

Figures 6.4 and 6.5 show the numerically calculated n_p and ϕ as functions of τ. Both figures are taken from ref. [4]. Note that ϕ reaches a maximum when $n_p = n_t$.

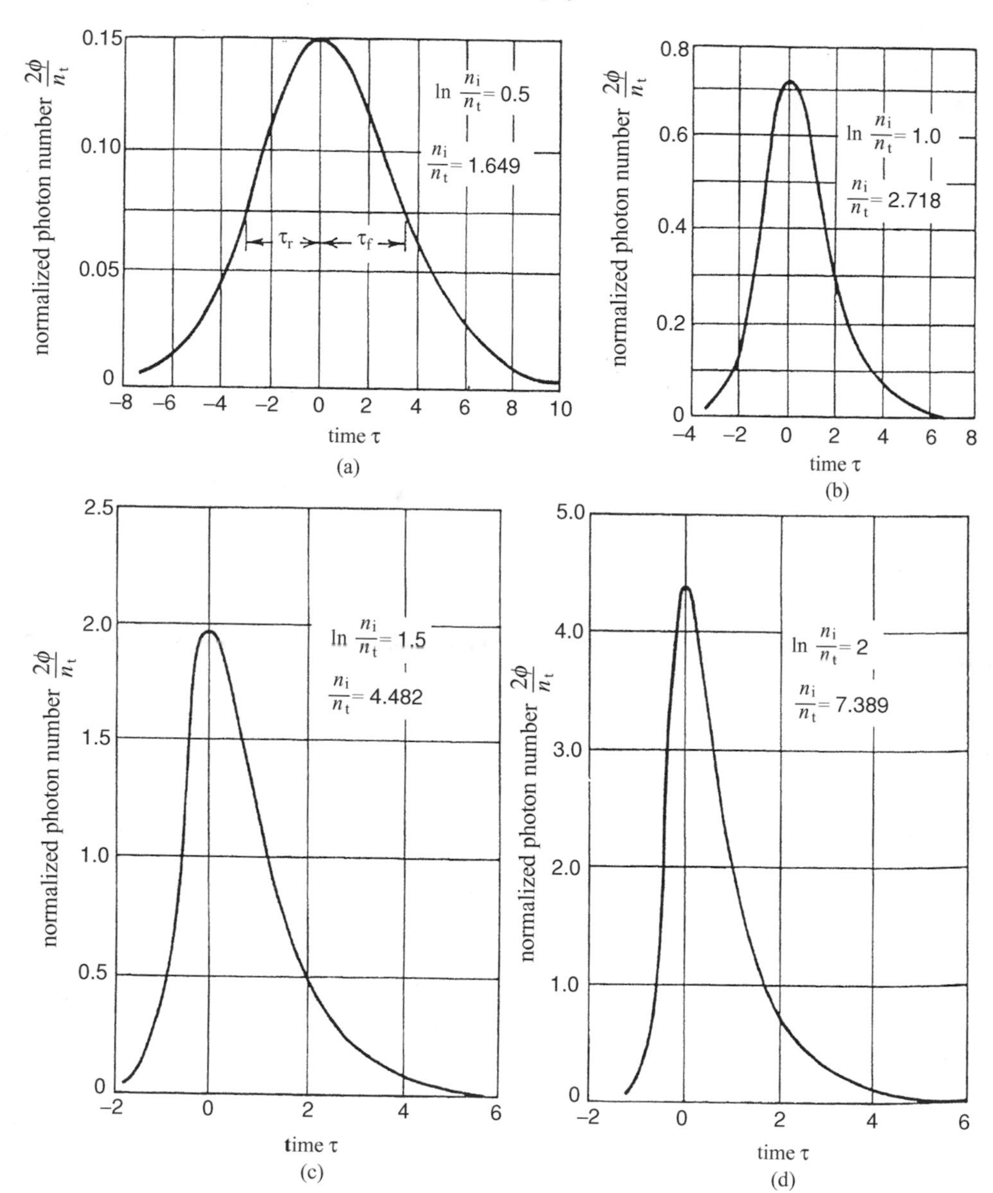

Figure 6.5. Normalized photon number versus time in a Q-switched giant pulse for various n_i/n_t. The pulse width of the total photon numbers ϕ depends on the initial population n_i. The larger the n_i/n_t, the sharper the pulse, and the higher the peak ϕ. Time is measured in units of photon lifetime τ. This figure is taken from ref. [4] with permission from the American Institute of Physics.

Note also the importance of having a large n_i/n_t ratio in order to achieve a large output and a sharp pulse. If we divide Eq. (6.10) by Eq. (6.11), we obtain

$$\frac{d\phi}{dn_p} = \frac{n_t}{2n_p} - \frac{1}{2}, \tag{6.12}$$

and, by integration,

$$\phi - \phi_i = \frac{1}{2}\left[n_t \ln\frac{n_p}{n_i} - (n_p - n_i)\right].$$

Mathematically, n_i and ϕ_i are integration constants; they are the initial values at $t = 0$. We will assume that $\phi_i = 0$. At $t \gg t_c$, again $\phi = 0$ and $n_p = n_f$, where n_f is the final population inversion after the transient. Thus, from Eq. (6.12) we obtain

$$\frac{n_f}{n_i} = \exp\left(\frac{n_f - n_i}{n_t}\right). \tag{6.13}$$

We note that the fraction of the energy initially stored in the inversion that is converted into laser oscillation energy is $(n_i - n_f)/n_i$. Figure 6.6 (taken from ref. [4]) plots the energy utilization factor as a function of n_i/n_t; it approaches unity as n_i/n_t increases. If we neglect cavity losses other than the transmission, the instantaneous power output will be given by $P = \phi h\nu/t_c$. We can find the maximum of P by setting $\partial P/\partial n_p = 0$; it occurs at $n_p = n_t$, as shown in Fig. 6.4. When $n_i \gg n_t$,

$$P_{max} \approx \frac{n_i h\nu}{2t_c}. \tag{6.14}$$

In cavities that have significant internal losses L_i, the output power P_o is related to P by

$$(P_o)_{max} = \frac{T}{T + L_i} P_{max}.$$

In continuous wave (CW) lasers, only the lower order modes will have the larger t_c or the smaller γ_t. Thus, the output radiation will consist primarily of the superposition of lower order oscillating modes. Most commonly, we would like to have only one transverse order mode to oscillate. Sometimes, only one oscillating mode, i.e. one transverse and one longitudinal order mode, is desirable. It is interesting to note here that we have not even discussed the characteristics of Q-switched lasers in terms of individual modes of the laser cavity. When $n_i \gg n_t$, there are many oscillating modes. It is more meaningful to discuss the total energy of all the modes than the energy in each mode. However, modes closer to the center frequency of the atomic transition will have a much larger intensity because of the larger γ. For these reasons, the Q-switched laser does not have a single-frequency output. It is used primarily in applications where a lot of pulsed power is required,

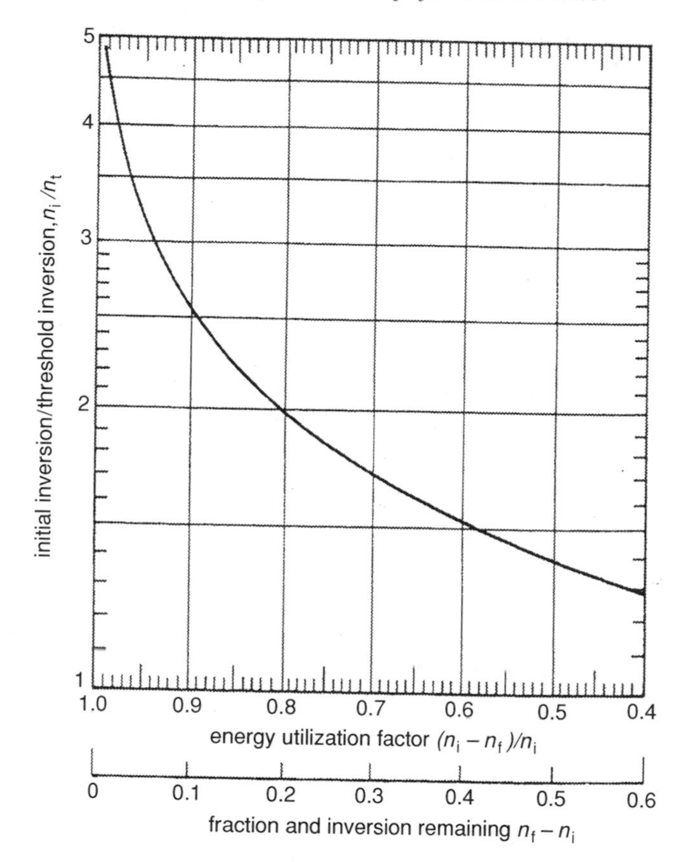

Figure 6.6. Energy utilization factor $(n_i - n_f)/n_i$ and residual inversion after the giant pulse. The fraction of energy stored in the total inverted population used to generate the pulse $(n_i - n_f)/n_i$ drops dramatically from unity to a low value as the n_i/n_f ratio decays to less than 2. When the n_i/n_t ratio drops, a larger fraction of the population inversion, n_f/n_i, remains after the pulse. This figure is taken from ref. [4] with permission from the American Institute of Physics.

without any precise control of its phase and frequency characteristics. The analysis of the Q-switched laser is interesting academically, because it demonstrates how a gross analysis of ϕ can be handled very simply by rate equations without any explicit information about the diffraction loss, the amplitude and the phase of the modes.

6.6 Mode locked laser oscillators

It is well known in Fourier analysis that when there are a number of Fourier terms with identical amplitude and phase, as well as equal frequency spacing between adjacent terms, the summation of all the terms will have a periodic time variation of sharp pulses. In other words, if we simply add the fields of a number of oscillating

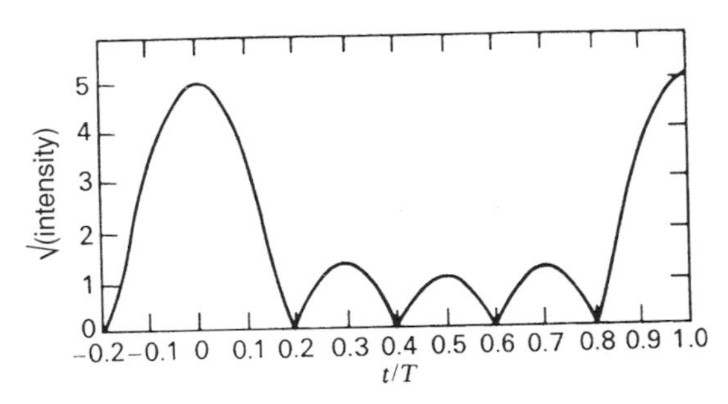

Figure 6.7. Theoretical plot of the time variation of the total optical field of five modes with equal amplitude and equal frequency spacing, $2\pi/T$, locked together. This figure is taken from ref. [3] with copyright permission from John Wiley and Sons.

modes with identical amplitude and phase, as well as equal spacing in frequency, we obtain mathematically

$$E = \pm \sum_{-(N-1)/2}^{+(N-1)/2} A e^{j(\omega_0 + n\omega + \phi_n)t} = e^{j\omega_0 t} e^{j\phi_n} A \frac{\sin(N\omega t/2)}{\sin(\omega t/2)}, \qquad (6.15)$$

where A and ϕ_n are the amplitudes and phases of all the modes, ω_0 is the center frequency and ω is the frequency spacing between adjacent modes. $E(t)$ is now periodic in $T = 2\pi/\omega$. The power, which is proportional to E^*E, is proportional to $\sin^2(N\omega t)/\sin^2(\omega t)$. Therefore, we can say:

(1) The total power is emitted in the form of a pulse train with $T = 2\pi/\omega$.
(2) The peak power is N times the average power. N is the total number of modes.
(3) The peak field amplitude is N times the amplitude of a single mode.
(4) The individual pulse width, defined as the time from the peak to the first zero, is $\tau = T/N$. Thus the pulse can be very narrow with large N.

Figure 6.7 (from Section 20.2 of ref. [3]) illustrates the time variation of the amplitude of five equally spaced modes locked together.

6.6.1 Mode locking in lasers with an inhomogeneously broadened line

When we analyzed the resonant modes in a cold cavity without the χ' of the quantum mechanical transition in Chapter 2, we found that the longitudinal modes of the same transverse order are spaced equally apart in frequency, with $\Delta\omega = 2\pi/T = \pi c/nD$ (or $T = 2Dn/c$), where n is the refractive index without the χ' contribution from the quantum mechanical transition and D is the length of the optical cavity. In inhomogeneous broadened lines, we could have many oscillating modes. The total number of oscillating modes N is approximately $\Delta\nu(2nD/c)$, which is equal to the line width $\Delta\nu$ of the quantum mechanical transition divided by the frequency spacing of the longitudinal modes. However, the longitudinal mode frequency spacing is no longer equal because of the mode pulling effect of the χ' contribution from the quantum mechanical transition. There is also no specific fixed phase relationship among the oscillating modes. Now we wish to show that by actively modulating the gain of the medium at a frequency close to $\Delta\omega$, we can induce the modes to oscillate with equal frequency spacing and fixed phase through non-linear interactions. Similar analysis shows that phase modulation can also achieve mode locking in inhomogeneous broadened lines.

It is interesting to analyze the mode locking in an inhomogeneously broadened line as follows (see Section 20.3 of ref. [3]). In a laser, let E be the total electric field that satisfies the wave equation. We can express the imaginary part of the χ as a conductivity term in the wave equation. There are two parts of the conductive term. The first part is the unmodulated gain or loss of the laser media; the second part is the modulated gain, used to achieve mode locking. Therefore, we obtain

$$\nabla^2 \underline{E} - \mu\sigma(\underline{r}, t)\frac{\partial E}{\partial t} - \mu\varepsilon\frac{\partial^2 E}{\partial t^2} = 0, \qquad \sigma(r, t) = \sigma_0 + \sigma_{\mathrm{m}} \cos\omega_{\mathrm{m}} t f(\underline{r}), \tag{6.16}$$

where σ_0 is the distributed equivalent gain or loss of the laser and σ_{m} is the amplitude of the gain modulation at frequency ω_{m} and with spatial variation $f(\underline{r})$. E can be expressed as a superposition of normalized "cold" cavity modes (see Eq. (2.3)),

$$E = \sum_s A_s(t)\, E_s(\underline{r}) e^{j\omega_s t}, \tag{6.17}$$

where

$$\nabla^2 E_a(\underline{r}) + \omega_a^2 \mu\varepsilon E_a(\underline{r}) = 0,$$

$$\int_V E_a(\underline{r})^* \cdot E_b(\underline{r})\, dv = \delta_{ab}.$$

Substituting Eq. (6.17) into Eqs. (6.16), we obtain

$$\sum_s E_s \frac{dA_s}{dt} e^{j\omega_s t} \approx -\sum_s A_s \frac{\sigma(\underline{r}, t)}{2\varepsilon} E_s e^{j\omega_s t}, \tag{6.18}$$

where d^2A_s/dt^2 terms are neglected because A_s is a slowly varying function and ω_s/ω_{s+n} is approximated by unity because the oscillating modes are all within the line width of the atomic transition. Utilizing the orthonormal properties of the modes and multiplying both sides by E_a, we obtain

$$\left.\begin{aligned} e^{j\omega_a t}\frac{dA_a}{dt} &= -e^{j\omega_a t}\frac{\sigma_0}{2\varepsilon}A_a - \sum_s S_{as}\frac{\sigma_{\text{m}}\cos\omega_{\text{m}}t}{2\varepsilon}A_s e^{j\omega_s t}, \\ S_{sa} &= \int_V f(\underline{r})E_s^* E_a dv. \end{aligned}\right\} \tag{6.19}$$

When $\sigma_{\text{m}} = 0$,

$$A_a = A_a(0)\, e^{-\frac{\sigma_0}{2\varepsilon}t}.$$

Thus, ε/σ_0 is the photon decay or rise time in the cavity. When there is steady state oscillation, gain always equals loss. Thus, $\sigma_0 \equiv 0$ and

$$\frac{dA_a}{dt} = -\sum_s \frac{S_{as}\sigma_{\text{m}}}{4\varepsilon}A_s(e^{j\omega_{\text{m}}t} + e^{-j\omega_{\text{m}}t})e^{j(\omega_s-\omega_a)t}.$$

Like any resonance response, A will only be affected significantly when the modulation frequency ω_{m} is close to $\pm(\omega_s - \omega_a)$. Let us consider the case where the modulation frequency is close to the frequency difference $\Delta\omega$ between adjacent longitudinal modes. Let

$$\Delta = \omega_{s+1} - \omega_s - \omega_{\text{m}} = \Delta\omega - \omega_{\text{m}} = \text{small frequency deviation.}$$

Then,

$$-\frac{dA_a}{dt} = \kappa A_{a+1}e^{j\Delta t} + \kappa A_{a-1}e^{-j\Delta t}, \tag{6.20}$$

where

$$\kappa = \kappa_{a,a+1} = \kappa_{a,a-1} = \frac{S_{a,a+1}\sigma_{\text{m}}}{4\varepsilon}.$$

When $\kappa = 0$, there is no coupling among adjacent modes. It happens when $\sigma_{\text{m}} = 0$ or when $f(\underline{r})$ is a constant.

In order to solve Eq. (6.20), we will make a substitution of variables. Let

$$C_a(t) = -je^{ja\Delta t}e^{-a\pi/2}A_a(t)$$

or

$$A_a(t) = jC_a(t)\,e^{-ja\Delta t}e^{ja\pi/2}.$$

Then

$$j\frac{dC_a}{dt} + a\Delta C_a = \kappa C_{a+1} - \kappa C_{a-1}. \tag{6.21}$$

At steady state, $dC_a/dt = 0$. The solution of the remaining difference equation is well known:

$$C_a = I_a\left(\frac{\kappa}{\Delta}\right),$$

where I_a is the hyperbolic Bessel function of order a. For $\kappa/\Delta \gg 1$,

$$\left.\begin{aligned}
I_a(\kappa/\Delta) &= \frac{1}{\sqrt{2\pi(\kappa/\Delta)}},\\
A_a(t) &= \frac{j}{\sqrt{2\pi(\kappa/\Delta)}} e^{-ja(\Delta t - \frac{\pi}{2})},\\
E(\underline{r}, t) &= \sum_s \frac{j}{\sqrt{2\pi(\kappa/\Delta)}} e^{j(\omega_0 + s\,\omega_\mathrm{m})t} e^{js\pi/2} E_s(\underline{r}),\\
\omega_s &= \omega_0 + s\Delta\omega.
\end{aligned}\right\} \tag{6.22}$$

There are four significant conclusions that can be drawn from this solution.

(1) All longitudinal modes have the same transverse variation of E_s.
(2) The frequencies of adjacent modes are locked to $\omega_0 + s\Delta\omega$ by ω_m.
(3) When $\kappa/\Delta \gg 1$, all modes have equal amplitude.
(4) The phase of the modes is fixed at $s\pi/2$.

Therefore, we have just demonstrated that active mode locking can be achieved by gain modulation, despite the mode pulling effect of χ'. Physically, this means that with gain modulation we are not only forcing all the modes to oscillate with equal frequency spacing, but also that power is transferred from one mode to its adjacent modes so that they oscillate with equal amplitude and fixed phase. A similar situation occurs with phase modulation. From Eq. (6.15), it is interesting to note that the pulse width of mode locked lasers will depend on the number of oscillating modes N.

6.6.2 Mode locking in lasers with a homogeneously broadened line

We have shown earlier that there can only be one CW oscillating mode in a homogeneously broadened line at steady state. Let us now put a shutter inside the laser cavity such that: (1) the shutter is open for a short period of time; (2) the shutter is closed for the rest of the time during one repetition period; (3) the shutter is open again at the end of the repetition period; (4) the repetition period is equal or close to the round trip propagation time of an optical wave inside the cavity (i.e. $2Dn/c$). Figure 6.8(b) demonstrates the transmission of such a shutter as a function of time. Only the specific summation of longitudinal modes that are consistent with this

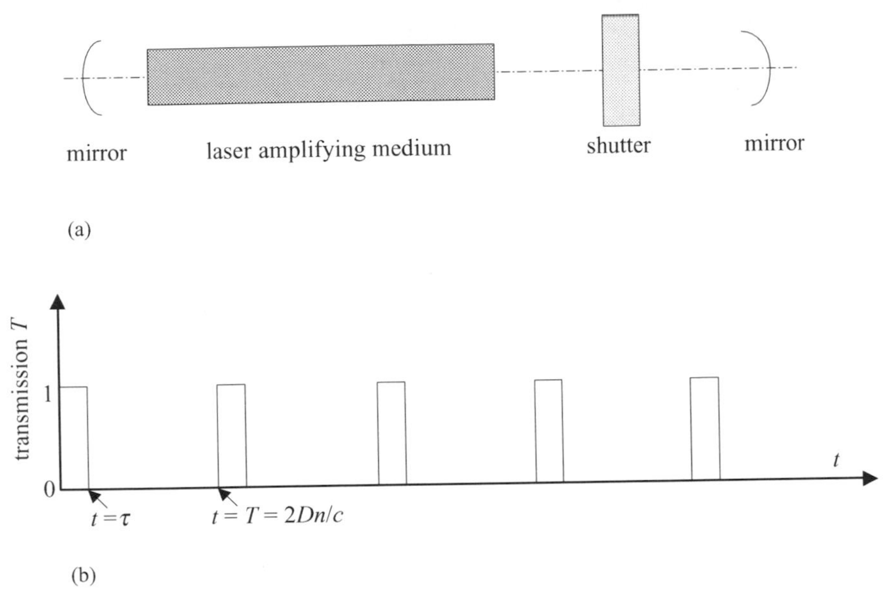

Figure 6.8. Mode locking by an intra-cavity shutter. (a) Laser cavity containing a shutter for mode locking. (b) Shutter time variation.

modulation will oscillate. Thus, a locking of longitudinal modes is achieved in a homogeneously broadened line. From another point of view, the modulation transfers continuously the power from the high gain modes (i.e. the oscillating modes) to the low gain modes (i.e. the non-oscillating modes at frequencies that are separated from the oscillating modes close to the side band frequency). The power transfer makes all the longitudinal modes within the line width of the transition oscillate in phase. A time dependent analysis of the mode locking in a homogeneously broadened laser is given in Section 11.3 of ref. [3].

6.6.3 Passive mode locking

The effect of a periodic gate can also be provided by the insertion of a saturable absorber gate in the optical path. A saturable absorber is usually made of a material that is transparent to intense radiation and opaque to weak radiation. The saturable absorber gate will clearly "encourage" the laser to oscillate as a circulating pulse with round trip transit time of $2nL/c$ since this mode of oscillation will undergo smaller losses than any other combination of modes in which the energy is spread more uniformly. From another point of view, all oscillations start with noise. Certain combinations of noise modes are transmitted preferentially by the saturable absorber. The saturable absorber gate favors the situation where the peak noise intensity fluctuation in the resonator travels and is amplified around the cavity. After a round trip, the amplified noise pulse will experience even less attenuation at the gate, thereby eventually creating the pattern of a circulating mode locked pulse. If the saturable absorber has a recovery time of S seconds, then the gate is open within time $1/S$ of the strong pulse. Thus, the time duration of the mode locked pulse tends to be $1/S$.

6.7 Laser amplifiers

Theoretically, the analysis of the amplification of optical signals in media with homogeneous and modest gain is straightforward. According to Eq. (5.43), when there is sufficient population inversion the intensity I_ν of the incident wave (e.g. plane waves or Gaussian beams) will be amplified in the z direction according to

$$I_\nu (z) = I_0 e^{\gamma z}. \tag{6.23}$$

Similarly to Eqs. (6.3) and (6.4), and in the absence of any saturation effect, we obtain

$$\gamma = \frac{k\mu^2}{2n^2\varepsilon_0\hbar}\left(N_2 - \frac{g_2}{g_1}N_1\right)g(\nu) - \alpha. \tag{6.24}$$

Since γz appears in the exponential, the gain of the amplifier can easily be 10 or 20 dB. Here, we have assumed that the χ' associated with the gain γ is sufficiently small that the effect of χ' on the wave propagation is negligible. In many applications, reflections at the input and output ends of the laser amplifier are reduced to a very low value. Therefore, in Eq. (6.23), we have assumed the reflections to be zero, or $T = 1$. However, for high gain laser amplifiers, even a small amount of reflection could trigger oscillation or a non-uniform gain profile.

In a medium with spatially homogeneous and modest gain γ, the spatial field profile of the incident wave is not altered by the gain. In the case of a Gaussian beam, it will be amplified as a Gaussian beam. However, the gain profile is frequently not homogeneous. An inhomogeneous gain profile could be caused by the variation of either the material index or the population inversion (e.g. due to non-uniform pumping). In those circumstances, whether the gain and its associated χ' variation will affect the lateral profile of the wave propagating through the medium needs to be considered. In the case of a material that has an index variation that supports a well guided waveguide mode, γ and χ' usually will not affect significantly the profile of a guided wave mode. Equation (6.23) is again applicable for each individual mode. However, Eq. (6.24) needs to be modified to take into account the overlap between the gain and the mode. For example, γ can be calculated by the perturbation analysis given in Section 4.1.2. The properties of semiconductor laser amplifiers are covered in Section 7.7.

Solid state laser amplifiers have been used in many applications in thc form of fibers, channel waveguides and bulk media. γ is usually independent of polarization in fibers and bulk media. It is dependent on polarization and mode order in channel waveguides or birefringent media. γ is clearly wavelength dependent. The transition is frequently inhomogeneously broadened, so that $g(\nu)$ is given by the inhomogeneously broadened line shape. When there is even a slight reflection of the waves in a high gain amplifier, the effective wavelength variation of the output could be significantly narrower than the wavelength dependence of γ.

Saturation of γ may occur in laser amplifiers in two different ways. (1) When I_s is large, χ'' saturates. The saturation is different for homogeneously and for inhomogeneously broadened lines. Details of the saturation effect of $[N_2 - (g_2/g_1)N_1]$ have been discussed in Section 5.4. Note that the wavelength bandwidth of the amplifier can be affected by the saturation mechanism. (2) When the pump power used to create the gain (or the population inversion) is very large, the pumping effect may be saturated. For the case of optical pumping, such as that used in erbium doped fiber amplifiers, the population of the upper levels that produce the N_2 may be saturated because of the large W_i. If we assume that the transitions used for optical pumping are homogeneously broadened lines, then Eq. (5.46b) can be used to describe the saturation effect. For the case of a single pump transition, the saturation of $[N_2 - (g_2/g_1)\, N_1]$ may then be represented by

$$\Delta N_{a0} = \Delta N_{\text{no pump sat}} \frac{1}{1 + (I_p/I_{sp})}.$$

I_p is the pump intensity and I_{sp} is the saturation parameter of the pump transition. In reality there may be more than one pump transition. Thus one should calculate the saturation effect for each pump transition separately and obtain the total ΔN_{a0} as the sum of the ΔN contributions from each one of the pump transitions. For the sake of brevity, the saturation expression for a single transition is used for multiple transitions where I_p is the averaged pump intensity per transition and where I_{sp} is the averaged saturation parameter for all the transitions.

When the two saturation effects are combined, we obtain

$$\begin{aligned} \Delta N_a &= \Delta N_{\text{unsat}} \frac{1}{1 + \left(\dfrac{I_\nu}{I_{s\nu}}\right)} \frac{1}{1 + \left(\dfrac{I_p}{I_{sp}}\right)} \\ &\approx \Delta N_{\text{unsat}} \frac{1}{1 + \left(\dfrac{I_\nu}{I_{s\nu}}\right) + \left(\dfrac{I_p}{I_{sp}}\right)}, \end{aligned}$$

for a moderate degree of saturation.

Laser amplifiers are used in two different ways: (1) as a power amplifier such as in the Nd/glass power amplifier, and (2) as a signal amplifier for communication networks. In case (1), the main considerations of the amplifier design are the mode control of the amplified radiation, the saturation effect in pulsed applications (e.g. pulse shaping similar to those occurring in Q switching) and the potential damage of material by high intensity optical fields (especially from reflection or focusing effects). In case (2), the gain as a function of wavelength, i.e. the bandwidth, is an important concern, especially in WDM (wavelength division multiplexed) applications. In addition, the noise of laser amplifiers is important for all communication applications; this will be discussed in the Section 6.8. See Chapter 4 of ref. [5] for more discussions on optical laser amplifiers.

6.8 Spontaneous emission noise in lasers

Spontaneous emission occurs in all media. It is a phenomenon that can be understood in two different ways as follows. (1) Phenomenologically, the blackbody radiation is spontaneous emission in thermal equilibrium. Therefore, one can find out the spontaneous emission rate from the known blackbody radiation intensity. This is the approach that we will use in this section. (2) In the quantized field theory (see Section 5.6 of ref. [3] and Section 5.7 of ref. [6]), both the radiation field and the atomic particles are quantized. The radiation field is expressed as a superposition of modes and the number of photons in each mode, while the ψ of atomic particles are expressed as summations of energy eigen states with a probability meaning for the coefficient of each state. The interaction between the radiation field and the atomic particles is expressed as the annihilation and creation of photons in each mode, while changes in the coefficients of the atomic energy states signify the change in energy of the atomic particles. The downward transition of the atomic particle can be induced by any radiation mode (i.e. the emission), even when there is no photon in that mode. This means that, in this case, radiation is emitted into all the modes in the absence of photons in the radiation field. This is the theoretical basis for spontaneous emission. In addition, radiation is emitted (i.e. created) or absorbed (i.e. annihilated) from those modes that have photons. This is the basis for induced transition.

In the phenomenological analysis, the effect of spontaneous emission might be included in the rate equation analysis of the lasers. In the case of laser oscillators, the spontaneous emission produces phase noise that is the origin of the minimum line width of laser output. It also produces an amplitude fluctuation that is the origin of the relative intensity noise. In laser amplifiers, it adds an additional noise component to the signal.

6.8.1 Spontaneous emission: the Einstein approach

Let us assume that there is a spontaneous emission transition probability, A, with which an atomic particle in a higher energy state will transfer into a lower energy state. There is no spontaneous absorption. We could evaluate A by considering the spontaneous emission at thermal equilibrium. This is known as Einstein's approach. A derivation of spontaneous emission probability A in semiconductors similar to the derivation here is given in ref. [7]. A derivation of spontaneous emission probability via the quantized field theory is given in ref. [6].

In Section 5.2, we showed that the induced transition probability from state k to state m for a monochromatic linearly polarized radiation is

$$W_{mk} = \sqrt{\frac{\mu_0}{\varepsilon}} \frac{\mu^2 I}{2\hbar^2} g(\nu).$$

For a broadband radiation with power $I(\nu)$,

$$W'_{mk} = \sqrt{\frac{\mu_0}{\varepsilon}} \frac{\mu^2}{2\hbar^2} \int I(\nu) g(\nu - \nu_{km})\, d\nu = \sqrt{\frac{\mu_0}{\varepsilon}} \frac{\mu^2}{2\hbar^2} I(\nu_{km}).$$

For non-polarized radiation, the I is not polarized in the y direction. If we take one-third of the $I(\nu)$ to be in the direction of the dipole moment, we obtain

$$W''_{mk} = \frac{1}{3} \cdot \frac{n\mu^2}{2c\hbar^2} I(\nu_{km}). \tag{6.25}$$

In thermal equilibrium, the radiation density is given by the blackbody formula,

$$\rho(\nu) = \frac{8\pi n^3 h \nu^3}{c^3} \left(\frac{1}{e^{h\nu/kT} - 1} \right), \tag{6.26a}$$

and

$$I = \frac{c}{n} \rho(\nu) = \frac{\rho(\nu)}{\sqrt{\mu_0 \varepsilon}}. \tag{6.26b}$$

The number of particles making the transition from state E_m to state E_k must equal the number of particles making the inverse transitions. If $E_m > E_k$, then the downward transition will have both induced and spontaneous transitions. The balance of downward and upward transitions can be expressed as

$$N_k [W''_{km} + A] = N_m W''_{km},$$

$$\frac{N_k}{N_m} = e^{-\frac{h\nu_{km}}{KT}}.$$

Therefore,

$$A = \frac{1}{t_{\text{spont}}} = \frac{N_m - N_k}{N_k} W''_{km} = \frac{8\pi^2 n^3 \nu^3}{3\varepsilon \hbar c^3} \mu^2. \tag{6.27}$$

$1/A$ is known as the spontaneous emission lifetime, t_{spont}. It is interesting to note that the spontaneous emission creates the blackbody radiation. Blackbody radiation at RF frequencies is the thermal noise. Since A is proportional to μ^2, t_{spont} is sometimes used to measure and represent the matrix element squared, μ^2:

$$\mu^2 = \frac{3\varepsilon\hbar c^3}{8\pi^2 n^3 \nu^3 t_{\text{spont}}}. \tag{6.28}$$

6.8.2 Spontaneous emission noise in laser amplifiers

The spontaneous emission in laser amplifiers will degrade the signal to noise ratio of the amplified signal. This is an important issue in communication.

In order to understand the analysis of amplifier noise, we shall clarify first our understanding of three important terms.

(1) The *spontaneous emitted power per unit volume*. This is equal to the population of the upper laser level times the energy of the photon and the spontaneous emission probability. The spontaneous emission probability is given in Eq. (6.27).

(2) The *amplified spontaneous emission per mode received by the detector*. The signal radiation and the spontaneously emitted radiation are both amplified by the laser amplifier. Let the laser signal radiation be in the form of a specific mode. The amplified field pattern of the signal will be in that mode. However, the spontaneous emission is distributed into all the modes. Since the amplification of various modes is different, the amplified spontaneously emitted power has its own intensity distribution and wavelength dependence. The amount of the signal and spontaneous emission noise which will be received by the detector will depend on the spatial and wavelength filtering characteristics of the receiver (i.e. how many modes and how wide a bandwidth the receiver will accept). It is customary to present the discussion on amplifier noise based on spontaneous emission received just in the mode of the signal and within a 1 Hz wavelength bandwidth. This is the minimum spontaneous emission that will be seen by the detector without any significant reduction of the signal.

(3) The *effect of the noise created by the signal and the spontaneous emission per mode*. The laser amplifier (seen by the detector with appropriate filters) will have an output power per mode, measured in terms of n photons per second. It consists of both the signal and the spontaneous emission (per mode) after amplification. There will be fluctuations of n; $\langle n \rangle$ is the mean number of output photons per second; $\langle n^2 \rangle$ is the variance of the radiation (i.e. the fluctuation of n^2). Noise characteristics are determined by $\langle n^2 \rangle - \langle n \rangle^2$. For applications using amplifiers, the important quantity is the noise figure, which is the ratio of signal/noise at the input to signal/noise at the output.

Let us next consider a signal which is in the fundamental Gaussian mode propagating from $z = 0$ to $z = l$ in an amplifier with power gain G. In order to receive minimum noise without reducing the signal, the receiver should detect only the radiation within the beam divergence of the fundamental Gaussian beam. In the following we will specify first the configuration of the spatial filter to be used with the amplifier. Then we will calculate the spontaneous emission noise power detected by the receiver in this amplifier configuration. Finally we will calculate the effect of the noise (i.e. the noise figure) in the receiver.

From Eq. (2.13), the signal radiation beam-width, θ, of a Gaussian beam is

$$\theta_{\text{beam}} = \frac{\lambda/n}{\pi\omega_0}. \tag{6.29a}$$

It corresponds to a solid angle Ω_{beam} centered about the z-axis, where

$$\Omega_{\text{beam}} = \frac{\lambda^2}{\pi n^2\omega_0^2}. \tag{6.29b}$$

Let us consider the case where there is negligible beam divergence within the amplifier; then the spot size A_a of the Gaussian beam at the end of the amplifier is approximately $\pi\omega_0^2$. Let the receiver be configured in such a way that it receives only the noise contained within the surface area A_a and the beam divergence Ω_{beam} of such a Gaussian beam.

Let the amplifier have length l and transmitted surface area limited to A_a at the output end, where $A_a = \pi\omega_0^2$. We shall now calculate the noise power detected by the receiver in such an amplifier configuration.

For spontaneous emissions expressed as a summation of plane waves, only those plane waves amplified and propagating in any angle θ within Ω_{beam} will be sensed by the detector. From Eq. (6.4d), the gain coefficient γ_a for linearly polarized radiation, neglecting any propagation loss or end reflections, is

$$\gamma_a' = \frac{k_0\mu^2}{2n^2\varepsilon_0\hbar}\Delta N_a g(\nu) = \frac{3\lambda^2}{8\pi n^2\, t_{\text{spont}}}\Delta N_a g(\nu).$$

For randomly polarized radiation, the gain needs to be averaged over all polarizations. Thus,

$$\gamma_a = \frac{\lambda^2}{8\pi n^2 t_{\text{spont}}}\Delta N_a\, g(\nu), \tag{6.30}$$

$$\frac{dI}{dz} = \gamma_a I.$$

For radiation within a cone $d\Omega$, propagating at an angle θ with respect to the z axis, as shown in Fig. 6.9, its power will come from the accumulated amplified spontaneous emission in small volumes dV. The power that will reach the end surface within the area A_a, in the direction θ, is the total integrated and amplified spontaneous emission power within the cylinder in Fig. 6.9. For high gain amplifiers within the frequency range from ν to $\nu + d\nu$, the total random polarized

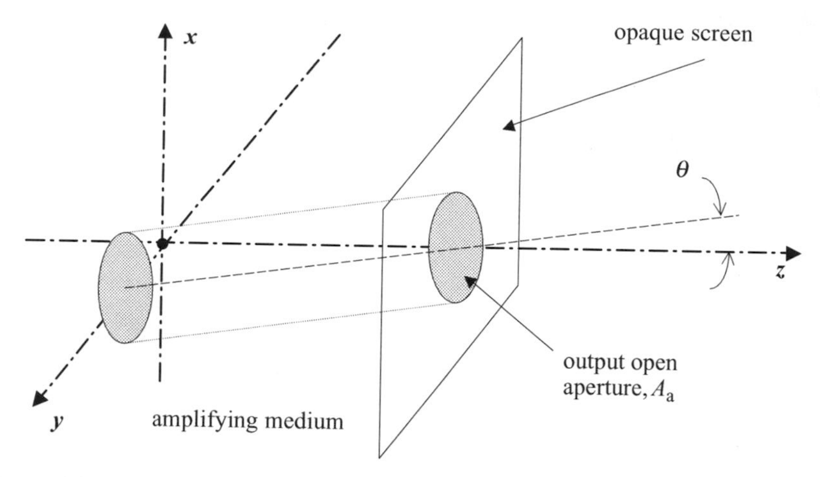

Figure 6.9. Model used to calculate the spontaneous emitted noise through the output aperture in an amplifier.

spontaneously emitted power, $N(\theta)h\nu$, within A_a at any angle θ and into a solid angle $d\Omega$ is:

$$N(\theta)h\nu = A_a \int_0^l dz \, \frac{dN}{dV} \, e^{\gamma_a (l-z)/\cos\theta},$$

where

$$\frac{dN}{dV} = h\nu \frac{N_2 g(\nu)}{t_{spont}} \frac{d\Omega}{4\pi} d\nu.$$

The evaluation of this integral yields

$$\begin{aligned} N(\theta) &= A_a \cos\theta \, \frac{[G-1] N_2 \, g(\nu)}{\gamma_a t_{spont}} \frac{d\Omega}{4\pi} d\nu, \\ &= 2\, d\nu \, \frac{N_2}{N_2 - \dfrac{g_2}{g_1} N_1} \, [G(\theta) - 1] \, \frac{A_a \cos\theta \, d\Omega}{(\lambda/n)^2} \\ &= 2d\nu \frac{N_2}{N_2 - \dfrac{g_2}{g_1} N_1} \, [G(\theta) - 1] \, \frac{\cos\theta \, d\Omega}{\Omega_{beam}}. \end{aligned} \tag{6.31}$$

Equation (6.30) has been used to express $\gamma_a t_{spont}$, and the gain of the amplifier G in the θ direction is

$$G(\theta) = e^{\gamma_a l/\cos\theta}.$$

If a polarization filter is used to filter out the polarization perpendicular to the signal polarization, we will reduce $N(\theta)$ by 2. For small θ, $\cos\theta \equiv 1$. When we sum all the $N(\theta)$ within the solid angle Ω_{beam} given in Eq. (6.29b), we obtain the noise power per mode, $N_0 h\nu$:

$$N_0 h\nu = h\nu\, d\nu\, \frac{N_2}{N_2 - N_1(g_2/g_1)} [G - 1]\,. \tag{6.32}$$

The expression $N_2/(N_2 - N_1(g_2/g_1))$ in the above equation is called the population inversion factor η_a. A different receiver may sense noise power from m modes, then the total noise power is $m_n N_0 h\nu$. The derivation for Eq. (6.32) follows a similar derivation to that given in Section 21.1 of ref. [3].

Knowing the noise power received by the detector, we can now calculate the effect of the noise from $\langle n^2 \rangle$ and $\langle n \rangle^2$. The variance of the total number of photons per second at the output of the amplifier, n, has been worked out statistically [8]. For an incident laser power P, an equivalent of m_n modes of noise received by the detector and wavelength bandwidth $\Delta\nu$,

$$\langle n^2 \rangle = G\left(\frac{P}{h\nu}\right) + 2m_n(G-1)\eta_a\Delta\nu + 2G\left(\frac{P}{h\nu}\right)(G-1)\eta_a$$
$$+ 2m_n(G-1)^2\eta_a^2\Delta\nu. \tag{6.33}$$

The first term is the shot noise of the signal radiation; the second is the shot noise of the beat spontaneous emission; the third is the signal–spontaneous noise; and the fourth is the spontaneous–spontaneous beat noise.

For strong incident radiation and large G, the first and third term dominate. Therefore, the noise figure F of the amplifier is given by

$$F = \frac{(P/N)_{\text{in}}}{(P/N)_{\text{out}}} \approx 2\eta_a. \tag{6.34}$$

For an ideal amplifier, $\eta_a \approx 1$ and the theoretical limit of F is 3 dB. In this discussion we have assumed that there are no reflections at the input and output ends of the amplifier. We have also neglected propagation loss inside the amplifier.

6.8.3 Spontaneous emission in laser oscillators

Spontaneous emission is the source that initiated the oscillation in the specific mode when the gain exceeds all the losses (including the outputs). Eventually the amplitude of the oscillating mode is limited by non-linear saturation.

Spontaneous emission causes intensity fluctuation of the laser oscillator, called the relative intensity noise (RIN).

When spontaneously emitted radiation with random phase is mixed with the radiation of the oscillating mode, it diffuses the phase, which results in a degradation of coherence. This is the basic reason for having only a finite line width, $\Delta\nu_{\text{osc}}$, for the laser oscillator. The relation governing the line width with respect to the cavity resonance line width, the output power of the laser and the population inversion of the laser was first reported by Schawlow and Townes [9]. This calculated $\Delta\nu_{\text{osc}}$ is the theoretical limit of the oscillation line width. The actual line width of laser oscillators is frequently broader because of other fluctuations not considered in the theoretical analysis.

The answers for both the RIN and the oscillator line width came from the solution for the optical electric field inside the cavity with gain, saturation and spontaneous emission,

$$\nabla \cdot \nabla \underline{e}(\underline{r}, t) - \mu\sigma \frac{\partial \underline{e}}{\partial t} - \mu\varepsilon \frac{\partial^2 \underline{e}}{\partial t^2} = \mu \frac{\partial^2}{\partial t^2}[\underline{P} + \underline{p}], \tag{6.35}$$

with

$$\underline{e}(\underline{r}, t) = \sum_m E_m(t)\underline{e}_m(\underline{r}),$$

$$\underline{P}(\underline{r}, t) = \sum_m P_m(t)\underline{e}_m(\underline{r}),$$

and

$$\underline{p}(\underline{r}, t) = \sum_m p_m(t)\underline{e}_m(\underline{r}).$$

Here, $\underline{P}$ is the instantaneous induced polarization of the laser transition, and $\underline{p}$ is the instantaneous polarization from the spontaneous emission. Equation (6.35) can be simplified by the orthogonality properties of the modes. The resultant equation for each mode, E_n, is

$$\ddot{E}_n + \frac{1}{\tau_{\text{p}}}\dot{E}_n + \omega_n^2 E_n = -\frac{1}{\varepsilon}(\ddot{P}_n + \ddot{p}_n),$$

where τ_{p} is the photon lifetime in the passive resonator, ω_n is the resonance frequency of the nth mode and

$$E_n = [E_{n0} + \delta(t)]\, e^{j[\omega_n t + \phi(t)]}. \tag{6.36}$$

E_{n0} is the average amplitude of the electric field, δ is the real amplitude deviation and ϕ is the instantaneous phase. The intensity fluctuation consists of a calculation of $\langle\delta(t)\delta(t+\tau)\rangle$, while the frequency spectrum, i.e. $\Delta\nu_{\text{osc}}$, depends on the calculation of $\langle\dot{\phi}(t_1)\,\dot{\phi}(t_2)\rangle$.

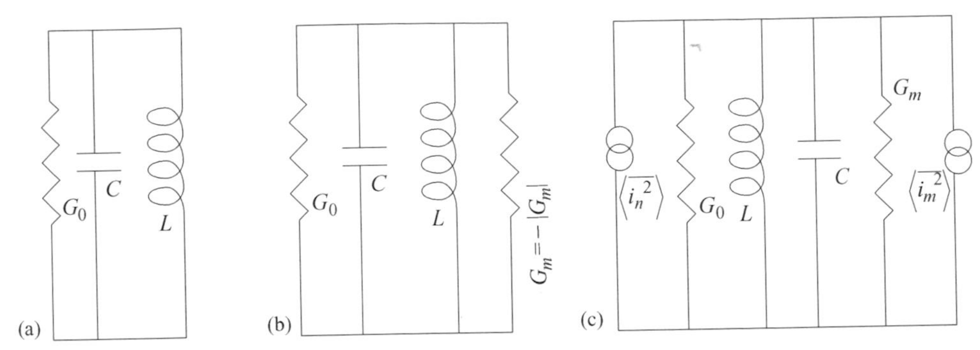

Figure 6.10. Equivalent circuits of an oscillating laser cavity. (a) Equivalent circuit of a resonant mode. (b) Equivalent circuit of a resonant mode with gain. (c) Equivalent circuit of a resonant mode with noise source and gain.

6.8.4 The line width of laser oscillation

Obtaining the solution of Eqs. (6.35) and (6.36) is a lengthy process. There is a more conventional method of calculating the line width, $\Delta\nu_{osc}$, which takes advantage of our knowledge of the noise and the circuit representation of a resonator. In this section, we analyze the Q factor of the resonance circuit, represent the thermal and the spontaneous emission noise by current sources and determine the $\Delta\nu_{osc}$ from the Q factor. This line width is known as the Schawlow–Townes relation.

The equivalent circuit and the Q of the resonance mode

The laser oscillator without gain is modeled frequently as a parallel RLC resonance circuit, as shown in Fig. 6.10(a). L and C are the equivalent inductance and capacitance of the cavity resonance, respectively. In this case, all the passive losses (including the output) of the resonant cavity are represented by the conductance G_0. The resonance frequency ω_0 is $1/\sqrt{LC}$. From Eq. (2.9), we know that the quality factor, Q_0, representing the passive losses (consisting mostly of the output) of the resonator is

$$Q_0 = \frac{2\pi D}{\lambda[1 - r_1 r_2 e^{-\alpha D}]}.$$

On the other hand, $Q_0 = 1/(G_0\omega_0 L)$ from the analysis of the equivalent circuit. Therefore, G_0 is related to cavity parameters as follows:

$$\frac{\omega_0 C}{G_0} = \frac{2\pi D}{\lambda[1 - r_1 r_2 e^{-\alpha D}]} = Q_0 = \frac{\nu_0}{\Delta\nu_0}, \tag{6.37}$$

where $\Delta\nu_0$ is the line width of the passive resonance.

Similarly, the effect of the gain on the resonance can be represented as a negative conductance G_m in the equivalent circuit, as shown in Fig. 6.10(b).

Circuit representation of the thermal noise

In Eq. (6.32), if we use the equilibrium relationship, $(g_2N_1)/(g_1N_2) = \exp(h\nu/KT)$ at temperature T, and if we consider $G = 0$ for passive material with large l, the noise per pass, $N_0h\nu$, is now the description of a blackbody radiator,

$$N_0h\nu = h\nu\,d\nu\frac{1}{e^{h\nu/KT} - 1}. \tag{6.38}$$

It is well known that the Johnson thermal noise in an RLC circuit can be represented by a noise source $\langle i_n^2(\omega)\rangle$ in parallel with G_0 (see Section 21.2 of ref. [3]), where, within a bandwidth $\Delta\omega$,

$$\left\langle i_{\mathrm{N}}^2\,(\omega)\right\rangle = \frac{4h\nu\,d\nu\,G_0}{e^{h\nu/KT} - 1} \tag{6.39}$$

and

$$\frac{\left\langle i_{\mathrm{N}}^2\,(\omega)\right\rangle}{\Delta\omega} = \frac{2h\nu}{\pi(e^{h\nu/KT} - 1)}G_0.$$

Circuit representation of spontaneous emission noise with gain and the negative emission temperature

For a medium with gain, if we just consider the effect of the amplification (without passive loss), we again obtain from Eq. (6.32) the amplified $N_0h\nu$ per pass and the Q factor,

$$N_mh\nu = h\nu\,d\nu\,\frac{1}{1 - (g_2N_1/g_1N_2)}(G - 1)\,, \tag{6.40a}$$

$$Q_m = \frac{\omega_0 C}{G_m} = \frac{2\pi D}{\lambda[1 - e^{\gamma D}]}. \tag{6.40b}$$

The equivalent $N_{m,\mathrm{eq}}h\nu$ per pass per mode, evaluated at the beginning of the pass, is

$$\begin{aligned} N_{m,\mathrm{eq}}h\nu &= h\nu\,d\nu\frac{1}{1 - (g_2N_1/g_1N_2)}\cdot\frac{G - 1}{G} \\ &\approx h\nu\,d\nu\frac{1}{1 - (g_2N_1/g_1N_2)}. \end{aligned} \tag{6.41}$$

Since $e^{\gamma D} > 1$, G_m and Q_m are negative. If we designate $(g_2N_1/g_1N_2) = \exp(h\nu/KT_m)$, then T_m is an equivalent negative temperature representing the population inversion

$$\frac{1}{1 - (g_2N_1/g_1N_2)} = \frac{1}{1 - e^{(h\nu/KT_m)}}.$$

In terms of T_m, the results for N_0 and $N_{m,\mathrm{eq}}$ expressed in Eqs. (6.41), (6.37) and (6.38) are the same. Thus, the circuit and the noise of the passive RLC circuit can be extended in terms of G_m and T_m to cover the case with population inversion.

In short, the circuit representation of spontaneous emission with gain is similar to thermal noise. The noise created by spontaneous emission can be represented as a noise generator $\langle i_m^2\rangle$ in parallel with and next to G_m, as in Fig. 6.10(c). For spontaneous emission, the $\langle i_m^2(\omega)\rangle$ is represented as a noise source (see Section 21.2 of ref. [2]),

$$\frac{\langle i_m^2(\omega)\rangle}{\Delta\omega} = \frac{2h\nu}{\pi(e^{\hbar\omega/KT_m} - 1)} G_m. \tag{6.42}$$

Figure 6.10(c) shows the equivalent circuit representation, including the G_0 for loss and output and G_m for gain, as well as the thermal and the spontaneous emission noise source.

Line width of laser oscillation

Both $\langle i_N^2\rangle$ and $\langle i_m^2\rangle$ are incoherent noise sources; thus their powers add. This means that for the total cavity, including both amplification and passive loss, we obtain

$$\left.\begin{aligned} \frac{1}{Q} &= \frac{1}{Q_0} - \frac{1}{|Q_m|} = \frac{G_0 - |G_m|}{\omega_0 C}, \\ \Delta\nu_{\mathrm{osc}} &= \frac{\nu_0}{Q} = \frac{G_0}{2\pi C}\left(1 - \frac{|G_m|}{G_0}\right), \\ \frac{\langle I(\omega)^2\rangle}{\Delta\omega} &= \frac{2}{\pi} h\nu \left[\frac{|G_m|}{1 - e^{h\nu/KT_m}} + \frac{G_0}{e^{h\nu/KT} - 1}\right]. \end{aligned}\right\} \tag{6.43}$$

When the mode is below oscillation, $|G_m| < G_0$. As the mode approaches oscillation, Q increases, and $\Delta\nu_{\mathrm{osc}}$ decreases. This effect is called line narrowing, and is observed experimentally. When saturation occurs in an oscillating mode, $|G_m| \approx G_0$. $1/Q$ is approximately zero in the first-order approximation. However, the accuracy of our knowledge about G_0 and G_m is insufficient to calculate such a small $1/Q$ from Eqs. (6.43). On the other hand, the output of the laser can be measured, and it is related to the noise sources in Fig. 6.10(c) via the circuit shown in that figure. Therefore we can evaluate Q from the output power of the laser P_o as follows.

When the mode is oscillating well above threshold, the second term of $\langle I(\omega)^2\rangle$ in Eqs. (6.43), involving G_0, is small in comparison with the first term; it can be neglected. The total emitted power is contained in the power dissipated in G_0. If we assume the power of the losses is much smaller than the power transmitted as the output, the total emitted power of the oscillator is

$$P_o = G_0 \int_0^{\infty} \frac{\langle V(\omega)^2 \rangle}{\Delta\omega}\, d\omega,$$

$$\langle V(\omega)^2 \rangle = \frac{1}{4C^2} \frac{\langle I(\omega)^2 \rangle}{(\omega - \omega_0)^2 + \left(\frac{\omega_0}{2Q}\right)^2}$$

with

$$\omega_0 = \sqrt{\frac{1}{LC}},$$

$$P_o = \frac{\hbar G_0^2}{2\pi C^2} \cdot \frac{1}{1 - e^{(h\nu/KT_m)}} \cdot \int_0^{\infty} \frac{\omega\, d\omega}{(\omega_0 - \omega)^2 + (\omega_0/2Q)^2}$$

$$= \frac{\hbar G_0^2 Q}{C^2} \cdot \frac{1}{1 - e^{(h\nu/KT_m)}}. \tag{6.44}$$

Thus,

$$\Delta\nu_{osc} = \frac{\nu_0}{Q} = \frac{\hbar \nu_0 G_0^2}{P_o C^2} \cdot \frac{1}{1 - e^{(h\nu/KT_m)}}$$

$$= \frac{2\pi h \nu_0 (\Delta\nu_0)^2}{P_o} \cdot \frac{1}{1 - e^{(h\nu_0/KT_m)}}. \tag{6.45}$$

Here, $\Delta\nu_0$ is the full line width (at half maximum) of the passive cavity resonance, without amplification. We have utilized $Q_0 = \omega_0 C/G_0 = \nu_0/\Delta\nu_0$. The results of the simplified analysis have ignored the coupling of amplitude fluctuations to phase fluctuations, i.e. the modulation of the index of refraction of the gain medium by fluctuations in spontaneous emission. More exact analysis based on Eqs. (6.35) and (6.36) has shown [10]

$$\Delta\nu_{osc} = \frac{2\pi h \nu_0 (\Delta\nu_0)^2}{P_o} \frac{1}{1 - e^{(h\nu/KT_m)}} (1 + \alpha^2), \tag{6.46}$$

where α is the line width enhancement factor due to the change of the real part of the index by the imaginary part.

6.8.5 Relative intensity noise of laser oscillators

There are fluctuations of laser intensity caused by random spontaneous emissions. This fluctuation is known as relative intensity noise, defined as

$$rin = \left(\frac{\overline{\delta p_1^2}}{P_L^2} \right) \Big/ \Delta f. \tag{6.47}$$

Here $\overline{\delta p_{\mathrm{l}}^{2}}$ denotes the mean square value of the intensity fluctuation (i.e. $\langle \delta(t)\,\delta(t+\tau)\rangle$ in Eq. (6.36)), P_{L} is the laser power and Δf is the band width. The rin is known to be independent of P_{L}. The relative intensity noise of lasers is usually specified in terms of the RIN, in dB, where

$$\mathrm{RIN} = 10 \log_{10}(rin). \tag{6.48}$$

Both P_{L}^{2} and $\overline{\delta p_{\mathrm{l}}^{2}}$ will exhibit themselves as current squared in the load resistor after detection. Since the same detector and circuit will be used for P_{L}^{2} and $\overline{\delta p_{\mathrm{l}}^{2}}$, the ratio of $\overline{\delta p_{\mathrm{l}}^{2}}/p_{\mathrm{l}}^{2}$ is the same as $\overline{i_{\mathrm{rin}}^{2}}/i_{\mathrm{L}}^{2}$. Hence the relative intensity noise is represented as a current generator with a mean squared current

$$\overline{i_{\mathrm{rin}}^{2}} = rin \cdot \overline{i_{\mathrm{L}}^{2}} \cdot \Delta f. \tag{6.49}$$

The RIN spectrum is not flat as the spectrum for white noise. The RIN is frequency dependent. However, for simplicity, most link analyses assume that RIN is a constant within the band width of interest. The RIN also differs for diode and solid state lasers, and for single-mode and multimode lasers. For example, single-mode solid state lasers may have a RIN of -170 dB for $\Delta f = 1$ Hz, whereas diode lasers typically have a RIN of -145 dB for $\Delta f = 1$ Hz.

References

1 O. Svelto, *Principles of Lasers*, Chapters 9 and 10, New York, Plenum Press, 1998
2 A. E. Siegman, *Lasers*, Sausalito, CA, University Science Books, 1986
3 A. Yariv, *Quantum Electronics*, New York, John Wiley and Sons, 1989
4 W. G. Wagner and B. A. Lengyel, "Evolution of the Giant Pulse in a Laser," *Journal of Applied Physics*, **34**, 1963, 2044
5 S. Shimoda and H. Ishio, *Optical Amplifiers and Their Applications*, New York, John Wiley and Sons, 1994
6 W. S. C. Chang, *Quantum Electronics*, Section 5.7, Reading, MA, Addison-Wesley, 1969
7 S. L. Chuang, *Physics of Opto-electronic Devices*, Section 9.2, New York, John Wiley and Sons, 1995
8 S. Shimoda and H. Ishio, *Optical Amplifiers and Their Applications*, Section 2.2, New York, John Wiley and Sons, 1994
9 A. L. Schawlow and C. H. Townes, "Infrared and Optical Masers," *Physical Review*, **112**, 1958, 1940
10 C. H. Henry, "Theory of the Line Width of Semiconductor Lasers," *IEEE Journal of Quantum Electronics*, **QE-18**, 1982, 259

7

Semiconductor lasers

The general principles of amplification and oscillation in semiconductor lasers are the same as those in solid state and gas lasers, as discussed in Chapter 6. A negative χ'' is obtained in an active region via induced transitions of the electrons. When the gain per unit distance is larger than the propagation loss, laser amplification is obtained. In order to achieve laser oscillation, the active material is enclosed in a cavity. Laser oscillation begins when the gain exceeds the losses, including the output. However, the details are quite different. In this chapter, the discussion on semiconductor lasers will use much of the analyses already developed in Chapters 5 and 6; however, the differences will be emphasized.

In semiconductor lasers, free electrons and holes are the particles that undertake stimulated emission and absorption. How such free carriers are generated, transported and recombined has been discussed extensively in the literature, [1, 2, 3]. We note here, in particular, that free electrons and holes are in a periodic crystalline material. The energy levels of electrons and holes in such a material are distributed within conduction and valence bands. The distribution of energy states within each band depends on the specific semiconductor material and its confinement within a given structure. For example, it is different for a bulk material (a three-dimensional periodic structure) and for a quantum well (a two-dimensional periodic structure).

For lasers, we require information on how the free carriers undergo stimulated emission and absorption, and what the χ produced from such transitions is. From the quantum mechanical point of view, unlike dopants in insulating crystals or molecules (or atoms) in gaseous media, free electrons and holes in semiconductors are not individually localized and identifiable particles. Pauli's exclusion principle dictates that there can only be one electron (or one hole) per energy state. The occupation probability of any state by electrons (or holes) is governed by the Fermi–Dirac distribution. A stimulated emission must involve an electron in an energy state in the upper band, such as a state in the conduction band, making a transition to a vacant state in a lower band, such as an available hole state in the valence band, in response to incident radiation. In the language of semiconductor physics, free electrons in the conduction band are recombined with free holes in the valence band, producing stimulated emission. A similar reverse statement can be made for absorption. Since there are many energy states, there are many pairs of transitions that emit (or absorb) photons of the same energy. The exact average number of transitions that take place depends on the distribution of the energy states, i.e. the density of states, and the Fermi distribution. In terms of our description of χ and lasers presented in Chapters 5 and 6, the susceptibility produced by stimulated emission and absorption in semiconductors is similar to the susceptibility obtained in inhomogeneously broadened transitions. The total averaged susceptibility is obtained from an accumulation of contributions from individual transitions. The averaged χ'' comes from a balance of emission and absorption transitions.

At the steady state (but not at thermal equilibrium), there is a separate Fermi distribution for electrons in the conduction band and one for holes in the valence band. Each Fermi distribution varies according to its own quasi-Fermi level. For certain quasi-Fermi levels in the conduction band and the valence band, we have net emission, i.e. $-|\chi''|$. At some other quasi-Fermi levels we have net absorption, i.e. $+|\chi''|$. The quasi-Fermi levels are themselves controlled by the electron and hole densities in the material at that location. Thus, amplification or oscillation in semiconductor lasers is controlled by carrier injection into the active layer. The easiest way to inject the necessary carriers in order to achieve $-|\chi''|$ is to apply a forward bias to a p–n junction diode. Therefore the semiconductor laser is also known as a diode laser.

The easiest way to understand stimulated emission (and absorption) and the susceptibility produced by stimulated transitions of electrons and holes is to discuss them via a simple semiconductor laser made of a bulk homogeneous semiconductor. This is the main objective of the discussion presented in Section 7.1. In order not to distract our attention from the basic physics of semiconductor lasers, the discussion of the amplification of optical waves in semiconductors is deferred to Section 7.7. Instead, how a laser oscillates and how such a susceptibility saturates after oscillation are discussed in Section 7.2.

Simple lasers in homogeneous bulk semiconductors are no longer used much in present day applications. A modern semiconductor laser is much more complicated than that described in Section 7.1. Different material structures, such as heterojunctions and quantum well materials, can be grown epitaxially. Different cavity configurations, including structures to provide the necessary feedback, such as DFB (distributed feedback) edge or surface emitting cavities, can be employed to control the resonant mode and the output. Various device structures can be fabricated to concentrate and to confine the injected carrier in the desired "active" region. The manner in which injected electrons and holes will affect the quasi-Fermi levels and the process by means of which the injected carriers are confined in the active region are determined by the semiconductor and by the device structure. How the injected carrier densities will be created efficiently by the laser current is determined by the electrical design of the device. How the gain in the active layer can be best utilized to obtain specific laser characteristics will be a matter of the optical design of the laser, including the control of the material indices and thickness. A discussion on these topics will be presented in Sections 7.3 to 7.5.

Section 7.6 discusses the modulation of the intensity of the output of laser oscillators by current modulation. Section 7.7 discusses the semiconductor laser amplifier. Section 7.8 discusses the noise in semiconductor lasers.

To summarize, there are five major differences between semiconductor lasers and solid state and gas lasers. (1) The cavity modes (usually the guided wave modes) of edge emitting semiconductor lasers have transverse dimensions comparable to the emission wavelength. They are usually not the Gaussian modes discussed in Chapter 2. (2) The electron and hole densities in the active region of semiconductor lasers are controlled through a balance of injection and leakage (or decay) of carriers. Structures such as hetero-junctions and current barriers have been used to increase the carrier lifetime and to reduce the required injection current. (3) The three-dimensional bulk material and a two-dimensional quantum well have different densities of states. Since the Fermi level for a given carrier density is affected substantially by the density of states, quantum size-confined structures are used to achieve the required $|\chi''|$ while using as small a carrier density as possible. (4) Since current injection has a high speed of response, electrical modulation of the injection current is used to yield direct modulation of laser intensity. (5) There will be noise generated by the carriers in addition to the noise generated by spontaneous emission.

7.1 Macroscopic susceptibility of laser transitions in bulk materials

In the following discussion it is assumed that the reader is acquainted with the fundamental properties of semiconductors (see refs. [2] and [3]). The energy diagram of the conduction and valence bands and concepts such as electron, holes, the Fermi level and p–n junctions, are not reviewed here. Our analysis in Section 7.1 is concerned only with the stimulated emission and absorption processes and the susceptibility in bulk homogeneous semiconductor materials for a small electromagnetic field. In order to describe clearly the basic physics and the analysis of the susceptibility, saturation effects such as those described in Section 5.4 are not discussed here. It would be instructional to compare this discussion with the discussion on the unsaturated susceptibility given in Chapter 5.

When a laser oscillates, there is a strong electromagnetic field, and the susceptibility saturates. The laser oscillation and the saturation of χ will be discussed in Section 7.2. The saturation of χ in semiconductors is very different from the saturation of χ in gas and solid state lasers.

The discussion in Section 7.1 will be divided into four parts: (1) the energy states in bulk semiconductors, (2) the density of states, (3) the Fermi distribution and the current densities, and (4) the susceptibility and the induced transitions.

7.1.1 Energy states

Within the conduction band and valence band of a three-dimensional periodic crystalline bulk semiconductor medium, each energy state has a wave function of the form [4],

$$\Psi_C(\underline{r}) = u_{C\underline{k}} e^{j\underline{k}\cdot\underline{r}}, \tag{7.1}$$

where $u_{C\underline{k}}(\underline{r})$ has the periodicity of the crystalline lattice. The energy of electrons in the conduction band for a state with a given $\underline{k}$ (known as the parabolic approximation of the energy band structure) is

$$E(|\underline{k}|) - E_C = \frac{\hbar^2|\underline{k}|^2}{2m_e}. \tag{7.2a}$$

A similar expression is obtained for energy states in the valence band,

$$E_V - E(|\underline{k}|) = \frac{\hbar^2|\underline{k}|^2}{2m_h}. \tag{7.2b}$$

E_C is the bottom of the conduction band and E_V is the top of the valence band. The effective masses of the electrons and holes are m_e and m_h. Note: this $\underline{k}$ is not to be confused with the propagation constant, k, of the optical waves in Chapters 1 to 4. For this reason, the magnitude of $\underline{k}$ will be presented as $|\underline{k}|$ in this chapter.

7.1.2 Density of energy states

There are a large number of such states per unit energy range (or per unit $|\underline{k}|$ range), called the density of states. The resultant density of states per unit volume expressions (for bulk materials) in the conduction and valence bands are well known [4, 5],

$$\rho_C(E - E_C)dE = \frac{1}{2\pi^2}\left(\frac{2m_e}{\hbar^2}\right)^{3/2}(E - E_C)^{1/2}\,dE, \; E > E_C, \tag{7.3a}$$

$$\rho_C(|\underline{k}|)d|\underline{k}| = \frac{|\underline{k}|^2}{\pi^2}d|\underline{k}|$$

and

$$\rho_V(E_V - E)dE = \frac{1}{2\pi^2}\left(\frac{2m_h}{\hbar^2}\right)^{3/2}(E_V - E)^{1/2}\,dE,\ E < E_V, \qquad (7.3b)$$

$$\rho_V(|\underline{k}|)d|\underline{k}| = \frac{|\underline{k}|^2}{\pi^2}d|\underline{k}|.$$

Here, the band gap energy E_g is $E_C - E_V$. The effective masses of the electrons and holes are m_e and m_h, respectively. Note that the density of states as a function of $|\underline{k}|$ is the same for holes and electrons. This means that, for direct transitions with no change in $\underline{k}$, the density of induced transitions (as a function of $|\underline{k}|$) is the same as the density of either the upper or the lower energy states in Eqs. (7.3).

When one analyses the density of states in material structures with just two or less dimensional periodic variation in the crystal, such as in a quantum well structure, the dependence of the density of state on E will change. This point will be discussed in Section 7.3.

7.1.3 Fermi distribution and carrier densities

The probability of the occupation of the energy states by electrons at equilibrium obeys the Fermi statistical distribution [3],

$$f(E) = \frac{1}{e^{(E-E_F)/KT} + 1}, \qquad (7.4)$$

where E_F is the Fermi level, K is Boltzmann's constant and T is the absolute temperature on the Kelvin scale. At equilibrium, there is only one Fermi level E_F. However, in a quasi-equilibrium situation (such as in a forward biased p–n junction at steady state), the probability distributions of electrons in the conduction band and in the valence band have different E_F. When electrons and holes are injected into an active region, there is a quasi-Fermi level, E_{FC}, used for describing the steady state electron distribution in the conduction band, and a quasi-Fermi level, E_{FV}, for the valence band. The $f_C(E)$ and $f_V(E)$ describe separately the probability of occupying the state by electrons at E in the conduction band and in the valence band. The hole (i.e. the absence of electrons) distribution is $[1 - f_C(E)]$ in the conduction band and $[1 - f_V(E)]$ in the valence band. Figure 7.1 illustrates the Fermi levels, the band diagram of a direct semiconductor as a function of the $|\underline{k}|$ of the electronic (and hole) state, the Fermi levels, the f_C and the f_V. The Fermi levels are shown with $E_{FC} > E_C$ and $E_{FV} < E_V$, describing the occupation probability of the energy states in degenerate semiconductors.

How are the quasi-Fermi levels controlled by electron and hole densities? The total number of electrons per unit volume in the conduction band, n_C, is the

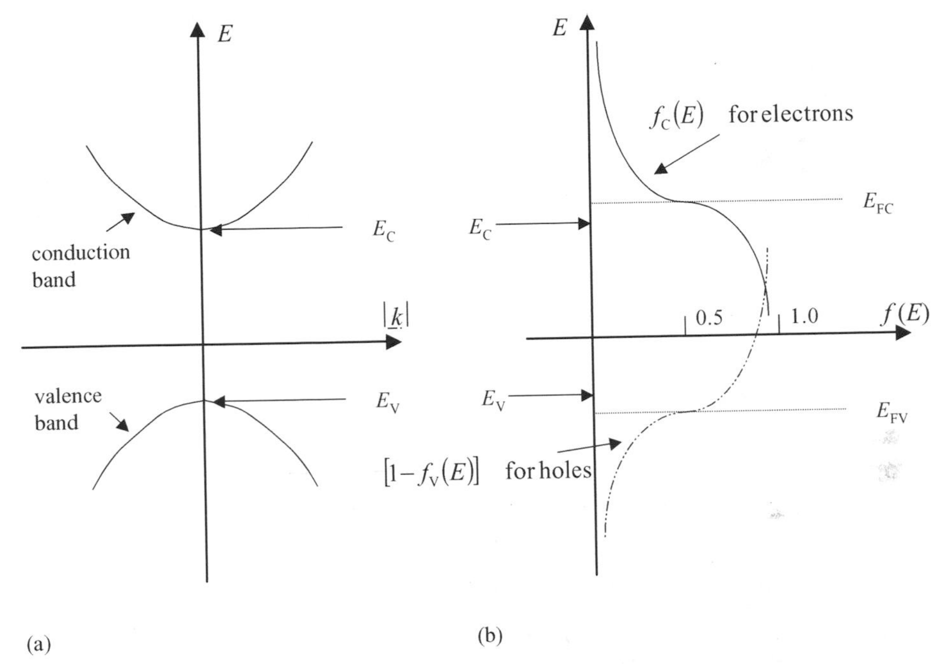

Figure 7.1. Energy diagram and Fermi distribution of electrons and holes in semiconductors. (a) Energy of electrons in the conduction band and holes in the valence band as a function of $|\underline{k}|$. (b) Fermi distribution of electrons and holes. The quasi-Fermi levels for electrons and holes, E_{FC} and E_{FV}, are shown here under current injection, with $E_{FC} - E_{FV} > E_C - E_V$.

integration of the product of $\rho_C(E)$ and $f_C(E)$. At 0 K and for electrons, according to Eq. (7.4), all the states above E_{FC} in the conduction band are empty and all the states below E_{FV} in the valence band are occupied. Thus,

$$n_C = \int_{E_C}^{\infty} \rho_C(E) f_C(E)\, dE.$$

At temperatures close to 0 K,

$$n_C \approx \int_{E_C}^{E_{FC}} \rho_C(E)\, dE = \frac{1}{2\pi^2}\left(\frac{2m_C}{\hbar^2}\right)\cdot\frac{2E_{FC}^{3/2}}{3}. \tag{7.5}$$

A similar expression for n_V is obtained for the holes. The significance of Eq. (7.5) is that the total number of injected electrons n_C in the conduction band (and the holes in the valence band, n_V) controls the value of quasi-Fermi levels, E_{FC} (and E_{FV}).

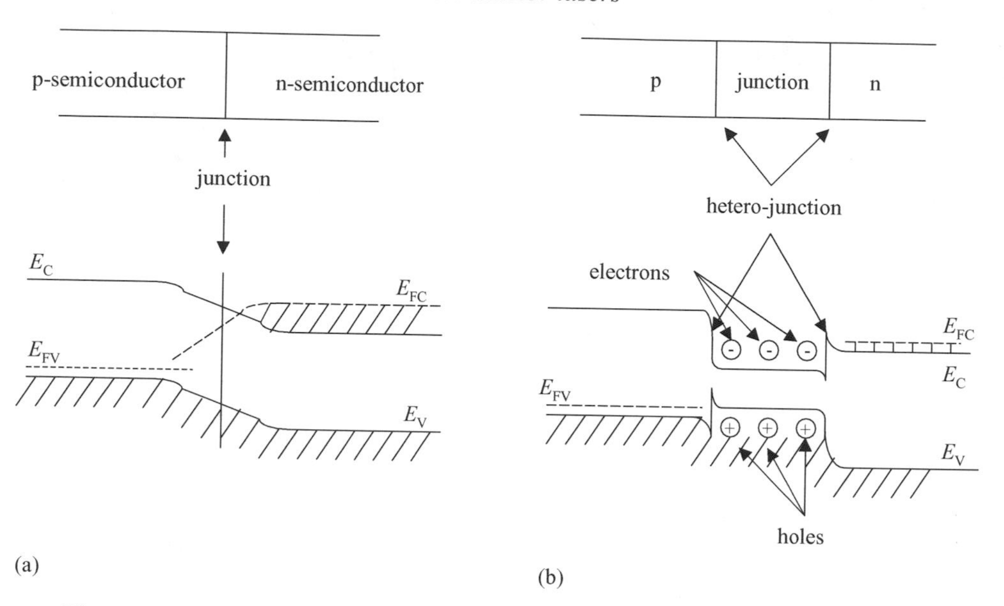

Figure 7.2. Energy band diagram and Fermi levels in a forward biased p–n junction. (a) Energy band diagram of a forward biased homo-junction. States in the shaded areas are most likely occupied by electrons. (b) Energy diagram of a forward biased double hetero-junction. States in the shaded areas are most likely occupied by electrons.

The carrier density, thus the Fermi level, is controlled by current injection. The Fermi level in turn controls the susceptibility (i.e. the gain of the laser material).

In a p–n junction, E_C and E_V change with position (see ref. [1] for more details). The values of E_{FC} and E_{FV} relative to E_C and E_V will also change as a function of position in a p–n junction. In quasi-equilibrium, E_{FC} and E_{FV} are independent of the position in the p and the n regions, outside the junction. Otherwise, current in the lateral directions, x and y, will flow. Figure 7.2(a) illustrates the band diagram and the quasi-Fermi levels in a forward biased p–n junction in a homogeneous medium, as well as the E_C and E_V, as a function of position. Figure 7.2(b) [6] illustrates the case for a p–n double hetero-structure diode in which the material in the active region has a band gap lower than the material for the p and n regions. As we will discuss in Section 7.5, the larger band gap is used to reduce the leakage of the carriers from the junction region.

7.1.4 Stimulated emission and absorption and susceptibility for small electromagnetic signals

Stimulated emission (or absorption) can take place only when the upper energy state in the conduction band is occupied by an electron (or empty) and when the lower electron energy state in the valence band is empty (or non-occupied by electrons, i.e. occupied by holes). Alternatively, we say that emission takes place when an electron and a hole recombine, and absorption takes place when the radiation generates an electron–hole pair. Thus, for a specific pair of energy states, the probability for net emission to take place, between an energy state E_2 in the conduction band and state E_1 in the valence band, is proportional to

$$f_C(E_2)[1 - f_V(E_1)] - f_V(E_1)[1 - f_C(E_2)] = f_C(E_2) - f_V(E_1). \quad (7.6)$$

For direct transitions, the $|\underline{k}|$ of the electrons and holes (generated or recombined) does not change, or $k_{\text{electron}} \approx k_{\text{hole}}$ with $\Delta k \approx 0$. Most semiconductor lasers use direct semiconductor transitions because direct transition probabilities between individual states are much larger than the indirect transitions that involve a change of $\underline{k}$. Let the photon energy of the radiation be $h\nu$; then $E_2 - E_1$ should equal the photon energy. The upper and lower levels, E_2 and E_1, for the energy states with eigen value $\underline{k}$ are given in Eqs. (7.2a) and (7.2b):

$$E_2 = E_C + \frac{\hbar^2 |\underline{k}|^2}{2m_e},$$

$$E_1 = E_V - \frac{\hbar^2 |\underline{k}|^2}{2m_h},$$

so

$$E_2 - E_1 = h\nu = E_g + \frac{\hbar^2 |\underline{k}|^2}{2m_r}, \quad (7.7)$$

where

$$E_g = E_C - E_V, \qquad \frac{1}{m_r} = \frac{1}{m_e} + \frac{1}{m_h},$$

where m_r is the reduced effective mass of the electron–hole pair and E_g is the energy of the band gap. This is a result of the parabolic approximation of the energy band diagram. There are many pairs of energy states that have the same $h\nu$ within a range $d\nu$.

Similar to the case of an inhomogeneous broadened line discussed in Section 5.4, the total susceptibility of the stimulated emission and absorption for a given $h\nu$ is obtained by integrating the χ_ξ of all the individual transitions for various ν_ξ values. The integrand, $\chi_\xi \, d\nu_\xi$, is the product of (1) the χ_ξ of the individual transition,

(2) the probability for allowing the transition as prescribed by the Fermi distribution in Eq. (7.6), and (3) the number of direct transitions per unit frequency ν_ξ range given by the density of states. From Eqs. (5.50) (see also ref. [5]), we obtain

$$\chi''_\xi(\nu, \nu_\xi)\, d\nu_\xi = \frac{\mu^2}{2\varepsilon_0 \hbar} \left\{ \frac{\frac{(\Delta\nu)}{2\pi}}{\left(\frac{\Delta\nu}{2}\right)^2 + (\nu - \nu_\xi)^2} \right\} [f_C - f_V] \frac{2|\underline{k}|m_r}{\hbar\pi}\, d\nu_\xi, \qquad (7.8)$$

where the following relations have been obtained from Eq. (7.7):

$$\nu_\xi = \frac{1}{h}\left(\frac{\hbar^2 |\underline{k}|^2}{2m_r} + E_g\right),$$

$$d|\underline{k}| = \frac{2\pi\, m_r}{\hbar\, |\underline{k}|}\, d\nu_\xi$$

and, from Eqs. (7.3a) and (7.3b),

$$\rho \frac{1}{V}\, d|\underline{k}| = \frac{|\underline{k}|^2}{\pi^2} \cdot \frac{2\pi\, m_r}{\hbar |\underline{k}|}\, d\nu_\xi$$
$$= \frac{2|\underline{k}|m_r}{\pi\hbar}\, d\nu_\xi.$$

Clearly, for small $\Delta\nu$, the $\{\ \ \}$ in Eq. (7.8) can be approximated by $\delta(\nu - \nu_\xi)$. Therefore, after integrating both sides of Eq. (7.8), we obtain

$$\left.\begin{aligned} \chi''(\nu) &= \frac{\mu^2 |\underline{k}| m_r}{\pi\, \varepsilon_0 \hbar^2} [f_C(h\nu) - f_V(h\nu)] \\ &= \frac{\mu^2}{2\pi\, \varepsilon_0 \hbar} \left(\frac{2m_r}{\hbar}\right)^{3/2} [f_C(h\nu) - f_V(h\nu)] \sqrt{\nu - (E_g/h)}, \\ \chi'(\nu) &= \frac{\mu^2 |\underline{k}| m_r}{\pi\, \varepsilon_0 \hbar^2} \int \left\{ \frac{\frac{(\Delta\nu)}{2\pi}}{\left[\frac{\Delta\nu}{2}\right]^2 + (\nu - \nu_\xi)^2} \right\} [f_C - f_V] \frac{(\nu_\xi - \nu)}{\frac{\Delta\nu}{2}}\, d\nu_\xi \end{aligned}\right\}. \qquad (7.9)$$

Here, the relation between χ'_ξ and χ''_ξ is obtained from Eqs. (5.41) and (5.49). Alternatively, one can obtain χ' from χ'' from the Kramers–Kronig relation presented in Eqs. (5.42).

7.1.5 Transparency condition and population inversion

In the 0 K approximation, f_C and f_V are either zero or unity:

$$f_C = u\,(E_2 - E_{FC}) = u\left(-E_{FC} + E_C + \frac{\hbar^2|\underline{k}|^2}{2m_e}\right), \tag{7.10a}$$

$$f_V = u\,(E_1 - E_{FV}) = u\left(E_V - \frac{\hbar^2|\underline{k}|^2}{2m_h} - E_{FV}\right) = u\left(E_2 - E_g - \frac{\hbar^2|\underline{k}^2|}{2m_r} - E_{FV}\right). \tag{7.10b}$$

Here, u is the unit step function that equals unity for positive arguments and zero for negative arguments. For a given $h\nu$ (i.e. $E_2 - E_1$) of the radiation, the $|\underline{k}|$ value is given by Eq. (7.7). In order to obtain gain, we need a negative χ'', or $f_C = 0$ ($E_2 < E_{FC}$) and $f_V = 1$ ($E_1 > E_{FV}$), or $E_2 - E_1 < E_{FC} - E_{FV}$. Similarly, for absorption we need positive χ'', or $E_2 > E_{FC}$, and $E_1 < E_{FV}$.

At other temperatures [7], we also have gain when

$$E_g \le E_2 - E_1 \le E_{FC} - E_{FV}. \tag{7.11}$$

Therefore, $E_{FC} - E_{FV} = E_2 - E_l$ is known as the transparency condition of semiconductor lasers. This condition is equivalent to that required for population inversion in solid state and gaseous lasers. E_{FC} (and E_{FV}) is determined by the density of electrons n_C (and holes n_V) in the conduction (and the valence) band. Equation (7.5) showed this relationship. In other words, there is an n_C (and n_V) required for achieving transparency.

Whenever the gain per unit length is larger than the residual propagation loss per unit length, there will be laser amplification. However, in order not to distract our attention away from learning the basic physics, our discussion on semiconductor laser amplification will be deferred to Section 7.7. First we will discuss the laser oscillation.

7.2 Threshold and power output of laser oscillators

Like all lasers, laser oscillation in a given cavity mode begins when the intensity gain of the mode (due to the $|\chi''|$ of the induced transition) exceeds the total loss, including the internal loss and the output coupling. At threshold, the required gain of the intensity of the oscillating mode, $\Gamma\gamma_t$, satisfies the condition:

$$e^{\Gamma\gamma_t L}\,\mathrm{Re}^{-\alpha_i L'} = 1,$$

or

$$\Gamma\gamma_t L = \alpha_i L' + \ln\left(\frac{1}{R}\right). \tag{7.12}$$

Here, γ_t is the required gain of the active medium at threshold, α_i is the propagation loss coefficient of the oscillating mode, L' is the propagation length per pass of the oscillating mode in the cavity, L is the propagation length per pass of the oscillating mode in the active medium, R is the equivalent intensity reflection coefficient for reflectors. For lossless reflectors, $1 - R$ is the effective transmission to the output. Γ is the optical filling factor of the active medium. It is the ratio of the optical energy of the mode in the active medium to the total optical energy of the mode in the cavity. It varies according to the configuration of the resonant cavity. A more explicit discussion of Γ will be given in Section 7.4.

The required γ_t is obtained by injecting carriers into the active layer in order to obtain the required $[f_C(h\nu) - f_V(h\nu)]$. The charge neutrality condition requires that $n_C = n_V$. The Fermi levels are determined by the carrier densities n_C and n_V. Therefore, there is a required $n_{C,t}$, corresponding to the γ_t at the threshold of oscillation.

For the current I injected into the laser, a portion of it, η_i, will be channeled into the active region and will emit photons. The current I is the integration of the current density (i.e. the current per unit area in the xy plane) J over the area of the device in the xy plane. The remainder, $1 - \eta_i$, of the injected carriers is lost as heat, electrical dissipation, carriers diffused into adjacent regions and carriers that are non-radiatively combined. Under the steady state condition and just before oscillation, the rate at which carriers are injected into the active layer equals the electron–hole decay rate. Thus we obtain

$$\eta_i \frac{J_t}{q} = \frac{\delta\, n_{C,t}}{\tau}. \tag{7.13}$$

Here, J_t is the threshold injection current density (number of injected carriers per unit area), δ is the thickness of the active layer, and τ is the lifetime of the holes and electrons in the active layer.

In all lasers, as we learned in Chapter 6, γ (i.e. n_C) is locked to its threshold value whenever the injection current exceeds its threshold at steady state. Saturation occurs because when the injected carrier density is above the threshold it will yield a larger photon density in the oscillating mode, and a larger photon density reduces the carriers via increased stimulated transitions in the active layer. The increase in carrier generation rate is balanced by the increased reduction of carriers. In other words, after oscillation begins, saturation will occur so that the saturated gain is always equal to the total losses, which include internal loss and output. As the laser current I exceeds its threshold, I_t, the total additional number of carriers recombined per unit time due to the stimulated transition will be proportional to $\eta_i(I - I_t)/q$. Thus the power emitted by stimulated emission is

$$P_e = \eta_i (I - I_t) \frac{h\nu}{q}. \tag{7.14}$$

Part of this power is dissipated inside the laser resonator and the rest is coupled out through the output end. We obtain the output power of a semiconductor laser as

$$P_o = P_e \frac{(1/L') \ln(1/R)}{(1/L') \ln(1/R) + \alpha_i}. \tag{7.15}$$

The external differential quantum efficiency, η_{ex}, is the ratio of the increase in photon output rate (that results from an increase in the injection) to the increase in the injection rate. It is

$$\eta_{ex} = \frac{d\left(\frac{P_o}{h\nu}\right)}{d\left(\frac{I}{q}\right)} = \frac{q}{h\nu} \cdot \frac{dP_o}{dI}. \tag{7.16}$$

Figure 7.3 shows the curve of light output as a function of drive current. The kink represents the beginning of the oscillation threshold. Note that the change of P_o is directly proportional to the change of I above the threshold. This means that the modulation of the light output will be directly proportional to I without much non-linear distortion. For this reason, there has been a great deal of interest in direct modulation of semiconductor lasers.

It is also interesting to keep in mind that when the gain saturates after oscillation, the carrier density in the active region is clamped to $n_{C,t}$. In other words, the quasi-Fermi levels are locked to their threshold values. This is what we mean by saturation. Such a saturation process is different from the saturation processes of solid state and gas lasers.

7.2.1 Light emitting diodes

It is interesting to note that, at injection current density smaller than J_t, a laser functions as a light emitting diode (LED). Let us define a radiative efficiency η_r as the fraction of the injected carriers that recombine through the radiation process with respect to the total injected carriers that decayed or leaked out through all mechanisms. Let us also define an optical collection efficiency η_c as the fraction of the optical power collected as the output of an LED with respect to the total optical

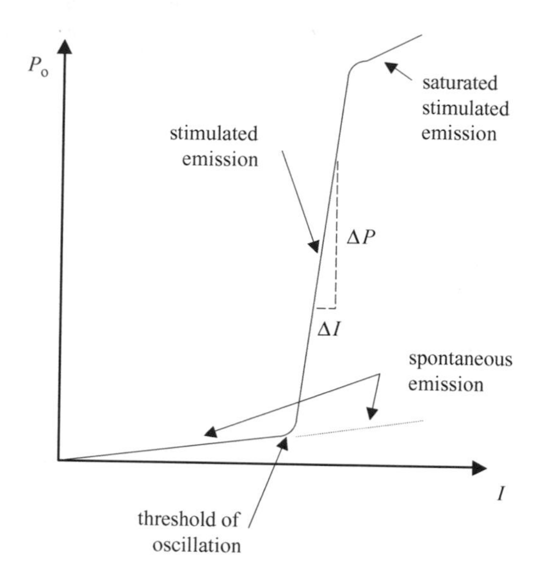

Figure 7.3. Output optical power versus the injection current for a diode laser. Below the threshold of oscillation there is spontaneous emission. The device is a light emitting diode (LED). Above threshold the power output due to stimulated emission is linearly proportional to injection current until saturation occurs.

power emitted in an LED. Then the output power of an LED is

$$P_{\text{LED}} = \eta_c \eta_r \frac{h\nu}{q} I \,. \tag{7.17}$$

7.3 Susceptibility and carrier densities in quantum well semiconductor materials

In order to understand how the susceptibility is affected by the material structures that have various densities of states, the susceptibility of a quantum well structure is presented here. A quantum well structure has the important property that its density of states is different from the density of states in a bulk material discussed in the previous section. For a given electron density, the quasi-Fermi level is higher in quantum well materials because of its smaller density of states. Thus, the susceptibility for a given carrier density is affected significantly by the density of states. A much more extensive discussion of quantum well lasers is given in ref. [8].

Quantum well structures are typically grown epitaxially on InP or GaAs substrates. Well layers usually have a smaller bandgap E_Γ, and barrier layers have a larger bandgap E_g, as illustrated in Fig. 7.4 for a single well among barriers. The growth direction is designated here as the z axis. The well thickness L_z is typically in the range of 50 to 150 Å in the z direction. The barrier thickness is typically less than 100 Å, just larger than the evanescent tail of the wave functions of the energy

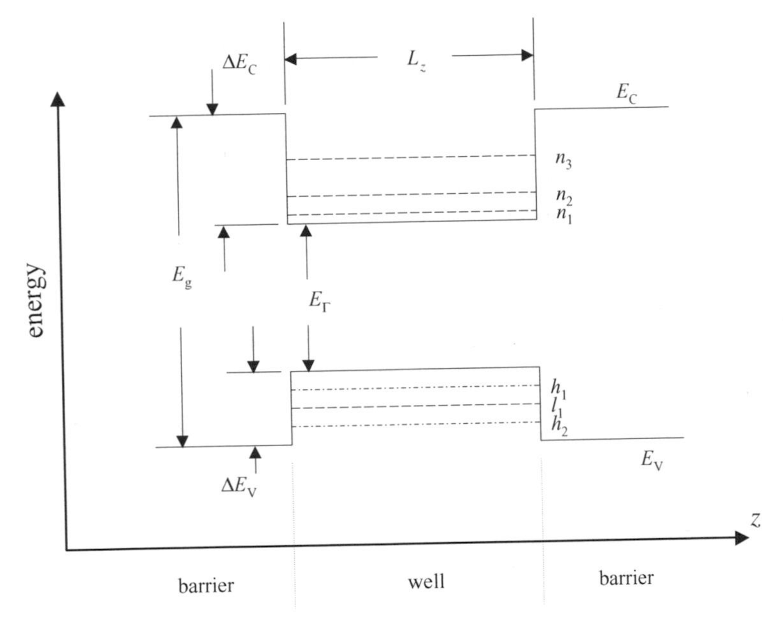

Figure 7.4. Energy band diagram and energy levels of electrons and holes in a quantum well. The thickness of the well is L_z. The well with an energy band gap E_Γ is sandwiched between barriers with a larger energy band gap E_g. The energy levels of the electrons are shown as n_1, n_2 and n_3. The energy levels of the heavy holes are shown as h_1 and h_2, while the first energy level of the light holes is shown as l_1.

states for quantum wells. In the lateral x and y directions, the material thickness and composition are uniform. For lattice matched well and barrier layers, the materials in the quantum wells have the same periodic crystalline variation in the xy directions as the host lattice.

The total discontinuity of the bandgap energy ΔE_g at the well–barrier interface, $E_g - E_\Gamma$, comprises ΔE_C of the conduction band and ΔE_V of the valence band.

7.3.1 Energy states in quantum well structures

The total energy of any state is the sum of the energy in the z direction and the energy in the xy directions designated by $|\underline{k}|$. See ref. [8] for a more detailed discussion of the energy states of quantum well hetero-structures. The energy eigen states in the z direction (for both the conduction and valence bands) will be the quantum mechanical solutions for electron and hole states in their potential wells. The discrete energy levels of the electrons and holes in a well are illustrated in Fig. 7.4. The energies of electron states in the conduction band are designated as n_j. Furthermore, the quantum size effect separates the light-hole from the heavy-hole energy band in the valence band. Therefore, in the valence band, there are light-hole energy states, l_1, l_2,... and heavy-hole energy states, h_1, h_2, ... Because of the larger effective mass of the heavy holes, h_1 is usually closer to E_V than is l_1.

In the two-dimensional periodic structure in the lateral direction, the *xy* variations of the wave functions of energy states still have the same form given by Eq. (7.1), with $\underline{k}$ restricted to the *xy* directions. Therefore the total energy of the upper state in the conduction band, i.e. E_2, is

$$E_2(|\underline{k}|) = n_j + \frac{\hbar^2|\underline{k}|^2}{2m_e}, \quad E_2 \geq n_j. \tag{7.18}$$

For the lower energy state in the valence band,

$$E_1(|\underline{k}|) = h_j - \frac{\hbar^2|\underline{k}|^2}{2m_{hh}} = l_j - \frac{\hbar^2|\underline{k}|^2}{2m_{lh}}, \tag{7.19}$$

if $E_1 \leq h_j$ or l_j.

7.3.2 Density of states in quantum well structures

In the conduction band in quantum well structures, the number of energy states for a specific E_2 is determined from Eq. (7.18). For $n_2 > E_2 > n_1$, there is only one $|\underline{k}|$ value for each E_2. For $n_3 > E_2 > n_2$, there are two $|\underline{k}|$ values for each E_2, $|\underline{k}'| = (1/\hbar)\sqrt{2m_C(E_2 - n_2)}$ and $|\underline{k}''| = (1/\hbar)\sqrt{2m_C(E_2 - n_1)}$. In other words, for $n_3 > E_2 > n_2$, there will be two energy states involved in an induced transition. The energy states will be identified by their $|\underline{k}|$, $|\underline{k}'|$, $|\underline{k}''|$, etc. Similarly, there will be j different $|\underline{k}|$ values for $n_j < E_2 < n_{j+1}$. Following the same pattern, the number of states in the valence band is determined from Eq. (7.19).

In summary, the number of energy states (identified by the eigen values of $\underline{k}$ in the *xy* directions) per unit volume V in the quantum well structure is different than the number of states (identified by the $\underline{k}$ in three dimensions) per unit volume in the bulk material. The density of states per unit volume for a specific $|\underline{k}|$ is [8]:

$$\rho(|\underline{k}|)\, d|\underline{k}| = \frac{1}{L_z}\frac{|\underline{k}|}{\pi}\, d|\underline{k}|. \tag{7.20}$$

Equation (7.20) applies to all the energy states in the conduction and valence bands.

For each $|\underline{k}|$ value associated with the n_j,

$$\rho_C(E_2)\, dE_2 = \frac{1}{\pi L_z}\sqrt{\frac{2m_e(E_2 - n_j)}{\hbar^2}}\, d|\underline{k}| = \frac{1}{2\pi}\left(\frac{2m_e}{\hbar^2}\right)\frac{1}{L_z}\, dE_2. \tag{7.21}$$

The total $\rho(E_2)$ for $E_2 > n_j$ is the sum of such terms for each $|\underline{k}|$ and n_j that yields the E_2 in Eq. (7.18). Therefore, for an E_2 in the conduction band of a quantum well

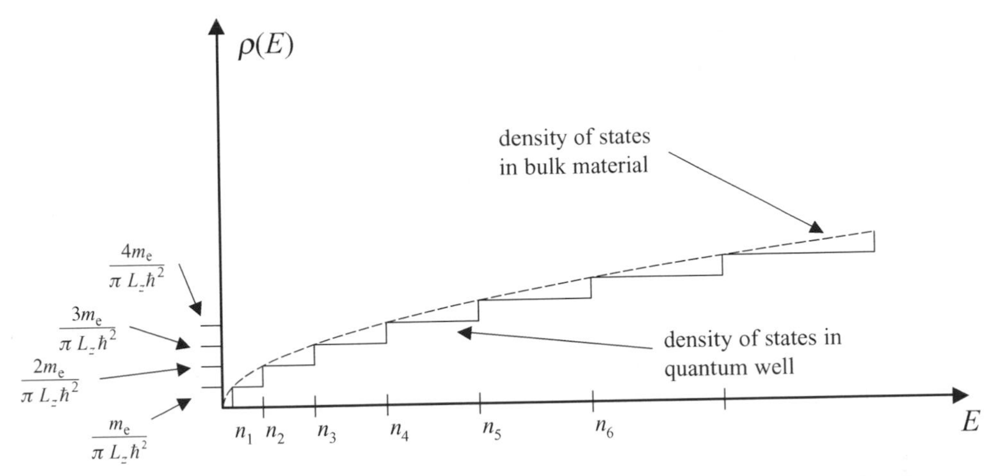

Figure 7.5. Density of electronic states in a two-dimensional quantum well compared with the density of states in a three-dimensional bulk material. The quantum well is assumed to have infinite ΔE_C. The density of states of a bulk material forms an envelope for the steps of the quantum well case.

structure, allowing different possible n_j [8],

$$\rho_C(E_2)\,dE_2 = \frac{1}{2\pi L_z}\left(\frac{2m_e}{\hbar^2}\right)\sum_j u(E_2 - n_j)\,dE_2. \tag{7.22}$$

Similar relations hold for E_1 in the valence band with respect to h_m and l_m energy levels in the z direction,

$$\rho_V(E_1)\,dE_1 = \frac{1}{2\pi L_z}\left\{\left(\frac{2m_{hh}}{\hbar^2}\right)\sum_j [u(h_j - E_1)] + \left(\frac{2m_{lh}}{\hbar^2}\right)\sum_j [u(l_j - E_1)]\right\} dE_1. \tag{7.23}$$

The effective mass of the heavy hole, m_{hh}, is different (usually larger) than the effective mass of the light hole, m_{lh}. Quite often, $n_1 < E_2 < n_2$, $E_1 > l_1$ and $h_1 > E_1 > h_2$, and therefore only the first terms of the series in Eqs. (7.22) and (7.23) are used. Figure 7.5 illustrates the density of states as a function of E_2 for the conduction band, in the bulk and in the two-dimensionally constrained quantum well materials. In this figure, the quantum well is assumed to have infinite ΔE_C.

7.3.3 Susceptibility

The susceptibility of a quantum well material can be calculated in a manner similar to the calculation of χ in Eqs. (7.8) and (7.9), using the density of states of two-dimensional periodic structures. Like the results obtained for the bulk media, the transparency condition is achieved when $E_{FC} - E_{FV} \geq E_2 - E_1 \geq n_1 - h_1$. Here $E_2 - E_1$ is the photon energy, and the highest energy level in the valence band is assumed to be h_1.

7.3.4 Carrier density and Fermi levels

In semiconductor lasers, charge carriers are injected into the active layer by the injected current in a forward biased p–i–n diode. The total number of electrons per unit volume in the conduction band is related to E_{FC} through the relation

$$n_C = \int_{E_C}^{\infty} \rho_C(E_2) f_C(E_2)\, dE_2. \tag{7.24}$$

Similarly, one can calculate E_{FV} due to the injected holes in the valence band. The number of injected carriers required to achieve the transparency condition or the threshold condition for laser oscillation is much less in quantum well materials than in the bulk, because of the big difference in the density of states. This effect is especially significant when L_z is less than 200 Å.

7.3.5 Other quantum structures

For unstrained quantum wells, heavy-hole transitions dominate because of the larger effective mass m_{hh} and smaller energy shift. Fields which have the electric field polarized perpendicular to the z direction, i.e. the TE polarized fields, have a larger gain. For strained quantum well layers, the presence of biaxial tension (or compression) alters the cubic symmetry of the semiconductor. The separations of both the heavy-hole and the light-hole band edge from E_V decrease under tension (or increase under compression). The degeneracy of the valence band edge for heavy-hole and light-hole bands is removed. Compressive strain yields a reduction in the hole effective mass and a reduction in the required carrier density to reach transparency and hence the oscillation threshold.

The density of states is further reduced in quantum wire and quantum dot structures, potentially yielding an even lower threshold of carrier density for oscillation.

7.4 Resonant modes of semiconductor lasers

Resonant cavities of semiconductor lasers are formed on material structures grown epitaxially on semiconductor substrates. They differ from cavities of solid state and gas lasers in several ways. (1) Their dimensions are different. Whereas cavities of solid state and gas lasers have reflectors with lateral dimensions of millimeters and cavity lengths of centimeters or meters, the cavities of semiconductor lasers have typical lateral dimensions of micrometers. The length of a typical long semiconductor laser is less than a few hundred micrometers. Instead of Gaussian mode analysis, guided wave analysis such as that discussed in Chapters 3 and 4 is used for analyzing edge emitting (i.e. in-plane) lasers. (2) Whereas the fields of oscillating modes are mostly contained within the gain region in solid state and gas lasers, the size of semiconductor laser resonant modes is often larger than the gain region. Attenuation of the mode outside the gain region may be high. Thus, optical confinement of the resonant mode is an important design consideration. (3) In some semiconductor lasers, resonance in the longitudinal direction can be obtained by distributed feedback as well as by end reflection. (4) Cavity design must be consistent with the design of material structure aimed at current confinement and reduction of carrier leakage.

Semiconductor laser cavity configurations must conform to the materials that can be grown by epitaxial growth technology, where the growth direction is designated as the vertical z direction in this chapter. There are two types of semiconductor lasers: the edge emitting lasers and the surface emitting lasers. (1) Edge emitting (or in-plane) lasers typically have a material structure that is a single transverse mode waveguide in the z direction. It consists typically of an optically active layer with gain and a high-index layer which serves as the core of the waveguide, surrounded by lower index cladding and contact layers. Cavity resonance is created by reflection of the guided mode between two ends in the longitudinal y direction. Material processing procedures such as photolithography, etching and regrowth are used to form a channel waveguide in the transverse direction x, perpendicular to the longitudinal direction y. Since the output beam is radiated from the end, this type of laser is called an edge emitting (or in-plane) laser. (2) Vertical cavity lasers typically have a bottom Bragg reflector, followed by the contact and active layers, then a top Bragg reflector or mirror. All layers are grown epitaxially in the z direction, which is also the longitudinal direction of the cavity. The resonance is obtained for TEM-like waves propagating back and forth in the longitudinal direction through the various layers and reflected by the top and bottom reflectors. The output beam is radiated either from the top reflector or from the bottom reflector (through the substrate). Thus these lasers are called vertical cavity surface emitting lasers (VCSELs). The index variation in the transverse direction is not strong enough to form a waveguide. Examples of both types of cavity are illustrated in Fig. 7.6.

7.4.1 Cavities of edge emitting lasers

Although the resonant modes of some lasers in earlier years were guided by the lateral variation of the gain generated from a non-uniform injected carrier density,

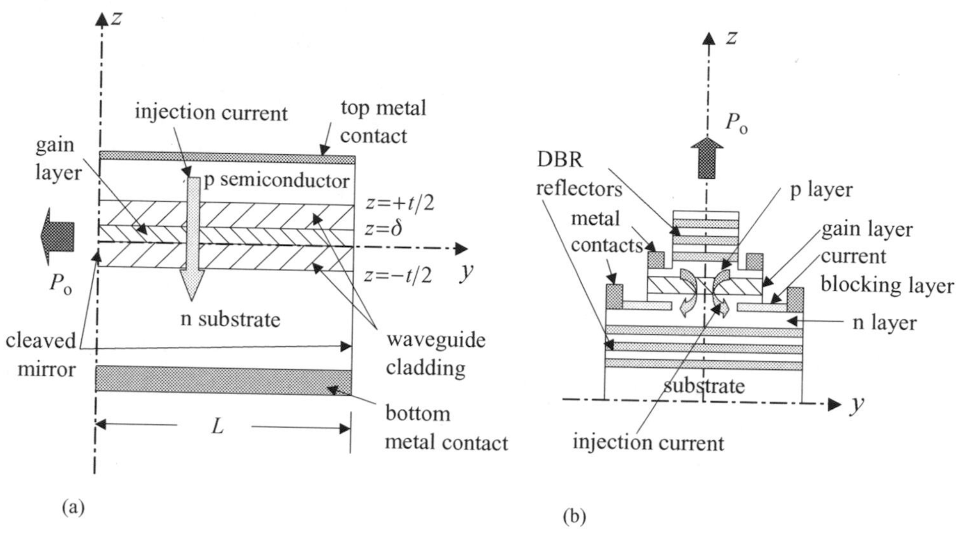

Figure 7.6. Cross-section of (a) an edge emitting (or in-plane) laser and (b) a vertical cavity surface emitting laser (VCSEL). (a) An example of an edge emitting (i.e. in-plane) laser. The gain layer is sandwiched between waveguide cladding layers (grown as a double hetero-structure) to form the high-index waveguide core for optical confinement in the z direction. Wave propagation and resonance are in the y direction. The p–n diode is formed by the p and n doping in various layers. The current is injected from the top and bottom metal contacts through the forward biased diode in the z direction. The higher band gap of the waveguide cladding layers serves the additional function of impeding carrier leakage from the gain layer. Optical and carrier confinement in the x direction may be achieved by etching and regrowth of, or diffusion in, the materials outside the channel region. (b) A vertical cavity surface emitting laser. The top and bottom Bragg reflectors, as well as the gain layer, are grown epitaxially on the substrate. The p–n diode is formed by the p and n doping in various layers between the metal ring electrodes. The current is injected through the forward biased metal ring contacts. The current blocking layer forces the injected current to be concentrated in the center region, coincident with the optical wave.

most semiconductor edge emitting lasers today have index variations in the lateral directions intended for control of the electromagnetic mode.

In the type of edge emitting laser illustrated in Fig. 7.6(a), the ternary and quaternary alloy layers of the hetero-structures are grown epitaxially on GaAs or InP substrate. The waveguide consists of a waveguide core (from $z = -t/2$ to $z = +t/2$) and an active layer with thickness δ within the waveguide, with $t \gg \delta$ (shown in Fig. 7.6(a) as located from $z = 0$ to $z = \delta$). The waveguide and the active layer have higher refractive indices than those of the surrounding contact layers. The active layer might just be a very thin quantum well layer. The indices of different layers are controlled by their composition. The lower the band gap, the higher the refractive index. The contact layers are chosen so that there is, in effect, a p–n diode in the z direction. Carriers are injected into the active layer from the top and bottom electrical contacts. In the other transverse direction x, not shown in Fig. 7.6(a), the material outside the channel waveguide core also has a lower index than the waveguide core materials, obtained by subsequent processing steps after the initial epitaxial growth.

The passive transverse modes of such waveguide structures have already been discussed in Chapter 3. They are identified by the effective index for the mode propagating in the longitudinal y direction and by the evanescent tails in the lower index regions in the z and x directions. Note that the size of the guided wave mode is usually much larger than the thickness of the active layer. Resonance is created by reflections of guided waves propagating in the y direction. Typically, the cavity length in the longitudinal direction is tens or hundreds of micrometers, whereas the transverse dimension of the guided wave mode is of the order of one micrometer or less.

The gain of the guided wave mode provided by the thin active layer could be analyzed by the perturbation analysis discussed in Chapter 4. Let the guided wave mode be $\underline{e}_j(x, z)$. Following Section 4.1.3, the χ'' provided by the active layer can be regarded as a perturbation $\Delta\varepsilon$ to the χ of all the layers that defined the mode. Let $\Delta\varepsilon$ be uniform within an active layer which extends from $z = 0$ to $z = +\delta$ and from $x = -w/2$ to $x = +w/2$. From Eqs. (4.6) and (4.7) and for the jth mode, with $a_j(y)\exp(-jn_{j,\text{eff}}k_0 y)$ variation in the longitudinal y direction, we have

$$\Delta\varepsilon = \varepsilon_0(-j\chi'')$$

and

$$\begin{aligned} \frac{da_j}{dy} &= -ja_j \left[\frac{\omega}{4} \int_{-w/2}^{w/2} \int_{0}^{+\delta} (\Delta\varepsilon)\, \underline{e}_j \cdot \underline{e}_j^* \, dz\, dx \right] \\ &= -j\Gamma\, a_j \left[\frac{\omega}{4} (\Delta\varepsilon) \int_{-\infty}^{+\infty} \int_{-\infty}^{+\infty} \underline{e}_j \cdot \underline{e}_j^* \, dz\, dx \right] \\ &= -\Gamma \left[\frac{\beta_j}{2(n_{\text{eff},j})^2} (\chi'') \right] \cdot a_j. \end{aligned} \tag{7.25}$$

$$\Gamma = \frac{\displaystyle\int_{-w/2}^{+w/2} \int_{0}^{+\delta} \underline{e}_j \cdot \underline{e}_j^* \, dz\, dx}{\displaystyle\int_{-\infty}^{\infty} \int_{-\infty}^{\infty} \underline{e}_j \cdot \underline{e}_j^* \, dz\, dx},$$

which is the optical filling factor. It is interesting to note that the gain for Gaussian modes in an unbounded medium in solid state and gas lasers is $-\omega\sqrt{\mu_0\varepsilon}\chi''/n^2$. According to Eqs. (5.43) and (7.25), the intensity of the jth guided wave will grow as $\exp(\Gamma\gamma y)$, where $\gamma = -(\beta_j\chi'')/n^2_{\text{eff},j}$. Clearly a negative χ'' (neglecting any propagation loss) will yield amplification of a_j as it propagates in the y direction.

In practice, $\Gamma \ll 1$ for many lasers. Therefore, epitaxial layers, such as a double hetero-structure (DH) with a low band gap material surrounded by higher band gap layers, are used to increase optical confinement, i.e. the Γ. This also stabilizes the mode with respect to small variation of χ'. Furthermore, DH layers outside the waveguide mesa in the x direction may be etched away and a layer of higher band gap material may then be regrown on it. Alternatively, instead of etching, the band gap of the material surrounding the mesa may be increased by implementing diffusion processes. The idea is to have a lower index material to surround the channel waveguide. These configurations are called the buried hetero-structure (BH) configurations. Figure 7.7 illustrates two examples of BH configurations. The channel waveguide in Fig. 7.7(b) is obtained by etching a V groove into the substrate, then following up with epitaxial growth.

The χ' created by current injection may, at times, also contribute to the perturbation of the lateral mode pattern in bulk lasers made of a homogeneous material. However, this contributes very little to any change of mode pattern in quantum well lasers because the quantum well layer is typically much less than 0.1 μm thick. Indeed, even the existence of the quantum well layers is usually neglected in the calculation of the field of guided wave modes.

Most commonly, the semiconductor waveguide is cleaved at both ends in the y direction. The cleavage provides an optically nearly perfect facet perpendicular to y. A dielectric discontinuity between the semiconductor and the air yields a 30% power reflection coefficient at each end over a broad band of wavelength. The length of the cavity L (see Fig. 7.6(a)) is the distance between the cleaved facets. Dielectric coatings may be applied to reduce or to increase the reflection, and to yield wavelength dispersion of the reflection coefficient. Since the $|\chi''|$ in the active region is larger for radiation polarized perpendicular to z than parallel to z, and since the propagation loss is lower for TE than for TM guided wave modes, the output is linearly polarized normally in the x direction for cavities oriented in the y direction. Oscillation of a mode begins when the gain of the mode exceeds the internal loss and the output coupling. Different longitudinal modes have similar gain and loss. Thus, cleaved cavity lasers are often multimode oscillators. The hopping of the oscillating longitudinal mode from one to another as a function of carrier injection creates undesirable characteristics, such as an increase in noise, poor oscillation wavelength stability and kinks in power output as the injection current increases.

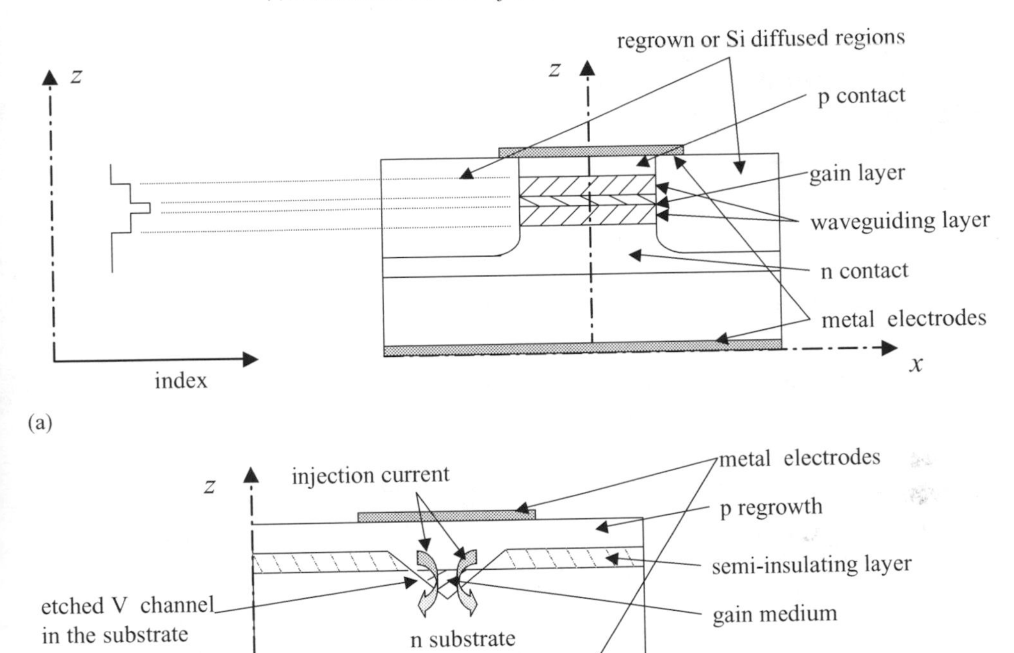

Figure 7.7. Two examples of BH edge emitting lasers. (a) Etched mesa or impurity induced disordered BH structure for current, photon and carrier confinement. The hetero-structure of the contact, gain and waveguiding layers is first grown epitaxially on the substrate. The waveguide and gain layers provide optical confinement of the guided wave mode in the z direction. Carrier confinement in the z direction is also provided by hetero-junctions. Regrowth of buried hetero-structure or diffusion is then used outside the channel waveguide for optical, current and/or carrier confinement in the x direction. (b) An etched channel substrate BH laser. A semi-insulating epitaxial layer is first grown on the substrate. A V groove is then etched so that the bottom extends into the conducting substrate. The double hetero-structure laser layer is then regrown. Due to the tendency of the regrowth to planarize, a thicker and separate active stripe is formed in the V groove where the current is constrained to flow. This figure is taken from Fig. 1.13 of ref. [6] by permission of John Wiley and Sons.

In order to have only one oscillating longitudinal mode, an interesting modification of the cleaved cavity is the use of a grating filter discussed in Section 4.2.1 to provide the desired reflectance at both ends and to select the resonant wavelength of the mode to oscillate. A grating at the end of the laser (for $y \geq L$ or $y \leq 0$) will yield a high power reflection coefficient within a narrow band width given in Eq. (4.14). The wavelength that will yield the maximum reflectivity is given by the Bragg condition in Eq. (4.9). The grating can be fabricated in materials which contain the evanescent field region of the guided wave mode. Lasers using grating reflectors are called distributed Bragg reflector (DBR) lasers.

A further modification of the DBR laser consists of introducing a continuous grating along the length of the cavity from $y = 0$ to $y = L$. This is intended to provide resonance in the y direction and to control the longitudinal mode [9]. This type is called a distributed feedback (DFB) laser. Theoretically, when there is a uniform grating of length L in the cladding layer, and when the waveguide is infinitely long in the y direction outside the grating region (i.e. the waveguide is not terminated by additional reflectors), Eqs. (4.12) (with appropriate boundary conditions) yields two independent equivalent longitudinal modes. In order to obtain only single-mode oscillation, this degeneracy can be removed by cleaving the waveguide at $y < 0$ and at $y > L$. The cleavage is used to control the phase of the reflected wave and to eliminate the oscillation of the second mode. However, the present cleavage process does not allow the position of the facet – and thus the phase of the reflected light – to be controlled precisely. Therefore, the long wavelength DFB lasers usually have a low-reflection coated front facet and a high-reflection coated rear facet to assure a single oscillating mode operation. The longitudinal mode degeneracy could also be removed by introducing a quarter-wavelength ($\lambda/4$) shift at the center of the device, without any cleaved facets.

7.4.2 Cavities of surface emitting lasers

Figure 7.6(b) illustrates an example of a vertical cavity laser. The longitudinal direction of the cavity is the vertical z direction of epitaxial growth. Resonance is obtained by a TEM-like wave propagating in the z direction through various layers and reflected by the Bragg reflectors at the bottom and on the top.

The index variation in the x and y directions is not strong enough to support a guided wave mode. Frequently, gain guiding determines the transverse beam size. Otherwise a mesa (i.e. a post) may be etched to control the transverse mode size. Sometimes a current blocking layer, such as that shown in Fig. 7.6(b), is used to direct the current just to the center region of the active layer to provide the gain more efficiently.

In order to provide effective coupling of its output to single-mode optical fibers, the lateral diameter of the optical output beam is typically of the order of 10 μm. The gain region overlaps substantially the mode in the transverse directions (i.e. $\Gamma \approx 1$). In such a cavity configuration, the mode propagating in the z direction is a TEM mode. The transverse variations of the cavity modes discussed in Chapter 2 are applicable here. However, in the longitudinal direction, the cavity length L in surface emitting lasers (including the effective penetration of the TEM waves into the Bragg reflectors) is typically just a few micrometers long. Within such a short distance and for a 10 μm mode size, the divergence of the beam and the diffraction loss per pass are negligible. Therefore, for the sake of simplicity, plane waves are commonly used to represent the fields inside the laser post for analyzing the resonance in the z direction. On the other hand, since the propagation distance in the gain medium is very short, very high reflectance with precisely controlled wavelength sensitivity is required for the end reflectors. For VCSELs, in terms of an analysis such as that given in Eq. (7.25), $L' \gg L$ ($L = \delta$, the thickness of the active layer in Fig. 7.6(b)), $\Gamma \approx 1$, and γ is related to the susceptibility by $\gamma = -\omega\sqrt{\mu_0\varepsilon}\chi''/n^2$.

The growth of periodic high- and low-index layers is carried out epitaxially to yield Bragg reflectors at the bottom and on the top. From Eq. (4.9), it is clear that the thickness of the high- and low-index layers (i.e. the periodicity) should be $\lambda_g/(2n)$, where λ_g is the desired free space wavelength of oscillation and n is the averaged index of refraction of the periodic layers. The magnitude and the wavelength range of the reflectivity will be controlled by the difference of the high and low index and by the number of layers. The lower the index difference, the larger is the number of layers required to achieve a high reflectivity, and the band width is narrower. The larger the number of layers, the more is precise control of the growth process required. For this reason, effective Bragg reflectors using a reasonably small number of layers can be obtained by means of AlAs and GaAs layers grown on GaAs substrate, while Bragg reflectors grown on InP substrates require many more layers. Naturally, λ_g needs to coincide with the wavelength range of large $|\chi''|$.

Additional micro-fabrication processing is sometimes used to define the lateral extent of the cavity, usually in the form of a circular laser post in the xy plane. Common techniques for current and gain confinement include proton implantation of the area outside the laser post, etching away the region outside of the laser post and regrowth of current blocking claddings around the etched laser post. For a given injected carrier density, the $|\chi''|$ is the largest for any radiation polarized in the xy plane. Since the laser cavity and the gain are symmetrical in the x and the y direction, the output can be linearly polarized in any direction in the xy plane. Additional optical elements must be added to remove this polarization degeneracy.

The primary advantages of a VCSEL oscillator are as follows. (1) The vertical cavity facilitates a circular low divergence beam, which can be coupled easily and efficiently with an optical fiber and bulk optics. (2) The emission wavelength is determined by the epitaxial growth rather than by micro-fabrication processes, and thus can be made with higher accuracy. (3) When the growth process can yield precisely controlled layer thickness as a function of x and y positions, large, monolithic arrays of single-wavelength lasers with distinct, equally spaced, wavelengths can be fabricated for wavelength division multiplexing (WDM) applications in optical fiber communication. (4) The volume of the active region is extremely small so that extremely low threshold current injection for laser oscillation can be obtained. (See ref. [10] for an extensive discussion on VCSELs.)

7.5 Carrier and current confinement in semiconductor lasers

For a given density of states, the electron and hole densities in the active region determine the unsaturated χ'', or gain, of the active region. Good current confinement and carrier leakage reduction would allow the desirable carrier density to be achieved with a small injection current. Under high injection levels, charge neutrality dictates that the electron density equals the hole density in the active region. Therefore, current confinement and carrier leakage in laser design could be discussed just in terms of electron densities. The discussion in this section applies to both edge emitting and surface emitting lasers.

Let the electron density in the active region be designated by n_{C}; then

$$\frac{dn_{\mathrm{C}}}{dt} = G - R_{\mathrm{recomb}}, \tag{7.26}$$

where G is the generation rate,

$$G = \frac{\eta_{\mathrm{i}} I}{qV}.$$

Here, I is the terminal injection current from the driver, q is the charge of an electron, V is the volume of the active region and η_{i} is the internal quantum efficiency, which is the fraction of the terminal current that generates carriers in the active region. R_{recomb} is the total rate of recombination of the electrons in the active region; it has four components:

$$R_{\mathrm{recomb}} = R_{\mathrm{stim}} + R_{\mathrm{spont}} + R_{\mathrm{leak}} + R_{\mathrm{nrad}}, \tag{7.27}$$

i.e.

$$R_{\mathrm{recomb}} = R_{\mathrm{stim}} + \frac{n_{\mathrm{C}}}{\tau}. \tag{7.28}$$

R_{spon}, R_{stim} and R_{nrad} are the rates at which electrons and holes combine via spontaneous emission, stimulated transition and non-radiative recombination, respectively; R_{leak} is the rate at which electrons are leaked to the areas outside of the active region; τ is the carrier lifetime in the absence of stimulated emission. Clearly, for a given I, n_{C} can be increased by increasing η_{i} and τ.

The common method of reducing the leakage (i.e. of increasing τ) is to use a hetero-barrier to impede the electrons from leaving the active region. The double hetero-structure illustrated in Figs. 7.6(a) and 7.7(a) serves the purpose of both optical confinement and barrier to impede carrier leakage.

An increase in η_i is achieved through the use of current blocking layers made from dielectric insulators, epitaxially regrown high band gap materials or impurity diffused materials. The objective is to channel the current efficiently into the active region. For example, the regrown or the Si diffuse layer of the BH structure shown in Fig. 7.7(a) serves the current confinement objective in addition to the objective of confining the optical guided wave mode. The objective of the semi-insulating layer in the device illustrated in Fig. 7.7(b) is to confine the current to the V groove region. The current blocking layer of the VCSEL illustrated in Fig. 7.6(b) is necessary to restrict the current to the area in the xy plane where amplification of the TEM wave is desired.

7.6 Direct modulation of semiconductor laser output by current injection

The discussions in the previous sections are valid under the steady state condition. For a sufficiently low frequency, when the steady state condition is satisfied, the laser output as a function of time is proportional to the input current, $I(t) - I_t$. Therefore, the output of a given laser oscillator can be modulated directly by the applied current. This is an attractive method for producing an intensity modulation of the signal at very low cost. Furthermore, the intensity will be linearly proportional to the current. However, in order to extend the modulation technique to a high frequency or to a short pulse, the steady state condition might not be satisfied. We must in that case examine the time dependent variation of n_C. For this reason, we will analyze the frequency variation of the laser output in response to an input current modulation at various frequencies.

Similar to Eqs. (6.1), the instantaneous rate at which n_C in the active layer and the number of photons in the oscillating mode N_p are increased or decreased in time is governed by the following rate equations [6]:

$$\begin{aligned}\frac{dn_C}{dt} &= \frac{\eta_i I}{qV} - \frac{n_C}{\tau} - A(n_C - n_{C,t})N_p,\\ \frac{dN_p}{dt} &= A\frac{V}{V_p}(n_C - n_{C,t})N_p - \frac{N_p}{\tau_p}.\end{aligned} \tag{7.29}$$

We assume that the oscillation of the mode has already been achieved. Therefore, the recombination rate due to spontaneous emission is negligible compared with the recombination rate of the stimulated emission. N_p is the number of photons per unit volume in the cavity for the oscillating mode; $n_{C,t}$ is the carrier density at threshold; V is the volume of injected carriers; V_p is the volume of the electromagnetic resonant mode in the cavity; A is the rate at which electrons and holes recombine, induced by the electromagnetic mode in the cavity; $1/\tau_p$ is the rate at which photons in the cavity mode decay, with or without significant stimulated transition (it includes the effect of passive cavity losses and output coupling); V/V_p is approximately the optical filling factor Γ. For the sake of simplicity, we have ignored in Eqs. (7.29) the spatial variations of these quantities.

In order to reconfirm the validity of Eqs. (7.29), we will first examine their solution at steady state. In the steady state and well above threshold, the time derivatives of n and N are approximately zero. We obtain from Eqs. (7.29),

$$\eta_i \frac{I_0}{qV} = \eta_i \frac{J_0}{\delta q} = \frac{n_{C0}}{\tau} + A(n_{C0} - n_{C,t})N_{p0} = \frac{n_{C0}}{\tau} + \frac{N_{p0}}{\Gamma \tau_p}. \tag{7.30}$$

The zero subscript stands for the steady state value of the variables. When $d/dt = 0$, the value of A, saturated by the oscillating mode, is $A = [\tau_p \Gamma (n_{C0} - n_{C,t})]^{-1}$. At threshold, N_{p0} is approximately zero. The results given in Eq. (7.30) are essentially the same as those obtained in earlier discussions in Section 7.2.

A prediction of the large signal characteristics for applications such as on–off modulation of laser radiation requires the time dependent solution of the rate equation. The rate equation given in Eqs. (7.29) is a non-linear equation; it is difficult to solve. However, if we consider the current driving the laser to be a DC current that establishes the steady state conditions, n_{C0}, N_{p0} and I_0, and a superposed small AC current used for AC small signal modulation of the DC output, then Eq. (7.29) can be linearized. Since a pulse is a superposition of Fourier components at different frequencies, results obtained in the small signal analysis can then be used as a reference for estimating the large signal behavior. Let

$$I = I_0 + i_1 e^{j\Omega t}, \tag{7.31a}$$

$$n_C = n_{C0} + n_1 e^{j\Omega t}, \tag{7.31b}$$

and

$$N_p = N_{p0} + p_1 e^{j\Omega t}. \tag{7.31c}$$

Substituting Eqs. (7.31) into Eqs. (7.29) and neglecting both the terms with $\exp(2j\Omega t)$ variation and the higher order terms, we obtain

$$\begin{aligned} j\Omega n_1 &= \frac{\eta_i i_1}{qV} - \left(\frac{1}{\tau} + AN_{p0}\right) n_1 - \frac{p_1}{\Gamma \tau_p}, \\ j\Omega p_1 &= A\Gamma n_1 N_{p0}. \end{aligned} \tag{7.32}$$

Hence,

$$p_1(\Omega) = \frac{-\dfrac{\eta_i i_1}{qV} A\Gamma N_{p0}}{\Omega^2 - \Omega_r^2 - j\Omega\gamma_s}, \tag{7.33}$$

$$\Omega_r^2 = \frac{AN_{p0}}{\tau_p},$$

$$\gamma_s = \frac{1}{\tau} + AN_{p0}.$$

Equation (7.33) shows that p_1 is approximately a constant for small Ω. p_1 reaches its peak value when $\Omega = \Omega_p$. For $\Omega > \Omega_p$, p_1 decreases proportionally to $1/\Omega^2$:

$$\Omega_p = \sqrt{\Omega_r^2 - \frac{\gamma_s^2}{2}} = \sqrt{\frac{AN_{p0}}{\tau_p} - \frac{\left(\frac{1}{\tau} + AN_{p0}\right)^2}{2}}. \tag{7.34}$$

A similar solution is obtained for n_1. Figure 7.8 illustrates the relative frequency response of the laser output $P_{ac}(\omega)/P_{ac}(0)$. It shows clearly that the band width for effective modulation, i.e. Ω_{3dB}, is determined by Ω_r. Usually, the first term under the square root is the dominant term in Eq. (7.34). Therefore, Ω_r is called the relaxation resonance frequency for semiconductor lasers. Ω_r is larger for larger AN_{p0}/τ_p, where A equals its saturation value, $[\Gamma\tau_p(n_{C0} - n_{C,t})]^{-1}$. In other words, one can extend to higher modulation frequency by using larger N_{p0} and smaller τ_p.

The modulation of electron and hole densities produced by the current modulation results in a modulation of χ' which affects the resonance frequency of the mode. In other words, the current modulation leads directly to a frequency modulation of the oscillating mode, known as chirping. Chirping has prevented the use of directly modulated lasers at data rates above 2 Gbits/s for long distance transmission of optical signals in fibers.

7.7 Semiconductor laser amplifier

Whenever there is gain, there is amplification of optical radiation in semiconductors. We have discussed laser amplification in solid state media in Section 6.7. In edge emitting laser amplifiers, the intensity of the jth guided wave mode, neglecting the propagation loss, will be amplified as $\exp(\Gamma_j \gamma y)$. The input incident radiation will excite guided wave modes plus various radiation modes. Here, different modes may have different optical filling factors because of the differences in the field pattern. Thus, different guided wave modes will have different gains. The output from the

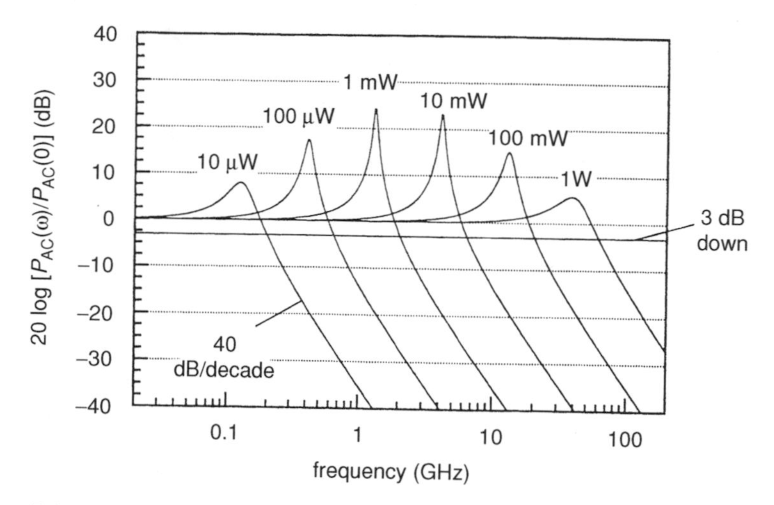

Figure 7.8. Frequency response of an idealized diode laser at various output power levels. The resonance peak and the damping of the resonance depend on the output power level. The resonance effect limits the highest frequency at which a diode laser can be modulated directly by current without serious distortion. Details about the active region and laser cavity are given in ref. [6], from which this figure is taken, with copyright permission from John Wiley and Sons.

laser amplifier will consist of the summation of all the amplified modes excited by the incident radiation. (See refs. [11] and [12] for an extensive discussion of semiconductor laser amplifiers.)

On comparing semiconductor amplifiers with the solid state amplifiers discussed in Sections 6.7 and 6.8.2, we note several differences. (1) The noise of semiconductor amplifiers will be larger for three reasons. First, it is more difficult to filter out in the receiver all the amplified spontaneous emission in unwanted modes. Secondly, carrier injection introduces another noise mechanism in addition to spontaneous emission. Thirdly, there is interference between the signal and the spontaneous emission components [12]. (2) Because of the large γ, large amplification can be achieved within a short distance of propagation in an edge emitting laser. In order to avoid the feedback due to reflections, it is necessary to reduce the reflection at the input and output ends of an edge emitting laser to a very low value. Multi-layer anti-reflection dielectric coatings have been used to yield less than 1% reflection. For these reasons, semiconductor in-plane laser amplifiers have not been commonly used in practice, as the erbium doped fiber amplifiers (EDFAs) have. However, semiconductor amplifiers can operate at wavelengths other than 1.55 μm, such as 1.3 μm, by tuning the semiconductor composition. (3) The gain of the semiconductor amplifier is polarization dependent since the susceptibility (i.e. the gain) is different between polarization directions parallel and

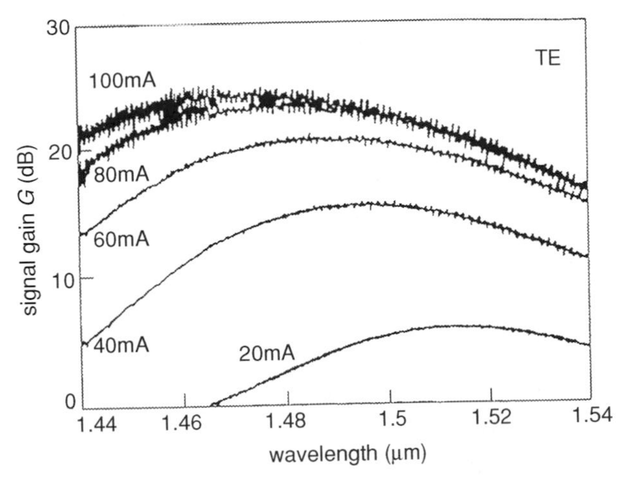

Figure 7.9. Amplifier gain versus signal strength for several current levels applied to a semiconductor amplifier. The figure is taken from ref. [11], with copyright permission from Kluwer.

perpendicular to the growth direction z. TE and TM modes also have different electromagnetic field patterns and propagation losses. The gain of the TE modes is typically bigger than that of the TM modes. (4) Because of the large line width of χ'' in semiconductors, the wavelength band width of a semiconductor amplifier is much wider than that of a solid state amplifier such as an EDFA. The amplifier gain versus signal wavelength for a semiconductor amplifier is illustrated in Fig. 7.9 [11].

Since the propagation length of the TEM wave in the active region in a surface emitting laser is very short, Fabry–Perot resonance is required to achieve significant overall gain. It is difficult to control the stability of the overall gain in a resonant cavity. The amplification will also occur in a very narrow wavelength range, unsuitable for applications such as WDM systems. Therefore, VCSELs have not yet been considered seriously for amplification of optical radiation.

The gain γ of a semiconductor amplifier will saturate at large optical intensity [11]. It can be analyzed as follows. The gain in the active region has been shown to be approximately proportional to the carrier density n_C in excess of the carrier density required for transparency n_{C0}, i.e. $g \propto (n_C - n_{C0})$. On the other hand, R_{stim} in Eq. (7.28) is proportional to n_C and the optical intensity I_{ph} in the laser. Thus, n_C in Eqs. (7.29) can be rewritten as

$$\frac{dn_C}{dt} = \frac{\eta_i I}{qV} - \frac{n_C}{\tau}\left(1 - \frac{I_{ph}}{I_s}\right),$$

where I_s is a proportionality constant. At steady state (i.e. $d/dt = 0$), the unsaturated $n_{C,unsat}$ at $I_{ph} = 0$ can be obtained from the above equation to be $\eta_i I\tau/qV$. Therefore,

$$n_C = \frac{n_{C,unsat}}{1 + \dfrac{I_{ph}}{I_s}} \tag{7.35}$$

or

$$\gamma = \frac{\gamma_0}{1 + \dfrac{I_{ph}}{I_s}},$$

where γ_0 is the unsaturated gain. Comparing the saturation of a semiconductor amplifier with the saturation of solid state and gas lasers, we see that it saturates like a homogeneous broadened transition.

7.8 Noise in semiconductor laser oscillators

Similar to solid state and gas lasers, spontaneous emission causes intensity fluctuations which yield relative intensity noise and frequency fluctuations, which yield a finite line width of laser oscillation. In addition, carrier fluctuations contribute to both the relative intensity noise (RIN) and the line width of oscillation $\Delta\nu_{osc}$. It is interesting to note that the fluctuation of carrier density not only modulates the gain, but also modulates the index of the active region, causing the resonant mode to shift back and forth in frequency. It has been shown that the shift in frequency is directly proportional to changes in carrier density [6].

In principle, the rate equations for the photon density and the carrier density, such as Eqs. (7.29), could include noise source terms for photons and carriers expressing quantum fluctuations [13]. For known auto- and cross-correlations of the Fourier components of these shot noise sources, the solution of such rate equations yields the noise power spectrum of the photon density and the RIN. Such an analysis [14] is too complex to be included here. However, the results from such an analysis can be summarized as

$$\text{RIN} = 2\tau_p^2 \left\{ \frac{C n_{C0}/\tau}{(\Gamma A\tau)^2} \frac{1}{N_{p0}^3} + \left(\frac{n_{C0}}{\tau} + \frac{1}{\Gamma A\tau\tau_p} \right) \frac{1}{N_{p0}^2} + \frac{1}{\tau_p N_{p0}} \right\}. \tag{7.36}$$

Here, N_{p0} is the photon density above threshold under steady state conditions, n_{C0} is the carrier density above threshold under steady state conditions, C is the spontaneous emission factor [13], $\Gamma \approx V/V_p$ is approximately the optical filling factor, and other symbols are explained after Eqs. (7.29). The first term originates from the beat noise between the signal and spontaneous emission; this term is proportional to the third power of the drive level $(I/I_t - 1)$. A reasonably good semiconductor laser typically has RIN $= -140$ dB/Hz. Other factors that may increase intensity noise include instability resulting from reflections, mode and polarization hopping, change of injection current, etc.

In semiconductor laser oscillators the line width is also broadened from the line width derived in Section 6.8.4, known as the Schawlow–Townes line width, by the line width enhancement factor α, which is the ratio of the change of the real part of the index of the active medium to the imaginary part. Equation (6.46) is applicable here. However, the spectral line width of semiconductor lasers is much larger than the line width of gas and solid lasers due to the much larger α caused by fluctuations of carrier densities [15]. Various ways of controlling the frequency modulation noise have been reviewed by Kourogi and Ohtsu [16].

References

1 S. M. Sze, *Physics of Semiconductor Devices*, New York, John Wiley and Sons, 1981
2 B. G. Streetman, *Solid State Electronic Devices*, Englewood Cliffs, NJ, Prentice-Hall, 1995
3 S. Wang, *Fundamentals of Semiconductor Theory and Device Physics*, Englewood Cliffs, NJ, Prentice-Hall, 1989
4 S. L. Chuang, *Physics of Optoelectronic Devices*, Section 2.2, New York, John Wiley and Sons, 1995
5 A. Yariv, *Quantum Electronics*, Chapter 11, New York, John Wiley and Sons, 1989
6 L. A. Coldren and S. W. Corzine, *Diode Lasers and Photonic Integrated Circuits*, New York, John Wiley and Sons, 1995
7 L. A. Coldren and S. W. Corzine, *Diode Lasers and Photonic Integrated Circuits*, Section 4.4, New York, John Wiley and Sons, 1995
8 J. J. Coleman, "Quantum-Well Heterostructure Lasers," in *Semiconductor Lasers*, ed. G. P. Agrawal, Woodbury, NY, AIP Press, 1995
9 N. Chinone and M. Okai, "Distributed Feedback Semiconductor Lasers," in *Semiconductor Lasers*, ed. G. P. Agrawal, Woodbury, NY, AIP Press, 1995
10 C. J. Chang-Hasnain, "Vertical-Cavity Surface Emitting Lasers," in *Semiconductor Lasers*, ed. G. P. Agrawal, Woodbury, NY, AIP Press, 1995
11 G. P. Agrawal, "Semiconductor Laser Amplifiers," in *Semiconductor Lasers*, ed. G. P. Agrawal, Woodbury, NY, AIP Press, 1995
12 S. Shimads and H. Ishio, *Optical Amplifiers and Their Applications*, Chapters 3 and 4, New York, John Wiley and Sons, 1994
13 K. Iga, *Fundamentals of Laser Optics*, Chapter 13, New York, Plenum Press, 1994
14 F. Koyama, K. Morita, and K. Iga, "Intensity Noise and Polarization Stability of GaAlAs-GaAs Surface Emitting Lasers," *IEEE Journal of Quantum Electronics*, **27**, 1991, 1410
15 C. H. Henry, "Theory of the line width of semiconductor lasers," *IEEE Journal of Quantum Electronics*, **18**, 1982, 259
16 M. Kourogi and M. Ohtsu, "Phase Noise and Its Control in Semiconductor Lasers," in *Semiconductor Lasers*, ed. G. P. Agrawal, Woodbury, NY, AIP Press, 1995

Index